The Sciences in Islamicate Societies in Context

This Variorum volume reprints ten papers on contextual elements of the so-called ancient sciences in Islamicate societies between the thirteenth and the seventeenth centuries. They address four major themes: the ancient sciences in educational institutions; courtly patronage of science; the role of the astral and other sciences in the Mamluk sultanate; and narratives about knowledge.

The main arguments are directed against the then dominant historiographical claims about the exclusion of the ancient sciences from the *madrasa* and cognate educational institutes, the suppression of philosophy and other ancient sciences in Damascus after 1229, the limited role of the new experts for timekeeping in the educational and professional exercise of this science, and the marginal impact of astrology under Mamluk rule. It is shown that the *muwaqqit*s (timekeepers) were important teachers at *madrasa*s and Sufi convents, that Mamluk officers sought out astrologers for counselling and that narratives about knowledge reveal important information about scholarly debates and beliefs. Colophons and dedications are used to prove that courtly patronage for the ancient sciences continued uninterrupted until the end of the seventeenth century. Furthermore, these papers refute the idea of a continued and strong conflict between the ancient and modern sciences, showing rather shifting alliances between various of them and their regrouping in the classifications of the entire disciplinary edifice.

These papers are suited for graduate teaching in the history of science and the intellectual, cultural and social history of the Middle East and for all readers interested in the study of the contexts of the sciences.

Sonja Brentjes is an historian of science with specialization in Islamicate societies, the late medieval Mediterranean and early modern Catholic and Protestant Europe. Currently, she is a visiting scholar at the Max Planck Institute for the History of Science in Berlin. Her latest books include: *Teaching and Learning the Sciences in Islamicate Societies, 800–1700* (2018) and *The Routledge Handbook of the Sciences in Islamicate Societies: Practices from the 2nd/8th to the 13th/19th Centuries* (2023), co-edited with Peter Barker (associate editor) and Rana Brentjes (assistant editor).

Also in the Variorum Collected Studies series

DONALD F. DUCLOW
Engaging Eriugena, Eckhart and Cusanus (CS1117)

VICTOR MALLIA-MILANES
The Winged Lion and the Eight-Pointed Cross
Venice, Hospitaller Malta, and the Mediterranean in Early Modern Times (CS1116)

SVETLANA KUJUMDZIEVA
Studies on Eastern Orthodox Church Chant (CS1115)

SONJA BRENTJES
Historiography of the History of Science in Islamicate Societies
Practices, Concepts, Questions (CS1114)

AVERIL CAMERON
From the Later Roman Empire to Late Antiquity and Beyond (CS1113)

ROBERTO TOTTOLI
Studies in Islamic Traditions and Literature (CS1112)

BRIAN CROKE
Engaging with the Past, c.250–c.650 (CS1111)

FELICE LIFSHITZ
Reading Gender
Studies on Medieval Manuscripts and Medievalist Movies (CS1110)

DOROTHEA MCEWAN
Studies on Aby Warburg, Fritz Saxl and Gertrud Bing (CS1109)

GAD G. GILBAR
Trade and Enterprise
The Muslim Tujjar in the Ottoman Empire and Qajar Iran, 1860–1914 (CS1108)

JEAN-CLAUDE HOCQUET
Le marchand et les poids et mesures (CS1107)

For more information about this series, please visit:
www.routledge.com/Variorum-Collected-Studies/book-series/VARIORUM

The Sciences in Islamicate Societies in Context

Patronage, Education, Narratives

Sonja Brentjes

VARIORUM COLLECTED STUDIES

Routledge
Taylor & Francis Group
LONDON AND NEW YORK

First published 2024
by Routledge
4 Park Square, Milton Park, Abingdon, Oxon OX14 4RN

and by Routledge
605 Third Avenue, New York, NY 10158

Routledge is an imprint of the Taylor & Francis Group, an informa business

British Library Cataloguing-in-Publication Data
A catalogue record for this book is available from the British Library

ISBN: 978-1-032-44496-3 (hbk)
ISBN: 978-1-032-44497-0 (pbk)
ISBN: 978-1-003-37245-5 (ebk)

DOI: 10.4324/9781003372455

Typeset in Times New Roman
by Apex CoVantage, LLC

VARIORUM COLLECTED STUDIES SERIES CS1119

CONTENTS

Introduction 1

1 'On the location of the ancient or 'rational' sciences in Muslim educational landscapes (AH 500–1100),' *Bulletin of the Royal Institute for Inter-Faith Studies*, *4*(1), 2002: 47–71. 3

2 'Shams al-Dīn al-Sakhāwī on *Muwaqqits*, *Mu'adhdhins*, and the Teachers of Various Astronomical Disciplines in Mamluk Cities in the Fifteenth Century,' in (eds.) Emilia Calvo, Mercè Comes, Roser Puig, Mònica Rius, *A Shared Legacy, Islamic Science East and West, Homage to Professor J.M. Millàs Vallicrosa*, Barcelona: Universitat de Barcelona, Publicacions i Edicions, 2008, 129–150. 27

3 'Ayyubid Princes and Their Scholarly Clients from the Ancient Sciences,' in Albrecht Fuess, Jan-Peter Hartung (eds.), *Court Cultures in the Muslim World: Seventh to Nineteenth Centuries*, SOAS/Routledge Studies on the Middle East, London: Routledge, 2010, 326–56. 48

4 'Patronage of the mathematical sciences in Islamic societies: structure and rhetoric, identities, and outcomes,' in Eleanor Robson, Jackie Stedall (eds.), *The Oxford Handbook of the History of Mathematics* (Oxford: Oxford University Press, 2008), 301–28. 81

5 'Courtly Patronage of the Ancient Sciences in Post-Classical Islamic Societies,' *Al-Qanṭara: revista de estudios árabes*, XXIX (2008), 403–436. 107

6 'The language of 'patronage' in Islamic societies before 1700,' *Cuadernos del CEMYR* 20 (2012), 11–22. 133

7 'The Study of Geometry According to al-Sakhāwī (Cairo, 15th c) and al-Muḥibbī (Damascus, 17th c),' in J. W. Dauben, S. Kirschner, A. Kühne, P. Kunitzsch and R. Lorch eds., *Mathematics Celestial and Terrestrial, Festschrift for Menso Folkerts zum 65. Geburtstag. Acta Historica Leopoldina* 54 (2008), 323–341. Halle/Saale: Deutsche Akademie der Naturforscher Leopoldina. 144

8 'On Four Sciences and Their Audiences in Ayyubid and Mamluk Societies,' in Syrinx von Hees ed. *Inhitat – The Decline Paradigm: Its Influence and Persistence in Writing Arab Cultural History*. Würzburg: Ergon, 2012, 139–172. 168

9 'Narratives of knowledge in Islamic societies: what do they tell us about scholars and their contexts?' *Almagest*, *4*(1), 2013: 74–95. 200

10 'Sanctioning knowledge,' *Al-Qanṭara: revista de estudios árabes*, 35(1), 2014: 277–309. 221

Index 247

INTRODUCTION

This volume unites ten papers on various contexts in which scholars of the mathematical sciences worked in Islamicate societies between the ninth and the seventeenth centuries. The ten papers address courtly patronage, education, narratives about the sciences and their origins and case studies on conflicts. Their main goal was to present concrete material invalidating fundamental historiographical convictions held by historians of science and historians of Islamicate societies since the nineteenth century. These positions stated that there were no noteworthy scientific activities in Islamicate societies after 1200 and no courtly patronage supporting them with the exception of Ulugh Beg (d. 1449) and that "the orthodoxy" rejected the ancient sciences and marginalized them which led to house arrests and edicts forbidding their teaching, that those sciences were excluded from the *madrasa*s and at best taught in private settings or clandestinely. These two broad themes are dealt with in six papers (2002, 2x2008, 2010, 2012, 2014). Specific subthemes of patronage and education are treated in two papers (2008, 2012) for the Ayyubid and Mamluk dynasties in Egypt and Syria. The focus is on the transmission of mathematical and astronomical knowledge at *madrasa*s in Cairo and its transfer to the muezzins at mosques and the place of astrology in Mamluk society. They argue against the belief that astrology was banned under the Mamluks and the science of timekeeping was not used in the mosques, addressing therewith central contextual issues of the astral sciences in Egypt and Syria between 1220 and 1515. These two papers underline my methodological position that working exclusively with scientific texts or instruments does not suffice to understand the positions of the sciences under concrete historical circumstances. The two remaining papers (2013, 2014) take another approach to the issues of exclusion and inclusion of the sciences into the socio-cultural practices, including the historical construction of memory, by analysing narratives about sanctions against the sciences and stories about their origins. They show that in contrast to religious positions, in the mathematical sciences as understood in premodern Islamicate societies there are no definitive sanctions against those sciences remembered in historical narratives, while stories about the origins of the sciences include at times traces of fascinating debates about divine right and power and indispensable human involvement in the development of knowledge. All ten

DOI: 10.4324/9781003372455-1

papers prove that limiting history of science in Islamicate societies to content studies loses sight of substantial parts of the factors which conditioned, enabled or constrained them and thus distorts our views of the achievements, problems and contradictions. The ten papers were written for different audiences with the aim to explain why and in which sense major historiographical positions with regard to the sciences and their socio-cultural contexts are contradicting the material found in the extant sources. Hence, positions, arguments and evidence overlap, at least partly, among several of them.

1

ON THE LOCATION OF THE ANCIENT OR 'RATIONAL' SCIENCES IN MUSLIM EDUCATIONAL LANDSCAPES (AH 500–1100)

"Knowledge is a door and its key is the question."
Muḥammad al-Muḥibbbī (d. AH 1111)[1]

The question being addressed in the present paper is this: Who made room for the ancient or 'rational' sciences in Muslim societies and under which terms was this space granted? I will discuss this question by studying the changing relationship between the religious and the ancient or 'rational' sciences and by investigating the forms of respectability granted to the latter.

While considering the evidence for the presence of the ancient or 'rational' sciences in Muslim educational landscapes, I will claim that the available sources are inadequate to provide us with reliable quantitative data about the role of the ancient or 'rational' sciences in the educational process. However, they can offer us important qualitative insights into the system. They show that the same discourse was applied to all types of sciences. Furthermore, they illustrate that the system of education was not stagnant, but fluid. The sources describe shifting borderlines between the major disciplinary units and point to changing alliances between some of the religious and some of the ancient or 'rational' sciences. Hence, they call into question the all too rapid and easy identification of the ancient sciences with the 'rational' sciences.[2] Since these two sets of sciences were apparently not identical, I use the term 'secular' sciences as an umbrella for both groups. I use the term in quotation marks to indicate that the 'rational' sciences also contained individual disciplines that belonged to the set of the religious sciences.

1 al-Muḥibbī, n.d., 4:40.
2 See the quotes given below from Makdisi, Berkey and Chamberlain, who all treat the 'rational' sciences as if they were identical to the ancient sciences.

DOI: 10.4324/9781003372455-2

Throughout this paper, I will insist that, when we study Muslim educational landscapes and ask which places were given to the ancient sciences and under which terms, we need to regard the biographical literature, as well as the various genres of writings on the classification of the sciences, as embodiments of these landscapes and as elements of their structural settings. From this perspective, the question of whether the ancient sciences remained alien or were alienated loses its meaning for, however they were evaluated by the individual authors of such texts, these disciplines are always mentioned in *ṭabaqāt* literature about jurists and in works on the classification of the sciences, although they appear less regularly in books focusing upon individuals who transmitted *ḥadīth* (sayings of the Prophet) or composed poetry.

The historical time-span examined in this paper begins in the late fifth century AH and stretches until the late eleventh century AH. The focus is upon the Ayyubid and Mamluk societies of Egypt and al-Shām, but other Muslim societies, such as the Ottoman, the Timurid and the Ilkhanid, are also discussed, mainly because they provide necessary points of reference and comparison.

Introduction

There are two major schools of thought among historians of science and historians of Muslim societies regarding the location of the ancient or 'rational' sciences in Muslim societies. One camp insists that there was no room for these sciences, that they remained alien to Muslim culture and that they stagnated in obscurity. A specific variant of this view has been formulated by George Makdisi in his influential book on the *madrasa* ('school'). He claims that the sciences of the ancient Greeks were excluded from the *madrasa*, where the legal and other religious sciences were taught according to formal curricula and where the philological disciplines were admitted as auxiliary studies.[3] His position has been adopted by

3 See Makdisi 1981, table of contents. Under chapter one, section two ("Typology of Institutions of Learning"), Makdisi differentiates between "pre-madrasa institutions" (section 1.II.2) and "the madrasa and cognate institutions" (section 1.II.3). He subdivides the pre-*madrasa* institutes into "institutions exclusive of the foreign sciences" (section 1.II.2.a) and "institutions inclusive of the foreign sciences" (section 1.II.2.b); in the latter section, he discusses libraries and hospitals. In section 1.II.3, he declares the emergence of the *madrasa* to be "a natural development of two previous institutions: the masjid, in its role as a college of law, and its nearby khan, as the residence of the law students in attendance" (27). He implies by this presentation that the *madrasa*, owing to its origin, belongs to the group of "institutions exclusive of the foreign sciences." In chapter two, on instruction, he treats "waqf and the dichotomous division of knowledge" (section 2.I.3). In the text, he writes:

> A striking feature of Muslim education in the Middle Ages was the dichotomy between two sets of sciences: the 'religious' and the 'foreign'. This dichotomy would not be so remarkable were it not for the fact that actual intellectual activity embraced the two sets, and scholarly production was prosperous in both. For a long time, this phenomenon has obscured our understanding of the true nature of the madrasa, an institution which, as a result, was readily assimilated to the university because it was assumed that all subjects were taught in it.

many historians of science in Muslim societies, even those who otherwise challenge the notion of the marginalization and alienation of the 'rational' sciences in Muslim societies, such as ʿAbd al-Hamid Sabra.[4]

The other camp, while not focusing primarily upon the ancient sciences, claims that the Muslim educational system was much less formalized than Makdisi and other scholars have suggested. In his study of the educational landscape in Mamluk Cairo, Jonathan Berkey follows A. Tibbawi to point out that the various kinds of teaching institutes "themselves played no actual role in Islamic education" since "Islamic law allowed no corporate identity to any particular institution, and no formal degree system was ever established."[5] He claims that "not all madrasas were exclusively or even principally educational institutions, and, as in earlier periods, much serious legal and religious instruction took place outside of madrasas."[6] With regard to the presence of the ancient sciences in Mamluk Cairo, however, Berkey merely follows Makdisi and others, writing that "certainly the curriculum offered in the madrasa concentrated on the traditional religious and legal sciences, to the exclusion of the so-called 'foreign' rational sciences inherited from the Hellenistic world."[7]

In another examination of Muslim educational landscapes, this time in Ayyubid and Mamluk Damascus, Michael Chamberlain insists that the Muslim discourse on education did not focus upon institutions when it came to students and studying. Students chose the teachers with whom they would read and discuss books and subsequently listed the names of those teachers in their *muʿjam* (register), not the names of the *madrasas* or any other *locales* that they may have visited. Chamberlain explicitly rejects the *madrasa* as the core institution of Muslim legal education, as well as the assumption that it possessed a formalized curriculum that

---

In reality, however, neither the madrasa nor its cognate institutions harboured any but the religious sciences and their ancillary subjects (77).

Then, he asks: "If such was the case, how is one to explain the flourishing of the philosophical and natural sciences?" (Ibid.). His answer counterpoises what he calls "the traditionalist forces represented by the madrasa and cognate institutions" with "the rationalist forces represented by the dār al-ʿilm and its cognates" (Ibid.). According to Makdisi, the major obstacle to accommodating the 'foreign' sciences was the *waqf*, which he describes as excluding "any and all things that were considered to be inimical to the tenets of Islam" (78). Although he concedes that the *waqf* did not completely succeed as an exclusionary force, he maintains that "the study of the 'foreign sciences' had to be pursued privately" (Ibid.). He even acknowledges that "there was nothing to stop the subsidized student from studying the foreign sciences," but specifies, at the same time, that this happened "unaided" or "in secret," "in the privacy of [teachers'] homes, or in the waqf institution, outside of the regular curriculum" (Ibid.). Nonetheless, Makdisi concludes that, in general, "the dichotomy between the two sets of sciences was maintained" and adds that this dichotomy "was matched by a dichotomy in the institutions of learning" (Ibid.).

4 See Sabra 1987; Sabra 1994; and Sabra 1996.

5 Berkey 1992, 16–17.

6 Ibid., 47.

7 Ibid., 7.

excluded the ancient sciences and *kalām* ('rational' theology).[8] He even writes that lecturers in Ayyubid and, later, Mamluk Damascus "taught the Hellenistic 'rational' sciences in madrasas as elsewhere."[9]

Berkey and Chamberlain conclude that the character of Muslim education has to be understood in terms of networks of individuals, rather than complexes of teaching institutions.[10] They stress that a consequence of this character was a relative freedom to teach particular disciplines and books. Furthermore, they point to the fact that *madrasas*, as well as other teaching institutes, were centres of activities that were not purely educational, such as worshipping God, housing the dead, hosting postal couriers, incarcerating prisoners, feeding the poor and needy, and accumulating wealth and reputation.[11] However, these conclusions remained without influence among historians of science.

The sciences of the ancients and the 'rational' sciences in Muslim educational landscapes

In this section, I will present results regarding the frequency of teaching associated with scholars mentioned in the *Ṭabaqāt al-shāfiʿiyya* of Ibn Qāḍī Shuhba (AH 779–851). I chose this biographical dictionary randomly in order to support my earlier point that any dictionary on Muslim jurists – in this case, adherents to the Shāfiʿī school of jurisprudence – contains material about the 'secular' sciences. My survey excludes the ninth century AH owing to the fact that coverage of it is incomplete, since Ibn Qāḍī Shuhba died in 851.

The survey (see Table 1.1) begins with the first explicit mention that Ibn Qāḍī Shuhba makes of a teaching institute (a *madrasa*), namely, biography number 233. This entry informs us about a scholar born in AH 414. Among the teaching institutes mentioned by Ibn Qāḍī Shuhba, we find the *madrasa*, the *jāmiʿ* (congregational mosque), the *dār al-ḥadīth* (house for teaching the sayings of the Prophet), the *qubba* (tomb) and the *ribāṭ* (Ṣūfī convent or house).

Table 1.1 Teaching activities and teaching institutes

	Entries	*A)*	*%*	*B)*	*%*
5th/6th	111			24	22.4
7th	152	84	55	49	32.0
8th	210	132	63	111	54.0

a) teaching activities; b) teaching institutes

8 Chamberlain 1995, 83–84.
9 Ibid., 84.
10 Berkey 1992, 20, 42–44. Chamberlain 1995, chapter four.
11 Berkey 1992, 47, 94, 188–193.

Although the number of scholars affiliated with a teaching institute doubled from the sixth to the seventh century AH, the figures derived from his book for the latter century imply that the act of teaching was deemed more laudable than association with a particular teaching institute or than the institute itself. During the eighth century AD, this trend continued upwards. Thus, while the number of scholars who held chairs in teaching institutes and received remuneration grew, the reputation for teaching as such, that is, teaching unconnected to or explicitly separated from the accumulation of worldly goods, continued to be ranked higher. This observation is confirmed by an aphorism coined by a Cairene jurist of the eighth century AH, Muḥammad ibn al-Ḥājj (d. AH 740). Ibn al-Ḥājj complained bitterly that "whereas before, a man spent his money in order to acquire knowledge, now he acquires money through his knowledge."[12]

Ibn Qāḍī Shuhba's rhetoric makes it clear that there were not yet sharp and rigid boundaries between teaching in an institute and teaching without remuneration, nor even between the various types of institutes themselves. Often only the proper name of an institute is given, without specifying which type is meant. Since several of these establishments possessed the same proper name, only a person initiated into the educational topography of the major towns of Egypt, Syria, Iraq, Iran and the Arabian Peninsula can follow, without difficulty, Ibn Qāḍī Shuhba's signposts through their educational landscapes.

When we inquire as to the location of the ancient or 'rational' sciences in this broad vista, it is easy enough to see that they are present in all three of the periods mentioned above (fifth/sixth, seventh and eighth centuries), albeit not very prominently placed. Ibn Qāḍī Shuhba's work indicates no clear case of affiliation between any one of these disciplines and a teaching institute, although such ties do exist with respect to *ḥadīth*, *tafsīr* (exegesis), *fiqh* (law) and the Arabic language. Furthermore, disciplines from the ancient sciences appear in the biographical entries much less frequently than do disciplines from the religious *or* philological sciences.

Table 1.2 shows that the 'secular' sciences are continuously present in the dictionary and that the frequency of references to them for the seventh and eighth centuries AH is surprisingly high. Although I did not play the same kind of number

Table 1.2 The 'secular' sciences in Ibn Qāḍī Shuhba's *Ṭabaqāt al-shāfiʿiyya*

	Entries	*A)*	*%*	*B)*	*C)*
5th/6th	111	4	3.7		
7th	152	29	22.0	4	7
8th	210	28	13.0	1	11

a) any 'secular' science; b) the ancient sciences *en bloc;* c) the 'rational' sciences

12 Quoted after Berkey 1992, 95.

games with other collections that I read through, the first point is confirmed by every one of them. All of the biographical dictionaries that I have examined contain a number of biographies that explicitly mention that their subject studied or taught one or more of the 'secular' sciences and wrote books in these fields.

I do not claim, however, that the percentages derived from Ibn Qāḍī Shuhba's book accurately reflect the historical distribution of the 'secular' sciences within the educational system. Nor do I claim that the increase and decrease in relative participation implied by the percentages express historical trends.

The reason for this cautious attitude becomes evident when we ask which books were read by students not explicitly linked with these sciences. Twenty-four students studied, for example, a book by Jamāl al-Dīn Abū ʿAmr ibn al-Ḥājib (d. AH 646) that is innocuously entitled *Mukhtaṣar*. This work treated law and contained chapters on logic, as well as on arithmetic. Students wrote commentaries upon each of these two chapters.[13] The book is directly related to a work by Sayf al-Dīn al-Āmidī (d. AH 631) on *uṣūl al-fiqh* (the foundations/fundamentals of law) which, according to Ibn Khaldūn (d. AH 789), was one of the works that discussed law from the perspective of *kalām* using tools taken from logic.[14] Thus, Ibn Qāḍī Shuhba's unspecific references to Ibn al-Ḥājib's work in other entries may obscure the fact that those who studied it did not stop when they reached its chapters on logic and arithmetic.

This view is supported by references to the *Mukhtaṣar* in the entries for three scholars who were active in various 'secular' disciplines. Quṭb al-Dīn al-Shīrāzī (AH 610 or 634–710) is described by Ibn Qāḍī Shuhba as the first commentator on the *Mukhtaṣar*. ʿAḍud al-Dīn al-Ījī (after AH 708–756) is mentioned as "the commentator" on the work. Al-Ḥasan ibn Sharafshāh (d. AH 715 or 718), a student of Naṣīr al-Dīn al-Ṭūsī (AH 599–672), who raised him to the highest position at Maragha, also commented upon it in an exposition of medium length.[15]

The authors of other works very often read and commented upon were Fahkr al-Dīn al-Rāzī (AH 544–606), Saʿd al-Dīn al-Taftāzānī (AH 712–791), ʿAḍud al-Dīn al-Ījī and al-Sayyid al-Sharīf al-Jurjānī (AH 741–816). All of these scholars are well known for their treatment of aspects of logic, metaphysics, natural philosophy, astronomy and/or mathematics in various of their works. Thus, while Ibn Qāḍī Shuhba does not explicitly link the overwhelming majority of students with the 'secular' sciences, several or even many among them might nevertheless have devoted some of their training to the discussion of a set of ideas and problems taken from these disciplines.

Shifts of alliances and emphasis

Up until now, I have spoken in a general manner about the sciences of the ancients or the 'rational' sciences. When we look at the details, however, and ask which of

13 Ibn Qāḍī Shuhba AH 1399/ AD 1979-AH 1400/ AD 1980, 2:247.
14 Ibn Khaldūn 1980, 2:29–30.
15 Ibid., 2:278 and 312; 3:33.

the disciplines subsumed under these categories are specifically mentioned in the biographies, we discover substantial differences of scale between individual fields. Logic and arithmetic are most often mentioned as fields of study and teaching.[16] Medicine appears less frequently.[17] Geometry, astronomy, algebra and philosophy occupy a third rank, while the occult disciplines are almost completely absent.[18] When we add other biographical sources to the *Ṭabaqāt al-shāfiʿiyya* of Ibn Qāḍī Shuhba, the picture changes slightly. Works covering only one century and unconfined to a particular jurisprudential school show a higher percentage of medical studies than works of the *ṭabaqāt* genre. Astronomy and its related disciplines, such as *ʿilm al-mīqāt* (the science of timekeeping), algebra, geometry, philosophy and the occult sciences, also appear more often. The relative distribution of the various disciplines, however, seems almost unchanged, with one exception. In collections focusing upon the ninth, tenth and eleventh centuries AH, geometry, astronomy and philosophy seem to be present more often than they earlier were.[19]

Entries containing references to one or the other of the 'secular' sciences point to the emergence of various alliances between them and some of the religious sciences. These alliances provided new spaces for the insertion of the 'secular'

16 In Ibn Qāḍī Shuhba's collection, arithmetic is mentioned twice and logic once in the entries between AH 414/AD 1023 and AH 586/AD 1190. During the seventh century AH (thirteenth century AD), logic appears nine times and arithmetic six times. The entries for the eighth century AH (fourteenth century AD) mention logic eight times and arithmetic three times.

17 Medicine does not appear in Ibn Qāḍī Shuhba's work during the fifth/sixth centuries AH (eleventh/twelfth centuries AD), but is present six times during the seventh century AH, and once in the eighth century AH.

18 *Falsafa* occurs once in the fifth/sixth centuries AH. It appears three times in the seventh century AH, as does *ḥikma*, while *ḥikma* alone is present and then only once in the eighth century AH. Natural philosophy is mentioned twice in the seventh century AH. The mathematical sciences in general and geometry and algebra appear once each in the seventh century AH, as do astronomy, timekeeping and algebra in the eighth century AH. Astrology, horoscope making and agriculture are referred to once each during the seventh century AH, as are magic squares and letter divination in the eighth century AH.

19 Not only does the frequency of the mathematical and philosophical sciences increase in such late biographical dictionaries, but the way of talking about them changes as well. To a certain extent, a sort of turning back to the notion of 'the sciences of ancients' seems to have taken place. See, for example, the following quotations from the dictionaries of Shams al-Dīn al-Sakhāwī and Muḥammad al-Muḥibbī:

> Muḥammad ibn Aḥmad . . . Shams al-Dīn Abū ʿAbd Allāh al-Bisāṭī . . . did not cease his continuous engagement with the sciences. He demanded the unequivocal and the meaningful from them until he had reached the top [rank] in law, the two *aṣl*, Arabic language, *maʿānī* [a form of rhetoric], *bayān* [another form of rhetoric], logic, philosophy, algebra, medicine, astronomy, geometry and arithmetic. He became the *imām* of his era (al-Sakhāwī n.d., 7:5–6).
>
> Sayyid Muḥammad Kibrīt . . . took the mathematical, philosophical and natural philosophical sciences and the science of truth [Sufism] with the great *muḥaqqiq* [truth-seeker], ʿAbd Allāh ibn Walī al-Ḥaḍramī, a student of Quṭb al-ʿĀrif bi-llāh taʿālā Sayyid Ṣibghat Allāh ibn Rūḥ Allāh al-Sindī (al-Muḥibbī n.d., 4:28).

sciences and lent them prestige and respectability. Yet they also circumscribed the directions into which these sciences led and cut off routes that they might have followed under other configurations.

Arithmetic and *ʿilm al-farāʾiḍ* (the science for determining inheritance shares) became allied at an early date and this association developed further with the passage of time. Algebra was a party to this alliance as well, although not invariably. The position of geometry, however, remained fluid and was, at times, precarious, despite al-Ghazālī's obeisance before the certainty of its proofs. Some authors placed it alongside philosophy as a useless and even dangerous discipline, while others integrated it into the 'rational' sciences.[20]

A second partnership evolved between astronomy and *ʿilm al-kalām*. Over the centuries, their interpenetration achieved such a degree of stability that even writers defending the so-called Islamic astronomy integrated elements of ancient Greek astronomy (such as the sphericity of the earth) into their schemes upon the authority of such scholars as Fakhr al-Dīn al-Rāzī. Still other sciences also formed associations with certain 'secular' disciplines. Some biographers seem to include, for instance, the philological sciences among the 'rational' disciplines.[21]

These shifts and moves are visible in the biographical literature through changes in the terminology used to refer to the set of the 'secular' sciences and various of its members. Until approximately the late sixth or early seventh century AH, this set of disciplines was most often called *ʿulūm al-awāʾil* (the sciences of the ancients). The rival term, *al-ʿulūm al-ʿaqliyya* (the 'rational' sciences), appeared long before this time, but only dominated the rhetorical scenery from the seventh century onward.

Similar processes may be observed with regard to individual disciplines, such as philosophy, astronomy, or arithmetic. Such changes in terminology conceal changes in content and meaning. While *falsafa* characterizes theories derived from ancient Greek philosophy in most cases, *ḥikma* (wisdom/philosophy) can mean *falsafa* and, moreover, *ishrāq* (illumination) or *taṣawwuf* (Sufi doctrine) as well.

Equally profound as the terminological shift from *falsafa* to *ḥikma* was a second change connected to the disciplines identified as belonging to the 'rational' sciences. In biography number 603, Ibn Qāḍī Shuhba remarks that Taqī al-Dīn

20 Tashköprüzade, for instance, writes that the sciences of philosophy and geometry are distant from the sciences of the world to come and that only those who prefer this life to the next occupy themselves with them; see MS Sprenger 1823 (Staatsbibliothek, Berlin), f. 6a,15–16. On the other hand, Muḥammad al-Marʿashī claims that some of the 'rational' sciences belong to the useful disciplines (*al-funūn al-nāfiʿa*). The examples he lists under this rubric are *ʿilm al-kalām* and the mathematical sciences, including geometry, arithmetic and astronomy; see MS N. F. 405 (National Library, Vienna), f. 2b,19–3a,1 and f. 3a,4–5.

21 For example, Muḥammad al-Muḥibbī writes: "And after ʿAlāmat al-Dīn ʿĪsā ibn Muḥammad al-Jaʿfarī al-Maghribī had come to Mecca, he served him as a companion and read with him such 'rational' sciences as the two *aṣl*, logic, *maʿānī*, *bayān*, *badīʿ*, grammar and morphology"; see al-Muḥibbī n.d., 1:252. See, also, al-Sakhāwī n.d., 1:301, 330.

al-Subkī (AH 683–756) "had read the two *aṣl* [foundations/fundamentals] and the remainder of the 'rational' sciences with ʿAlī ʿAlāʾ al-Dīn al-Bājā, and logic and *khilāf* [legal controversy] with Sayf al-Dīn al-Baghdādī."[22] This passage suggests that the two *aṣl*, that is, *uṣūl al-dīn* (the foundations/fundamentals of religion) and *uṣūl al-fiqh*, were seen as parts of the 'rational' sciences. While I was initially rather surprised to discover this, I later observed that either the 'rational' sciences en bloc or logic alone were repeatedly mentioned alongside not only those disciplines closely connected to *ʿilm al-kalām*, but the two *aṣl* and *khilāf* as well.[23]

This is illustrated by the case of Shams al-Dīn al-Iṣfahānī (AH 616–688). He studied the 'rational' sciences with Tāj al-Dīn al-Urmawī (d. before AH 656) in Baghdad and *jadal* (dialectics) and *ḥikma* with Athīr al-Dīn al-Abharī (d. AH 662).[24] Al-Urmawī was one of Fakhr al-Dīn al-Rāzī's best students and excelled in the 'rational' sciences.[25] Fakhr al-Dīn al-Rāzī is praised by Ibn Qāḍī Shuhba as "the *imam* of his time in the 'rational' sciences and one of the *imāms* in the sciences of the *sharīʿa*."[26] Athīr al-Dīn al-Abharī was a student of Kamāl al-Dīn ibn Yūnus (AH 551–640) and the author of three mathematical works, seven astronomical books and a very influential *summaa* on philosophy.[27] Shams al-Dīn al-Iṣfahānī is characterized in Ibn Qāḍī Shuhba's entry as a scholar whose particular fields of study, teaching and writing were *uṣūl al-fiqh*, *khilāf* and logic, but who also excelled in *kalām* and *jadal*. This evaluation is shared by the authors of other biographical dictionaries.[28]

In Cairo, Shams al-Dīn al-Iṣfahānī successfully taught the two *aṣl* and the 'rational' sciences. Most of the nineteen of his students honoured with entries in Ibn Qāḍī Shuhba's work studied *uṣūl* with him. Three among them took logic or the 'rational' sciences as well.[29] Two are said to have combined *uṣūl* and the 'rational' sciences or logic when studying with him.[30] Al-Iṣfahānī not only combined the two *aṣl* and logic in his lecturing, but also wrote a textbook called *al-Qawāʿid fī l-ʿulūm al-arbaʿa*, which combined the two *aṣl*, logic and *khilāf*.[31] Two of his works are specifically devoted to logic.[32]

An unknown student of al-Iṣfahānī wrote two treatises upon the classification of the sciences. Both texts deal with several seriously disputed topics, one of them

22 Ibn Qāḍī Shuhba AH 1399/ AD 1979-AH 1400/ AD 1980, 3:48–49. Al-Bājā is praised by Ibn Qāḍī Shuhba as "*imām* of the two *aṣl* and of logic" (Ibid., 2:292).

23 See, for instance, Ibid., 2:245, 292, 248 and 303; 3:39 and 108.

24 Ibid., 2:259.

25 Ibid., 2:152.

26 Ibid., 2:81.

27 Matvievskaya and Rozenfel'd 1983, 2:388–390.

28 Ibn Qāḍī Shuhba AH 1399/ AD1979-AH 1400/ AD 1980, 2:261.

29 Ibid., 2:303 and 309; 3:108–109.

30 Ibid., 2:303; 3:108.

31 Ibid., 2:260–261.

32 One is called "Ghāyat al-maṭlab"; see Ibid., 2:261. The title of the other is "Nāẓir al-ʿayn"; see MS Wetzstein 51 (Staatsbibliothek, Berlin), f. 22b,12.

being the question of whether *ʿilm* (knowledge, science) can be defined at all and, if so, how. Another relates to whether all sciences have to consist of "*mawḍūʿ*, *mabādiʾ*, *masʾala* (object, principles, question)." A third concerns the licit or illicit status of individual sciences according to the assessments of different groups of religious scholars.

Here I will focus, however, upon three points made in the second treatise. The first is the information that the legal scholar, Nāṣir al-Dīn al-Bayḍāwī (d. AH 691), composed a work called *Miṣbāḥ al-arwāḥ fī l-manṭiq wa-uṣūl al-dīn*.[33] Next, the anonymous student tells us that, much like al-Bayḍāwī, Fakhr al-Dīn al-Rāzī and his followers had declared that knowledge of *ʿilm al-naẓar* (the science of theoretical speculation) was one of the conditions that had to be met by a *mujtahid* (spiritual guide). The *mujtahid* needed this knowledge in order to differentiate between correct and false thinking.[34] Third, this treatise makes it clear that the writings of Abū ʿAlī al-Ḥusayn ibn Sīnā on the classification of the sciences and on philosophy were carefully studied in late Ayyubid and early Mamluk Cairo.

These three points are of the utmost importance for our topic. They suggest that an important process took place between the fifth and seventh centuries AH: the forging of an alliance between two major religious disciplines, *uṣūl al-dīn* and *uṣūl al-fiqh*, and some of the 'secular' sciences – in particular, logic, but possibly some of the other philosophical disciplines as well. The treatises bind this development into a chain linking Fakhr al-Dīn al-Rāzī directly with Nāṣir al-Dīn al-Bayḍāwī and indirectly with Ibn Sīnā. The attribution of this process to Fakhr al-Dīn al-Rāzī and Nāṣir al-Dīn al-Bayḍāwī is supported, to some extent, by Ibn Qāḍī Shuhba, who declares that al-Bayḍāwī united the 'rational' and the 'traditional' sciences.[35]

In his famous work, *al-Muqaddima*, Ibn Khaldūn describes this alliance between the two *aṣl* and logic in slightly different terms. According to him, *uṣūl al-fiqh* took over elements from logic once Ashʿarites and Muʿtazilites applied themselves to the latter discipline.[36] He confirms that this process started in the fifth century AH. Major religious scholars were involved in it, among them Imām al-Ḥaramayn Abū l-Maʿālī al-Juwaynī (AH 419–478), Abū Ḥāmid al-Ghazālī (d. AH 505), ʿAbd al-Jabbār (d. AH 415) and Abū l-Ḥusayn al-Baṣrī (d. AH 436).[37]

Ibn Khaldūn's tale endorses the report by Shams al-Dīn al-Iṣfahānī's student that new developments took place in the late sixth and in the seventh centuries AH.

33 MS Wetzstein 7 (Staatsbibliothek, Berlin), f. 3a,6–7.

34 Ibid., f. 32a,4–7. Specifically:

> And we said in it that the *imām*, Fakhr al-Dīn al-Rāzī, and those who followed him, such as the judge, Nāṣir al-Dīn al-Bayḍāwī, listed among the conditions for the *mujtahid* that he knows, in addition to that, the science of theoretical speculation [or 'verification'] in order to [be able to] differentiate between false and correct thinking.

35 Ibn Qāḍī Shuhba AH 1399/ AD1979-AH 1400/ AD 1980, 2:220.

36 Ibn Khaldūn 1980, 3:28 (chapter 5, section 13 on *uṣūl al-fiqh*).

37 Ibid., 3:28–29.

Two scholars, Fakhr al-Dīn al-Rāzī and Sayf al-Dīn al-Āmidī, abridged the works of their predecessors and put particular emphasis upon verification and argumentation, with slight differences in their respective approaches.[38] From there, lines may be drawn to Tāj al-Dīn al-Urmawī, Nāṣir al-Dīn al-Bayḍāwī, Jamāl al-Dīn Abū ʿAmr ibn al-Ḥājib and other scholars of the seventh century AH.[39] Thus, due to a tradition inaugurated, to some extent, by Fakhr al-Dīn al-Rāzī and carried on by his direct and indirect students, legal scholars aspiring to become *mujtahid*s had to face the challenge of familiarizing themselves with some of the 'secular' sciences. This challenge was not confined to northern Iran/Afghanistan and Syria plus Egypt. As is indicated by Ibn Khaldūn's report on their spread to northern Africa, as well as his own reading list, it also was faced in some of the lands of Western Islam (Tunis, Bougie, Tlemcen) and became an acceptable part of their educational landscapes despite Ibn Khaldūn's lamentations about the decline of "scientific instruction" in the Maghreb.[40]

Ibn Khaldūn treats in his *Muqaddima* two further alliances between the religious and 'secular' disciplines. The first occurred between logic and *ʿilm al-kalām*. This merger was appreciated by Ibn Khaldūn himself and, or so he claims, by "a large number of scholars (who) followed in their steps [that is, the steps of al-Rāzī, al-Ghazālī and al-Bayḍāwī] and adhered to their tradition."[41] The second alliance occurred between *kalām* and philosophy. Ibn Khaldūn attributed this admixture, to which he clearly objected, to Nāṣir al-Dīn al-Bayḍāwi and unnamed later Persian scholars who did this "in all their works."[42]

Hence, it comes as no surprise that the Ottoman scholar, Tashköprüzade (AH 901–967), declared in his treatise, *al-Risāla al-jāmiʿa li-waṣf al-ʿulūm al-nāfiʿa*, that "the sciences of logic" were "*mabādiʾ al-uṣūlayn* [sic] [the principles (of the sciences) of the foundations/fundamentals]" and that he further insisted that these sciences were indispensable for the confirmation of the necessary, that is, for the proof of God's existence.[43]

No surprise, too, that the two *aṣl* as well as *al-ʿilm al-ilāhī* (the divine science, that is, metaphysics) appear as elements of the 'rational' sciences in the late ninth and tenth centuries AH. Abū Zakariyyāʾ al-Anṣārī (d. AH 926), a legal scholar of the Shāfiʿī rite, composed a rather meagre treatise called *al-Luʾluʾ al-naẓīm fī rawm al-taʿallum wa-l-taʿlīm* on the classification of the sciences. He declares the universe of Muslim education to be composed of four classes of sciences – the *sharʿī* (Sunni legal), *adabī* (belletristic), *riyāḍī* (mathematical) and *ʿaqlī* (rational). The last class of the 'rational' sciences comprises ten members: *manṭiq* (logic), *jadal*, *uṣūl al-fiqh*, *uṣūl al-dīn*, *al-ʿilm al-ilāhī*, *al-ʿilm al-ṭabīʿī*

38 Ibid., 3:29.
39 Ibid.
40 Ibid., 2:428–429.
41 Ibid., 3:52.
42 Ibid., 3:53.
43 MS Sprenger 1823 (Staatsbibliothek, Berlin), f. 5b,12–14.

('natural philosophy'), *al-ṭibb* (medicine), *al-mīqāt*, *al-nawāmis* (the laws or the talismans?) and *al-kīmiyā'* (alchemy/practical chemistry).[44]

It is no wonder, then, that Muḥammad al-Muḥibbi had no qualms when writing, in his biographies of eleventh-century AH scholars, that a certain savant had "composed useful works in which the discourse on the 'rational' and the 'traditional' [sciences] was united."[45] Thus, while boundaries demarcating distinct disciplines continued to separate the 'rational' from the 'traditional' sciences, the process of teaching and explaining the former in textbooks reduced the number of barriers and integrated them under a common umbrella. This common umbrella is often characterized as *fā'ida* (benefit) or *naf'* (use), two terms heavily saturated with Muslim values concerning virtuous behaviour and, occasionally, as *baḥth* (organized study).

The respectability of the 'secular' sciences

When I said in the first section of this paper that the 'ancient' or 'rational' sciences were always present in Muslim educational landscapes, albeit not prominently so, I did not mean to imply that their lack of prominence reflected a lack of respectability. The respectability of these disciplines is expressed, first and foremost, by their constant presence in the encyclopaedias and in the *marātib al-'ulūm* (ranks of the sciences) literature. It is also reflected in the inclusion of major representatives of the ancient sciences, such as Abū Naṣr al-Fārābī (about AH 257–339), Abū Ḥayyān al-Tawḥīdī (AH 310/320-?414), Ibn Sīnā (d. AH 429), 'Abd al Laṭīf al-Baghdādī (AH 558–629), Naṣīr al-Dīn al-Ṭūsī, Quṭb al-Dīn al-Shīrāzī and Abū l-Fidā' (AH 672–732), in biographical dictionaries of various kinds. Third, the *ṭabaqāt* literature further demonstrates that all of the Sunni schools of jurisprudence, as well as the Sufi orders, accepted these disciplines into their educational worlds. And, fourth, the respectability of the so-called 'rational' sciences is signified by their linkage, on equal footing, with the religious sciences in such expressions as "*al-'ulūm al-naqliyya wa-l-'aqliyya* (the traditional and the rational sciences)."

Moreover, social sanction *is* also attested by the esteem that a certain discipline and mastery over it might lend to a particular scholar. Shams al-Dīn al-Sakhāwī (d. AH 902), for instance, speaks of Shihāb al-Dīn ibn al-Majdī (AH 767–850) as "foremost among men in the branches of arithmetic, geometry, astronomy, the calculation of inheritance shares and the science of time, without dispute."[46] Muḥammad ibn Sulaymān, called al-Kāfiyajī (born before AH 790–879), is

44 MS Sprenger 1056 (Staatsbibliothek, Berlin), f. 31b.

45 al-Muḥibbī n.d., 1:171.

46 al-Sakhāwī n.d., 1:300. Al-Sakhāwī was a disciple of Ibn Qāḍī Shuhba; see *Encyclopaedia of Islam*, 2nd ed., 3:814. He had a biting tongue and criticized some of his contemporaries sharply for real or invented misdeeds or ideas not shared by the biographer; see *Encyclopaedia of Islam*, 2nd ed., 8:881–882.

described as "the first of the era, the jewel of the time and the fame of the period" in the 'rational' sciences, including mathematics and philosophy.[47] And it is said about a certain Aḥmad ibn Muḥammad al-Makhzūmī (AH 789–827) that "he excelled in law, the calculation of inheritance shares, arithmetic and astronomy, and [displayed] good behaviour in affairs."[48]

Yet another indication of the respectability of the ancient or 'rational' sciences may be found in the professional careers followed by scholars who had studied one or more of these disciplines and the social status that they attained. After their studies were completed or in conjunction with teaching duties, students of the 'rational' sciences (including mathematics, astronomy, or philosophy) faced no obstacles to becoming judges, even at the highest levels, and could be appointed as the chief judge of a town, a province or even an entire country. Similarly, they could also rise to head teaching institutes and might ultimately amass as many appointments as any of their colleagues who had confined themselves to the study of religious disciplines.

The substantial number of students linked to specific teachers of particular 'rational' disciplines, who are frequently mentioned in various biographical entries, attests to the respect accorded to them and to a broad acceptance of the act of studying such topics with them. Above, I have already referred to Shams al-Dīn al-Iṣfahānī and his students. Other examples are ʿIzz al-Dīn ibn Jamāʿa (AH 749–819) and Shihāb al-Dīn ibn al-Majdī. Ibn Jamāʿa was praised by the biographers as "shaykh of Egypt in the 'rational' sciences" and as "miracle of miracles in the knowledge of the literary sciences, the 'rational' sciences and the two *aṣl*."[49] In addition, he had a profound knowledge of medicine and surgery.[50] It is said that he composed treatises in twenty different sciences, among them *adab*, *raml* (sand divination), *ḥadīth* and *kalām*.[51] According to Ibn Qāḍī Shuhba, he taught *ḥadīth* and *uṣūl*.[52] In Shams al-Dīn al-Sakhāwī's biographical dictionary, *al-Ḍawʾ al-lāmiʿ*, he appears, moreover, as a teacher of the 'rational' sciences.[53] Ibn al-Majdī was *muwaqqit* (timekeeper) at the Azhar mosque in Cairo and the

47 al-Sakhāwī n.d., 7:261.

48 Ibid., 2:135.

49 Ibn Qāḍī Shuhba AH 1399/ AD1979-AH 1400/ AD1980, 4:61–62.

50 Ibid., 4:62.

51 Ibid., 4:62–63.

52 Ibid., 4:64, 128 and 144.

53 Al-Sakhāwī reports, for instance, about Aḥmad ibn Muḥammad Shihāb al-Dīn al-Qāhirī, called Ibn al-Qurdāḥ (born after AH 780), that he served as a companion to ʿIzz al-Dīn ibn Jamāʿa in the study of such arts as music (al-Sakhāwī n.d., 2:142). Concerning Muḥammad ibn Aḥmad Shams al-Dīn Abū ʿAbd Allāh al-Bisāṭī (AH 760–796), he says that he had travelled to Cairo in order to study with Nūr al-Dīn al-Jalawī al-Maghribī al-Mālikī, serving him as a companion for ten years in *fiqh*, the 'rational sciences and other disciplines. When his teacher fell ill, he recommended him "to read the 'rational' [sciences] with ʿIzz al-Dīn ibn Jamāʿa. Thus, he served him in what he read from the sciences, the 'rational' ones and the 'traditional' ones" (al-Sakhāwī n.d., 7:5).

head of the teachers at the *madrasa* al-Jānibākiyya al-Dawādāriyya.[54] He wrote three mathematical and 21 astronomical treatises that still exist.[55] According to al-Sakhāwī, Ibn al-Majdī taught at least 20 students in arithmetic, geometry, algebra, astronomy, timekeeping, astronomical instruments and philosophy.[56] Some of his students later became teachers of the mathematical and astronomical disciplines as well. One of them, al-Sayyid ʿAlī al-Faraḍī, was even given the epithet "tilmīdh [student of] Ibn al-Majdī."[57]

The lasting impact that the 'secular' sciences had upon the scholarly mentality of the Muslim world and preserved well into the early modern period is evinced by other epithets, such as Imam Ibuqrāṭ or Shaykh Uqlīdis. Praising Muslim scholars as the Hippocrates, Plato, Galen, or Ptolemy of their age was a none too rare element of educational rhetoric.[58] Not only medieval Muslim, but ancient Greek representatives of mathematics, astronomy, medicine and philosophy were accepted as authoritative sources of information, argumentation and proof, even in works not devoted to these sciences. This attitude was not limited to the early centuries of Islam.

Indeed, the evidence for the overall respectability of the 'rational' sciences may be detected in works by their reputed enemies, such as Jalāl al-Dīn al-Suyūṭī (AH 849–911). One of his texts treats the history of Egypt, devoting two sections to the ancient sciences. The first section occurs in the context of Egypt's pre-Islamic history and consists of a brief survey of sages, philosophers and physicians who allegedly worked in Egypt, at least for a while. In it, we find such names as Hermes Trismegistos, Pythagoras, Plato and Empedocles.[59]

The second section is embedded in a very long series of chapters on the intellectual history of Islam in Egypt and is called "Dhikr man kāna bi-miṣr min arbāb al-maʿqūlāt wa-ʿulūm al-awāʾil wal-ḥukamāʾ wa-l-aṭibbāʾ wa-l-munajjimīn."[60] The section embraces five folios and is thus much longer than others on such figures as Ḥanbalī legal scholars, historians, preachers, or storytellers. It is almost as long as the sections on the Ḥanafī and Mālikī jurists.[61]

54 Ibid., 1:301. Ibn al-Majdī was installed in this position upon the order of Sultan al-Ashraf. He introduced Sufism into the *madrasa*, an act made possible by the great age of its benefactor, Jānibāk al-Dawādār (d. AH 833), who relied heavily upon the scholar and his fame. Berkey presents the Jānibākiyya as one of the numerous examples of the terminological fluidity that marked the educational landscape of Mamluk Cairo. According to him, the Jānibākiyya was called a *madrasa* by some Mamluk authors and a *jāmiʿ* by others. See Berkey 1992, 49.

55 Matvievskaya and Rozenfel'd 1983, 2:490–492.

56 al-Sakhāwī n.d., 1:310, 317, 376 et al.; 2:142, 183 et al.; 7:48, 115, 116, 155 et al.

57 Ibid., 1:82 et al.; 7:115 et al.

58 Al-Sakhāwī, for instance, calls Quṭb al-Dīn al-Rāzī, "Aflāṭūn zamānihi (the Plato of his time)"; see Ibid., 2:197. Al-Muḥibbī quotes his father as saying about the head-physician of Damascus, Ibn al-Ghazāl al-Ḥimṣi, that he was "Ibuqrāṭ waqtihi wa-zamānihi wa-Jālīnus ʿaṣrihi [the Hippocrates of his time and era and the Galen of his century]"; see al-Muḥibbī n.d., 4:299.

59 MS Petermann I 185 (Staatsbibliothek, Berlin), f. 21a–22b.

60 Ibid., 191b, 3–4.

61 Ibid., 156a–171b, 196a–198a.

The earliest of the scholars mentioned in this section are figures from the late second century AH and the last is al-Suyūṭī's own teacher, al-Kāfiyajī. Al-Kāfiyajī taught the two *aṣl*, *tafsīr*, grammar and other philological disciplines, logic, astronomy, geometry, philosophy, dialectics, spherical trigonometry and optics, including burning mirrors. He also was proficient in law and medicine.[62] Al-Suyūṭī's chapter on the scholars of the ancient and 'rational' sciences in Egypt contains the names of three Christian and 12 Muslim physicians, one astrologer, five scholars who engaged in the sciences of the ancients (*hay'a* or mathematical cosmography, *handasa* or geometry, *ḥikma*, *manṭiq*, *riyaḍiyya*, *ṭibb* and *falsafa*), nine scholars simply linked to the 'rational' sciences, eight scholars who combined the two *aṣl* with one or more of the ancient or 'rational' disciplines, four scholars who were famous for their knowledge in *uṣūl*, *kalām* and *fiqh*, and five scholars who simply knew many sciences, were highly educated, or had an excellent knowledge of the philological disciplines.

Thus, al-Suyūṭī's description reflects the amalgamation between the two *aṣl* and logic, as well as some of the other 'rational' sciences. Furthermore, it shows that *'ilm l-kalām*, too, was tentatively regarded as a 'rational' science. It also implies that the borderlines between the various classes of the sciences had indeed become fluid, allowing the philological sciences to be included among the 'rational' sciences. Moreover, al-Suyūṭī specifies that, even in the eighth century AH, individuals were engaged in *falsafa*, medicine was taught in the Ibn Ṭūlūn mosque, and scholars that he mentions solely because of their fame in the 'rational' science were teaching at various institutes.[63] The following statement by al-Suyūṭī about Shaykhzāde Khuzbānī (d. AH 808) is one of those that come closest to indicating that such disciplines were indeed taught within the physical confines of a teaching institute:

> He was excellent in the 'rational' [sciences] and *hay'a*, *ḥikma*, logic and Arabic. He (composed] treatises and was able to solve difficulties. [Sulṭān] Barqūq demanded him from the ruler of Baghdad in order to install him as the head of the *shaykh*s of the Shaykhūniyya in the place of al-Kilistānī.[64]

Thus, the fame of scholars well-versed in the 'rational' sciences – and expressly in astronomy, *ḥikma* (philosophy?) and logic – was enough to move Mamluk sultans to write urgent letters in order to acquire their services from neighbouring rulers or simply command their transfer from one town to another. Contrary to past practice, the Mamluk sultan in question did not wish to install the scholar in his court, but to appoint him over a teaching institute and its faculty.

62 al-Sakhāwī n.d., 7:261.

63 MS Petermann I 185 (Staatsbibliothek, Berlin), ff. 194a,19, 24–25; 194b,9–10, 28–29; 195a, 3–5.

64 Ibid., f. 195a, 3–6.

Where were the 'secular' sciences taught or studied?

Berkey has already pointed out that biographical dictionaries very seldom indicate where teachers taught, whether by reading texts with their students or by sharing their own thoughts through conversations or lectures.[65] Thus, it is extremely difficult to find an unequivocal statement to the effect that a certain 'rational' discipline was studied or taught at a specific locale. Other scholars have taken this vagueness to mean that these sciences were confined to private houses and banned from *madrasa*s and cognate teaching institutes.[66] Yet, there are almost no clear references in the dictionaries to support this position. Indeed, it seems apparent that the locus of teaching simply did not matter nearly as much as the teaching itself. As a consequence, when we try to assess the places available for the 'secular' sciences in Muslim educational landscapes, it is almost irrelevant to ask whether these sciences were taught in a teaching institute, a private house, or a garden.

In several cases, however, biographers write that a subject was taught by an individual living at a teaching institute. Moreover, a particularly compelling example is the case of the *madrasa* al-Jānibākiyya al-Dawādāriyya. Its head was Ibn al-Majdī, who transformed it into a Sufi convent and who taught, as was mentioned above, a number of 'secular' disciplines. Moreover, a certain Muḥammad ibn Aḥmad Shams al-Dīn al-Ṭūlūni (born AH 828) studied logic, philosophy and astronomy, as well as religious and philological disciplines, with another inhabitant of the Jānibākiyya, Mullā ʿAlī, while it was a Sufi convent.[67] The teaching of these disciplines by Ibn al-Majdī and Mullā ʿAlī in a *madrasa* turned into a Sufi convent may represent the extension of a trend already in place during the eighth century AH.

Berkey makes reference to the teaching of *ʿilm al-mīqāt* at Cairo's greatest *madrasa*, which was established by Sultan Ḥasan in the second half of the eighth century AH. The school financed 506 students, among them 400 who studied law according to one of the four major Sunni rites. Additionally, there were paid classes in *tafsīr*, *ḥadīth*, *uṣūl al-fiqh*, Arabic language, medicine and *ʿilm al-mīqāt*. The last class was rather small; according to Berkey, only six students attended it.[68] Since Berkey's source was the foundational deed, however, this number only indicates the number of students planned and provided for by the founder.[69] Actual participation was not necessarily limited to these six students, as Berkey has shown for other classes and subjects.[70]

Study and instruction in the 'rational' sciences were stable features of Muslim educational landscapes. Their embeddedness in these landscapes was ensured

65 Berkey 1992, 88.
66 Makdisi 1981, 77–78.
67 Ibid., 7:113.
68 Berkey 1992, 69.
69 Ibid., 67–69.
70 Ibid., 201–216.

by the networks of relations, by the evolving chains of teachers and by the submission of the pursuit of these disciplines to the very rules and norms that were applied to the religious and the philological sciences. Central elements of these rules and norms were *qara'a* (to read), *ḥafaẓa* (to memorize), *lāzama* (to serve as a companion), *a'āda* (to tutor) and *durūs* (lectures). There is ample evidence for the application of most of these terms to the 'secular' sciences.

Students were the companions of teachers of logic, philosophy, algebra, the mathematical sciences, timekeeping, music, medicine, or the 'rational' sciences in general.[71] In the eleventh century AH, the appropriation of such mentorship by the 'rational' sciences became so widespread that the author of a biographical dictionary, Muḥammad al-Muḥibbī, could proudly claim that his father, too, had taken in companions in the two *aṣl*, *bayān* and logic.[72] Texts on geometry, astronomy, or logic could be studied by *qirā'a*, in other words, upon the students' personal responsibility under the guidance of a tutor or professor.[73] Texts on arithmetic are reported to have been studied by *samā'*, that is, by means of a teacher's voice.[74] Texts in medicine, logic, or algebra were memorized, as were the Qur'ān, books on law, or treatises on grammar.[75] The term *durūs* may also be found in connection with the 'rational' sciences. 'Izz al-Dīn ibn Jamā'a held lectures (*durūs*) on the 'traditional' and the 'rational' sciences and Ibn al-Majdī lectured (*durūs*) upon the calculation of inheritance shares and upon arithmetic.[76]

These patterns were not limited to Cairo, Damascus, Aleppo, or Baghdad. The 'rational' sciences were studied and taught in the same style in Tunis, Jerusalem, Safed, al-Ghazza, Ramla, Mecca and Sana'a, and in small provincial towns, such as Asyut.[77] Even philosophy had a place in the holy cities of Mecca and Jerusalem.[78]

Bones of contention: the sciences of the ancients, religion and faith

There can be no doubt that the position of the sciences of the ancients was heavily disputed in Muslim societies. At least, this is the overwhelming impression conveyed by both the biographical and the *marātib al-'ulūm* literature. Historical chronicles confirm this picture to some extent. The controversy, however, appears to have been largely confined to scholarly discourse. I know of no example where

71 al-Sakhāwī n.d., 1:25–26, 30, 75, 82 and 376; 2:7–8, 60, 142, 195 and 225; 7:5, 44, 174 et al.
72 al-Muḥibbī n.d., 4:212.
73 al-Sakhāwī n.d., 2:175.
74 Ibid.
75 Ibid., 2:225; 7:140. See, also, Berkey on the professorship of medicine at the Ibn Ṭūlūn mosque; Berkey 1992, 29.
76 al-Sakhāwī n.d., 1:251; 2:183.
77 Ibid., 1:363; 2:253; 7:60, 132, 149 et al.
78 Ibid., 7:60.

a scholar engaged in the ancient sciences was executed because of any alleged corruption of his faith, although Shiʿis, Sufis and even Sunnis were prosecuted for heterodox beliefs and were occasionally sentenced to death.

Once in a while, however, material punishments of a lower degree were inflicted: manuscripts were burnt and scholars were dismissed from offices, evicted from schools or homes, or expelled from towns. Occasionally, too, a particular ruler banned philosophy from a court-sponsored school or a religious scholar issued a *fatwā* (legal opinion) in order to suppress – with no lasting success – the teaching of logic and philosophy in a town.

Scholars who disliked, feared, or hated the ancient sciences for any one of a multitude of reasons created a reservoir of rhetorical and narrative tools in order to set safe boundaries to contain these disciplines and their adherents. Their biographical dictionaries preserved, for instance, a limited number of stories illustrating the destructive powers of the ancient sciences and the sad fates of their adherents. These authors coveted and nurtured such stories as rare treasures capable of impressing the minds and souls of believers.

By comparison, the number of scholars with training or even a reputation in the sciences of the ancients who rose to high posts – becoming, for example, *qāḍī al-quḍāt* (high judge), *kātib al-sirr* (private secretary to the ruler), or *nāẓir bayt al-māl* (exchequer) – or occupied teaching chairs or average legal positions is much higher and represents a steady level of achievement by men so educated. This development did not preclude, however, the appearance of fresh stories about the difficulties that adherents of the 'rational' sciences might encounter. When analysing such stories, one must consider carefully whether these troubles were directly or indirectly caused by the persecuted person's study or teaching of such disciplines. While there are cases in which difficulties do seem to have been caused by a scholar's connection with these disciplines, one may also find cases in which this was merely a pretence.

Although certain Muslim scholars, such as Ibn al-Ṣalaḥ (AH 577–643), al-Dhahabī (d. AH 748), or al-Suyūṭī, would have liked nothing better than to rid themselves of the treacherous, dangerous and barely comprehensible sciences of the ancients for once and for all, no Muslim dynasty, religious order (such as the Sufis), or social group (such as the *fuqahāʾ* or legal scholars) ever took concerted action to ban them in practice. However, al-Ghazālī's verdict against philosophy and his introduction of the category of the forbidden sciences were very influential in subsequent writings about the classification of the sciences. The same holds true for his other two categories, those of the useful sciences and the irrelevant sciences. The search for quotations from the Qurʾān or an appropriate *ḥadīth*, became standard means either to defend the permissibility of a specific discipline not explicitly mentioned by the great theologian and mystic or to elevate the status of one that was. Al-Ghazālī's views acquired a strong normative authority and were transformed into an informal institution regulating modes of speech, value systems and interpersonal relationships.

The central accusation uttered against the ancient sciences was their potential to corrupt faith and thus threaten the afterlife of the believer. The biographical literature overflows with remarks about the quality of belief (for example, excellent, good, indifferent, or corrupted) of individual scholars. An illuminating example is the way in which Abū Ḥayyān al-Tawḥīdī is presented. Various biographers suppressed his preoccupation with philosophy or explicitly credited him with sound faith. Others, such as Sibṭ ibn al-Jawzī (d. AH 654) and al-Dhahabī protested against such testimonials, with the former finding delight in the *bon mot* coined much earlier: "There are three heretics in Islam: Ibn al-Rawandī, Abū Ḥayyān al-Tawḥīdī (and) Abū al-ʿAlāʾ al-Maʿarrī."[79] Al-Dhahabī vehemently opposed the favourable assessments of some of his colleagues and maintained that Abū Ḥayyān "was a malicious enemy of God."[80]

Al-Dhahabī also rejected the standard view that Abū Ḥayyān's *nisba*, 'al-Tawḥīdī,' referred to the fact that his father had traded in a particular kind of date, called *tawḥīd*, in Baghdad. In his view, Abū Ḥayyān had adopted the *nisba* for the same reason that Ibn Tūmart had called his followers "*al-muwaḥḥidūn* (believers in the oneness of God)."[81] Thus, al-Dhahabī reminded his readers both of Abū Ḥayyān's leanings toward the teachings of the Sufis and of Ibn Tūmart's pretensions to be a *mahdī*, positions that al-Dhahabī himself heartily despised.

In addition to casting aspersions upon the sincerity of faith of individual scholars, other efforts were taken to domesticate the ancient sciences. One consisted of the constant repetition of stories about the *miḥna* ('inquisition'), the heroic resistance exercised against it by Aḥmad ibn Ḥanbal and a few other firm religious scholars, and the relationship between this event and the introduction of the ancient sciences by Caliph al-Maʾmūn. Lying at the centre of these stories was the issue of the createdness of the Qurʾān, a doctrine that the court and leading Muʿtazilites forced upon the subjects of the early Abbasid Caliphate.

For instance, Tāj al-Dīn al-Subkī (d. shortly after AH 769), a famous author of the Shāfiʿī rite and a son of Taqī al-Dīn al-Subkī, mentioned above, was initiated into the circle of stories about this unfortunate event by his equally famous teacher, al-Dhahabī, who praised Aḥmad ibn Ḥanbal for his steadfastness, which had saved the community. While al-Dhahabī had leanings toward a kind of theological anthropomorphism, al-Subkī was more inclined towards *kalām* and wrote, according to George Makdisi, his great biographical work on the scholars of the Shāfiʿī rite with the goal of persuading them to be more open toward Ashʿarī *kalām*.[82] Despite the substantial differences in religious and scholarly outlook between the two men, al-Subkī shared his teacher's belief that the doctrine of the

79 Quoted after Ibn Qāḍī Shuhba AH 1399/AD 1979-AH 1400/AD 1980, 1:179–180.

80 Ibid., 1:181.

81 Ibid. Ibn Qāḍī Shuhba describes al-Dhahabī's insistence upon Abū Ḥayyān's lack of faith as a "great exaggeration."

82 See *Encyclopaedia of Islam*, 2nd ed., 2:214–216 and 9:744–745.

createdness of the Qur'ān had been the direct result of al-Ma'mūn's interest in philosophy and the ancient sciences and of the efforts of the many scholars well-versed in these disciplines who came to his court.[83]

Al-Dhahabī and Tāj al-Dīn al-Subkī were neither the first nor the last religious scholars who told and retold this story to stir up the passions of believers and admonish them to follow the right path. I have taken the abridged version of this tale from a biographical dictionary on legal scholars of the Hanafi rite compiled by Taqī al-Dīn al-Ghazzī (d. AH 1005 or 1100).

Another, very similar effort to remind believers of their religious duties with regard to the ancient sciences was the repetition of derogatory stories surrounding the *Rasā'il Ikhwān al-Ṣafā'*. As late as the eleventh century AH, the *Rasā'il* were portrayed as an effort to buttress the Ismā'īlī cause by harmonizing religion with philosophy in order to raise, ultimately, the latter above revelation.[84]

A third variant of these efforts to maintain distrust and suspicion against the ancient sciences was the streamlining of incidents surrounding the difficult professional careers of otherwise famous and respected scholars. This was carried out in such a manner that accusations of intrigue due to greed and envy, allegations of arrogance and the disruption of acknowledged networks, and insinuations of courtly power games between overlords, 'vassals' and insignificant neighbours disappeared from most of the reports. The end result was that the wronged scholar was seen as a guilty party whose lack of belief was due to his occupation with the ancient sciences. The exemplary case for this reconstruction of Muslim educational history by some of its biographers is the story of Sayf al-Dīn al-Āmidī.[85]

Conclusions

As I have shown, the 'secular' sciences were present in Muslim educational landscapes between the fifth and eleventh centuries AH under different dynasties and in various geographical areas. The network character of this landscape granted scholars a substantial degree of freedom to choose what to study, with whom and where, as well as what to teach. The all-embracing rhetoric applied to describe these landscapes indicates that the forms essential to the study and teaching of the religious or philological sciences were applied to the 'secular' disciplines as well. Consequently, such forms of mentorship as living for shorter or longer periods of time with a teacher in his home, which was often a teaching institute, and serving him in preparing his lectures, readings and writings were not solely restricted to the religious and philological sciences, that is, those sciences which, according

83 "And he was from among those who cared for philosophy and the ancient sciences. He excelled in them. A group of their scholars flocked around him. And this led him to the claim of the createdness of the Qur'ān." Quoted after the repetition of Tāj al-Dīn al-Subkī's narrative by Taqī al-Din al-Ghazzī AH 1390, 1:346.

84 See al-Muḥibbī n.d., 4:7–8.

85 See Brentjes 1997, 21–33.

to George Makdisi, were the only ones allowed into the *madrasa*. Thus, although there were probably almost no endowed chairs for the 'rational' sciences – except medicine and timekeeping – before the Ottoman and Moghul dynasties, the hybrid character of the entire system, the extension of major elements of these sciences to *ḥisāb* (calculation, arithmetic), *hay'a* and *ḥikma* and the evolution of the *madrasa*s and the houses of *ḥadīth* or Qur'ān into spaces of birth, family and death makes meaningless any differentiation between teaching in a *madrasa* or a cognate institute and teaching in the teacher's house.[86] While the 'secular' sciences did form changing alliances with the religious disciplines, transforming themselves in the process, they remained stable, if minor, components of Muslim educational landscapes.

When we ask what caused their decline – a question looming large in discussions of the history of science in Muslim societies after the seventh century AH – the answer cannot be their exclusion from the *madrasa*, but rather the contrary.[87] Their inclusion in the religiously-dominated educational scenery and their partnership with the religious sciences granted them stable spaces for their existence and niches for their efflorescence.[88] It also subjected them to the same rules of behaviour as the religious sciences in order to be accepted, to impart respectability to students and teachers, and to be transmitted. Consequently, the same modes of writing epitomes, paraphrases, commentaries and supercommentaries applied to all of the sciences, as did the same modes of learning, whether by heart, with a tutor, through serving a master, or by reading authoritative texts, often only in extracts.

Novelty and innovation were not banned from Muslim educational landscapes. Their hybrid character meant, however, that these features were preferably sought and gained through interpersonal relationships. Thus, scholarly texts, whether by ancient Greek or medieval Muslim authorities, were increasingly supplemented and tentatively replaced by new texts derived from earlier ones, but written by the student's teacher or the teacher of his teacher. All of these elements worked together to set stable boundaries around the knowledge studied and taught, boundaries that appear to have constricted over time. They also contributed to the apparent ease with which the Ottoman system absorbed, for at least three and a half centuries, an amazing amount of knowledge from earlier Muslim societies, as

86 See also Berkey's description of the fluid terminology in Mamluk sources concerning the *madrasa* and other centres of education and worship. He concludes that there were no strict boundaries between the various kinds of institute, private houses included; see Berkey 1992, 45–60.

87 I do not claim, of course, that the embeddedness of the 'secular' sciences in the system of education in Muslim societies and their affiliations with religious sciences were the only or the most important causes for shifts in their cognitive contents, methods, goals, questions and modes of practice over time. Other factors drawn from such areas as politics, trade, technology, war, ecology and social structures also contributed to these processes.

88 Perhaps the strongest case for this claim is the efflorescence of astronomy in the form of timekeeping; see King 1983, 534–549.

well as Western Europe, particularly, in the latter case, in the fields of medicine, astronomy, astrology, history, geography and philosophy. Muslim educational landscapes were fluid, but stable, with regard to their discourse and to their ability to accommodate the 'secular' sciences. With regard to the new, early modern foreign sciences, they were stable, but flexible, until most of their centres broke apart in the course of the nineteenth century.

APPENDIX: ARABIC QUOTATIONS

Page 47: al-Muḥiibbī: العلم باب ومفتاحه المسألة. ١

Page ٥٧: al-Sakhāwī: رأس الناس في انواع الحساب والهندسة والهيئة والفرائض وعلم الوقت بلا منازع.

Page 57: description of al-Kāfiyajī: اوحد العصر ونادرة الزمان وفخر هذا الوقت.

Page 57: description of al-Makhzūmī: كان ماهراً في الفقه والفرائض والحسب والفلك حسن السيرة في القضاء.

Page 59: description of Shaykhzāde Khuzbānī: كان فاضلاً في المعقول والهيئة والخكمة والمنطق والعربية. وله تصانيف واقتدار حل المشكلات. طلبه برقوق من صاحب بغداد فولاه مشيخة الشيحونية عوضاً عن الكلستاني.

Note 19: al-Sakhāwī: محمد بن أحمد . . . الشمس ابو عبد الله البساطي . . . لم يزل يدأب العلوم يتطلب المنطوق منها والمفهوم حتى تقدم في الفقه والاصلين والعربية واللغة والمعاني والبيان والمنطق والحكمة والجبر والمقابلة والطب والهندسة والحساب. وصار امام عصره.

Note 19: al-Muḥibbī: السيد محمد كبريت . . . اخذ العلوم الرياضية والحكمية والطبيعية وعلم الحقيقة عن المحقق الكبير عبد الله بن ولي الحضرمي تلميذ لقطب العارف بالله تعالى السيد صبغة الله بن روح الله السندي . . .

Note 21: al-Muḥibbī: . . . ولما قدم العلامة عيسى بن محمد الجعفري المغربي الى مكة لازمه وقرأ عليه العلوم العقلية كالاصلين والمنطق والمعني والبيان والبديع والنحو الصرف . . .

Note 22: Ibn Qāḍī Shuhba: وقرأ الاصلين وسائر المعقولات على علاء الدين الباجى، والمنطق والخلاف على سيف الدين البغدادي . . .

Note 34: MS Wetzstein 7 (Staatsbibliothek, Berlin): . . . وذكرنا فيه ان الامام فخر الدين الرازي ومن تابعه كالقاضي ناصر الدين البيضاوي ذكر من شروط المجتهد ان يعرف مع ذلك علم النظر فيفرق بين الفكر الصحيح والفاسد.

Note 45: al-Muḥiibbī: ألف تأليف مفيدة جامعة فيها ابحاث عقليات ونقليات.

Note 62: al-Sakhāwī: . . . الاستاذ في الاصلين و لتفسير والنحو والصرف والمعاني والبيان والمنطق والهيئة والحكمة والجدل والاكر والمرايا والمناظر مع مشاركة حسنة في الفقه والطب . . .

Note 83: al-Ghazzī: وهو . . . ممن عُنِي بالفلسفة وعلوم الاوائل، ومهر فيها، واجتمع عليه جمعٌ من علمائها، وجَرهُ ذلك في القول بخلق القرآن.

Bibliography

Berkey, Jonathan. 1992. *The transmission of knowledge in medieval Cairo: A social history of Islamic education*. Princeton, NJ: Princeton University Press.

Brentjes, Sonja. 1997. *'Orthodoxy,' ancient sciences, power, and the madrasa ('College') in Ayyubid and early Mamluk Damascus*. Preprint 77. Berlin: Max Planck Institute for the History of Science.

Chamberlain, Michael. 1995. *Knowledge and social practice in medieval Damascus, 1190–1350*. Cambridge: Cambridge University Press.

al-Ghazzī, Taqī al-Dīn. AH 1390. *Ṭabaqāt al-ḥanafiyya*. Vol. 1. Beirut.

Ibn Khaldūn. 1980. *The Muqaddima*. Translated by Franz Rosenthal. 3 vols. 3rd ed. Princeton, NJ: Princeton University Press.

Ibn Qāḍī Shuhba. AH 1399/AD 1979-AH 1400/AD 1980. *Ṭabaqāt al-shāfiʿiyya*. Edited by Al-Hafiz 'Abdul 'Aleem Khan. 4 vols. Hyderabad: Osmania Oriental Publications Bureau.

King, David A. 1983. 'The astronomy of the Mamluks.' *Isis* 74: 531–555.

Makdisi, George. 1981. *The rise of colleges*. Edinburgh: Edinburgh University Press.

Matvievskaya, Galina Pavlovna, and Boris Abramovič Rozenfel'd. 1983. *Matematiki i astronomy musul'manskogo srednevekov'ya i ich trudy (VIII-XVII vv.)*. Vol. 2. Moscow: Izdatel'stvo nauka.

al-Muḥibbī, Muḥammad. n.d. *Khulāṣat al-athar fī aʿyān al-qarn al-ḥādī ʿashar*. 4 vols. Beirut: Maktabat Khayyat.

Sabra, Abd al-Hamid. 1987. 'The appropriation and subsequent naturalization of Greek science in medieval Islam: A preliminary statement.' *History of Science* 25: 223–243.

———. 1994. 'Science and philosophy in Islamic medieval theology: The evidence of the fourteenth century.' *Zeitschrift für Geschichte der Arabisch-Islamischen Wissenschaften* 9: 1–42.

———. 1996. 'Situating Arabic science: Locality versus essence.' *Isis* 87: 654–670.

al-Sakhāwī, Shams al-Dīn. n.d. *Al-Daw' al-lāmiʿ li-ahl al-qarn al-tāsiʿ*. 11 vols. Beirut: Dār maktabat al-ḥayāt.

2

SHAMS AL-DĪN AL-SAKHĀWĪ ON *MUWAQQITS*, *MU'ADHDHINS*, AND THE TEACHERS OF VARIOUS ASTRONOMICAL DISCIPLINES IN MAMLUK CITIES IN THE FIFTEENTH CENTURY

Dedicated to David A. King on the occasion of his retirement in 2007

In appreciation of David's contribution to the history of science in Islamic societies and of his intellectual generosity over many years I will take up in this paper some of his questions about the context of the sciences in certain Islamic societies and discuss them on the basis of sources other than those that he focused on. They complement the picture he has built, modify it and occasionally challenge it.

David has written about the new astronomical professional, the *muwaqqit*, who begins to be visible in the sources in the late thirteenth and early fourteenth century, and the work of these men, particularly in Mamluk and Ottoman Egypt and Syria.[1] Trying to contextualize the new profession David investigated the relation-

1 See, in particular, David A. King, "On the role of the muezzin and muwaqqit in medieval Islamic societies". In F. Jamil Ragep, Sally P. Ragep, with Steven J. Livesey (eds.), *Tradition, Transmission, Transformation: Proceedings of Two Conferences on Premodern Science Held at the University of Oklahoma*. Leiden, New York & Cologne: E.J. Brill, 1996, pp. 285–346; modified reprint: *In Synchrony with the Heavens: Studies in Astronomical Timekeeping and Instrumentation in Medieval Islamic Civilization. Volume One: The Call of the Muezzin*. Leiden, Boston: E.J. Brill, 2004, pp. 631–677. The page references in my paper are to the first version. See also the earlier paper David A. King, "The astronomy of the Mamluks". *Isis* 74 (1983), 531–555 reprinted in David A. King, *Islamic Mathematical Astronomy*. Aldershot: Variorum, 1986, item 3, which discusses the major Mamluk writers on astronomy and the types of astronomy they engaged with. It also notes their affiliation to mosques and madrasas as *muwaqqits* and, occasionally, as teachers.

DOI: 10.4324/9781003372455-3

ship between the *muwaqqit*s and the *mu'adhdhins* – asking which practices were used for determining the prayer times, the rules of the *fuqahā'* or the procedures of the *muwaqqits*.[2] He sought to determine the profile and practice of a *muwaqqit* in addition to establishing what kind of tables and instruments they have left us.[3] He also asked where in the Islamic world and when did *muwaqqit*s operate.[4]

While David focused primarily on texts written by *muwaqqit*s, the tables they devised and used and the instruments they constructed, which he called the primary sources, supplementing them by legal and other texts, I want to discuss his three questions from the angle of a different kind of source – biographical dictionaries of the educated elite during Mamluk rule.[5] I will limit myself here to information and views offered in one such dictionary, *al-Ḍaw' al-lāmiʿ li-ahl al-qarn al-tāsiʿ* (*The shining light on the people of the ninth century*) by Shams al-Dīn al-Sakhāwī (830–902/1427–1497), one of the most important *ḥadīth* transmitters and historians of the fifteenth century. The main reason for this focus is a point of methodology. I wish to show how our answers on questions that we ask are shaped by the genre of the sources we use, their specific character, the beliefs, methods and practices of their authors and our own beliefs, methods and practices. The persona of the *muwaqqit* in David's work is mainly presented as a well-defined astronomer "in the service of Islam" who was attached to mosques, produced a huge, impressive body of astronomical tables as well as other written and instrumental work in the domain of *ʿilm al-mīqāt*, which the *muwaqqit* taught to students who later often also became *muwaqqit*s. This new professional was, in David's opinion, however unable to influence the practice of prayer times of the *mu'adhdhin*, which remained guided by the views and positions of the *fuqahā'*. I wish to argue that when we look in a different kind of source, i.e., biographical dictionaries, this persona loses its clear contours, takes on different shades and appears much more to be a *mudarris* at a *madrasa* teaching a broad range of themes and treatises than being actively involved in regulating prayer times and creating or repairing instruments at mosques. This argument and its material

2 He claimed that the rules of the *fuqahā'* ("folk" or "ethno-" astronomy) generally prevailed over those of the *muwaqqit*s. King, *On the Role of the Muezzin*, p. 288. An outline of the content of this 'folk'-astronomy and David's views on its relationship to mathematical astronomy see King, *Synchrony*, pp. 465–475.

3 He briefly defined this profile as that of "the mosque official responsible for regulating the times at which the muezzin should perform." King, *On the Role of the Muezzin*, p. 286.

4 He answered this third question by stating that astronomical timekeeping "involving complicated – and sometimes highly sophisticated – procedures [. . .] seems to have been restricted mainly to Egypt (thirteenth century onwards), Syria, the Yemen, and also Tunis (fourteenth century onwards), and finally Istanbul (fifteenth century onwards). There is little evidence on the practice in the rest of the Islamic world, notably the Maghrib, Iraq, Iran, India, and Central Asia." King, *On the Role of the Muezzin*, p. 288. See also King, *On the Role of the Muezzin*, pp. 298–300. For the wealth of tables calculated for timekeeping and the determination of the direction of prayer direction and the mathematical methods developed in this context see King, *Synchrony*, pp. 1–456.

5 King, *On the Role of the Muezzin*, p. 288.

underpinning do not mean, of course, that teaching *'ilm al-mīqāt* was not an important component of the classroom work of this *muwaqqit-mudarris*. It rather means that from the perspective of the biographers it was not the most important element of this teaching. But not only is the ranking modified; the status of *'ilm al-mīqāt* itself shifts. It is not primarily the training of a body of professionals that the teacher of this discipline achieves. *'Ilm al-mīqāt* appears in al-Sakhāwī's dictionary rather as part of the general education given to members of the educational elite in the first place, but also to visitors in search of knowledge, well-off merchants and occasionally sons of the military elite. Biographical dictionaries yield a similar effect of displacement with regard to the *mu'adhdhin* and the spread of *'ilm al-mīqāt* towards the east. The *mu'adhdhin* moves away from his clear affiliation with or subordination under the *fuqahā'*, while *'ilm al-mīqāt* moves east as part of pilgrimage, commerce and travel for knowledge. In addition, biographical dictionaries point to phenomena less visible in the manuscripts and on the instruments. According to al-Sakhāwī, an exchange of different types of mathematical, astronomical and astrological skills and practices takes place in which Iranian scholars played an important role. In addition, some scholars from Anatolia participated in this transfer. Biographical dictionaries also add new information about the social ranking of those we consider mainly as *muwaqqits* and teachers of arithmetic and algebra among the educational elite at large and about their interaction with the ruling military aristocracy.

1 *Muwaqqits*

The entries in al-Sakhāwī's *al-Ḍaw'* range from one line merely stating the name and the *nisba al-mīqātī* to substantive descriptions of origin, family, education, reputation, social, professional and other activities. The briefest type are statements like 'Umar b. 'Abd al-Raḥmān al-Zawāwī al-Mīqātī (d. 885 h/1480–81).[6] One of the longest entries for a person professionally involved with *'ilm al-mīqāt* is that of Aḥmad b. Rajab b. Ṭaybughā, Shihāb al-Dīn, known as Ibn al-Majdī (767–850/1365–1447).[7] Ibn al-Majdī's biography and the information given about him in biographies of his students neatly depict the shift in persona I referred to in the introduction to this paper. Al-Sakhāwī reports about him that he mostly taught *'ilm al-farā'iḍ* and *ḥisāb*, then algebra and only then *'ilm al-mīqāt* writing profusely about these four disciplines. He is said to have taught occasionally *hay'a* and *handasa*, although he apparently did not write about these topics. But he wrote treatises on *ḥadīth* which he apparently never taught.[8] In addition, he also taught *fiqh*, *uṣū al-fiqh*, Arabic and even *ḥikma*. The enumeration of *ḥikma* as one among other topics of teaching in fifteenth-century Cairo challenges the widespread

6 Shams al-Dīn al-Sakhāwī, *al-Ḍaw' al-lāmi' li-ahl al-qarn al-tāsi'*. Bayrut, s.d., vol. VI, p. 90.
7 Al-Sakhāwī, *al-Ḍaw'*, vol. I, pp. 300–302.
8 Al-Sakhāwī, *al-Ḍaw'*, vol. I, p. 301.

assumption about the death of philosophy and related matters at least in the Arab-speaking parts of the Islamic world after al-Ghazālī's verdict on the four types of philosophical infidelity and subsequent declarations against non-religious fields of knowledge such as the infamous *fatwā* of Ibn al-Ṣalāḥ in Ayyubid Damascus against teaching anything but the religious sciences. The content of Ibn al-Majdī's teaching of *ḥikma* remains, however, opaque.[9] In *fiqh* or grammar, Ibn al-Majdī was famous for the quality of his reading of *al-Ḥāwī*, which he had learned by heart.

Members of all *madhhabs* and classes studied with Ibn al-Majdī because of the benefits he brought to them. Among those who served (*lāzama*) him were al-Sakhāwī's Shaykh Ibn Khiḍr and nine other scholars, only three of whom later also taught *al-farā'iḍ*, arithmetic, algebra, *'ilm al-mīqāt* and other disciplines. Al-Sakhāwī did not try to give a complete list of those who served Ibn al-Majdī. In other entries he names Ibn al-Majdī as served by scholars not mentioned in the biography dedicated to him.[10] This kind of incompleteness needs to be taken into consideration for any kind of information, i.e., silence in biographical dictionaries does not lend itself to infer, for instance, that a certain discipline or work was not taught.

Although the Mamluks did not sponsor astrology and astronomy officially at their courts, they did so individually. They also included some of these disciplines in their patronage for madrasas, mosques and Sufi convents. David has pointed to a certain Ibrāhīm al-Ḥāsib al-Malikī al-Nāṣirī who is supposed to have compiled an astrological work in Cairo around 1358 and has worked for Sultan al-Nāṣir Aḥmad.[11] Sultan al-Nāṣir Muḥammad b. Qalā'ūn consulted astrologers and geomancers in addition to doctors when he fell ill.[12] Ibrāhīm b. Muḥammad al-Qurashī from Ghazza, known as Ibn Zuqqā'a (1344–1407), a teacher of Ibn Ḥajar, did also some study of astrology, *'ilm al-ḥarf* and the usefulness of plants and herbs which he searched for by traveling across the countryside. He became a Sufi with considerable powers whose fame spread widely. He was invited repeatedly to the festival of Mawlūd. Several Mamluk sultans favoured him, among them Sultan al-Nāṣir Faraj who made him move to Cairo where he lived close to the Nile. "He became very close to al-Nāṣir so that (the Sultan) did not leave for travels unless (Ibn Zuqqā'a) cast the horoscope for him and he did not (miss) the time (Ibn Zuqqā'a) specified for him."[13] Sultan al-Mu'ayyad, however, proved hostile

9 This also applies to most of the entries in Sakhāwī's dictionary in which *ḥikma* is mentioned as part of the learning or teaching. In a few entries relating to Cairo works by Ibn 'Arabī, Suhrawardī and Mawlānāzāde are mentioned. In entries about Iranian and some Anatolian scholars, Ibn Sīnā, Naṣīr al-Dīn al-Ṭūsī, Athīr al-Dīn al-Abharī, Qutb al-Dīn Rāzī, Sa'd al-Dīn al-Taftāzānī, Ḥāfiz al-Dīn al-Taftāzānī, al-Sharīf al-Jurjānī and students of the last three named scholars appear. Al-Sakhāwī, *al-Ḍaw'*, vols. II, pp. 196, 197, 310; VI, pp. 184, 187, 190; VII, p. 261; VIII, pp. 127, 224 et al.

10 Al-Sakhāwī, *al-Ḍaw'*, vols. I, pp. 310–311, 376–377; II, pp. 6–7; VI, p. 82; X, p. 226.

11 King, *The Astronomy of the Mamluks*, pp. 535, 550.

12 King, *The Astronomy of the Mamluks*, p. 535.

13 Al-Sakhāwī, *al-Ḍaw'*, vol. I, p. 130.

against him due to his close connection with the previous sultan and dismissed him. Although al-Sakhāwī's formulation is somewhat ambiguous it could be that a part of al-Mu'ayyad's hostility against the Sufi was caused by his astrological counselling of Sultan al-Nāṣir Faraj.[14] Quoting from Ibn Ḥajar, al-Sakhāwī added that the sultan tried him for 'reprehensible' behaviour with many witnesses from among the eunuchs and others, but at the end decided to drop the case.[15] Some of the Mamluks had an undeniable interest in *'ilm al-mīqāt* and other disciplines and entertained relationships with scholars of these fields. Sultan Ḥasan donated a professorship and six studentships for the study of timekeeping at his *madrasa*.[16] The Mamluk Qujmas hired 'Alī b. 'Umar al-Maqsī "who was passionately fond of *al-mīqāt*" as his client for his expertise in *'ilm al-mīqāt*.[17] Ibn al-Majdī had the trust of Sultan al-Ashraf Barsbay who called on him for help in an affair that worried him and made him anxious and helpless. Ibn al-Majdī offered to pray for him and brought back as a good omen an inscription written at the side of the *miḥrāb* from the *madrasa* closest to the fortress. The sultan appointed Ibn al-Majdī as the head of the *Madrasa al-Jānibākiyya al-Dawādāriyya*. As a *mudarris*, Ibn al-Majdī turned the *madrasa* into a Sufi convent because of the testament of its donor. As a widespread habit since the thirteenth century, Ibn al-Majdī made sure that his son in law inherited his position as the head of the convent.[18]

Al-Sakhāwī's other entries confirm by and large the picture set in Ibn al-Majdī's biography. *Muwaqqits* were first and foremost teachers of *farā'iḍ* and *ḥisab*. The third place of their teaching activities is taken by algebra. Only then follows *'ilm al-mīqāt*. The *muwaqqit-mudarrisūn* all were teaching also other fields than mathematics and astronomy, in particular Arabic, grammar, *fiqh* and *ḥikma*. This contradicts in some sense the available evidence of treatises, which these teachers and *muwaqqits* left behind, because those treatises consist to a substantial degree of writings on *'ilm al-mīqāt*. But al-Sakhāwī's record is fairly consistent throughout the volumes of his dictionary for all of those he graced with the *nisbas muwaqqit* or *mīqātī*. A possible interpretation of this contradiction is to assume that in fifteenth-century Cairo and other Mamluk cities teaching focused more on reading and debating than on writing texts.

Only rarely did al-Sakhāwī consider it worthwhile to mention the engagement of those he labelled *muwaqqit* or *mīqātī* with practical affairs such as setting the time in a mosque or a *madrasa* or constructing instruments or scales. Examples can be found in the biographies of Nūr al-Dīn b. al-Naqqāsh, 'Abd al-Khāliq

14 Al-Sakhāwī, *al-Ḍaw'*, vol. I, p. 130.

15 Al-Sakhāwī, *al-Ḍaw'*, vol. I, p. 132.

16 Jonathan Berkey, *The Transmission of Knowledge in Medieval Cairo: A Social History of Islamic Education*. Princeton: Princeton University Press, 1992, p. 69.

17 Al-Sakhāwī, *al-Ḍaw'*, vol. V, p. 265.

18 David A. King, *A Survey of the Scientific Manuscripts in the Egyptian National Library*. Winona Lake, IN: Eisenbrauns, 1986, C62.

al-Ṣāliḥī and Ḥasan b. ʿAlī al-Takhawī al-Qāhirī.[19] While not everybody reported by al-Sakhāwī to have studied *mīqāt* became later a *muwaqqit* or *mīqātī*, he did not give such a *nisba* even to some of those whom he mentioned as having set the time in a mosque or erected sundials.[20]

Hence, in al-Sakhāwī's presentation, the *muwaqqit* does not appear to have been the result of a professional educational focus on *ʿilm al-mīqāt* and related disciplines. Rather, he describes the *muwaqqit* as only one facet of another persona, mostly that of a *mudarris*, but also that of an *imām*, a *khāṭib*, a *wāʿiẓ*, a *muḥtasib* or a physician.[21] The reason for this more complex persona, i.e., for the combination of several posts and professional obligations, is in all likelihood to be found in economic and social factors. As David has shown on the basis of *waqfiyya*s, the post of a *muwaqqit* was not well remunerated, although slightly better paid than that of a *mu'adhdhin*.[22] A *mudarris* in contrast often could reap a fair salary. This alone should have stimulated a healthy interest among those who worked as *muwaqqit*s and taught *ʿilm al-mīqāt* to get an appointment as a *mudarris*. But even the salary of one professorship often did not satisfy the *ʿulamā'* in Cairo, Damascus and other cities of Egypt, Syria and Palestine. Simultaneous holding of several professorships was widespread. It brought higher prestige and more income. It also allowed for establishing a personal network of younger scholars (students, friends, family members) who deputized for the holder of the chairs in the smaller and less well-paid *madrasas* for less than the stipulated salary. This network of patron-client relationships included marrying off one's daughter(s) or niece(s) to the most promising young scholars and choosing the successors for the held chairs, preferably the professor's son(s) or other younger male relative(s). *Muwaqqit*s were no exception of this widespread behaviour.[23] Ibn al-Majdī made his son in law successor of his post as head of the *Jānibākiyya Dawādāriyya*, ʿAlī b. ʿAbd al-Qādir al-Naqqāsh al-Mīqātī was "the leader (of the *muwaqqit*s?) at *al-Maqsī* Friday mosque, *al-Jamaliyya al-Ṣāḥibiyya madrasa* and others than the two, for instance the *Ashrafiyya* tomb of Sultan Īnāl, and taught the art (of *mīqāt*?) at several places" and Muḥammad b. Aḥmad al-Makhzūmī al-Qāhirī, known as Ibn al-Khashshāb, determined the prayer times at the *Ashrafiyya Barsbay madrasa*, the Friday mosque of *al-Ṣāliḥ* and at the *Manṣūriyya madrasa*.[24]

David raised two more issues, one referring to Tāj al-Dīn al-Subkī's (d. 771/1369) claim that the *muwaqqit*s were engaged in astrology and occult sciences, not in science, and hence contributed to lower the religious moral of the Muslim population

19 Al-Sakhāwī, *al-Ḍaw'*, vols. II, p. 25; III, p. 115; IV, p. 41.

20 Al-Sakhāwī, *al-Ḍaw'*, vols. IV, p. 41; V, p. 108; VI, p. 285; VIII, pp. 75–76.

21 Al-Sakhāwī, *al-Ḍaw'*, vols. II, p. 142; III, p. 150; IV, p. 192; V, p. 108; VI, p. 285; VII, p. 44; VIII, pp. 238–239; IX, p. 179; X, p. 95.

22 King, *On the Role of the Muezzin*, pp. 301–303.

23 Berkey, *The Transmission*, pp. 105–127.

24 Al-Sakhāwī, *al-Ḍaw'*, vol. V, p. 242.

of Damascus.[25] The other issue relates to those who produced the yearly ephemerides.[26] In al-Sakhāwī's dictionary only two of the scholars named *muwaqqit* are explicitly mentioned for having studied astrology and possessed great knowledge in astrological procedures and judgments, although we know from David's survey of the scientific manuscripts of the Dār al-Kutub in Cairo that at least Ibn al-Majdī also wrote about astrological topics. Moreover, astronomical handbooks (*Zīj*) as a rule always included some astrological tables. Hence, the scholars who wrote ephemerides and studied *Zījes* encountered astrology at least as a reading matter. Ibrāhīm b. Aḥmad al-Shīrāzī, the *muwaqqit*, was in Alexandria where the *ḥadīth* transmitter al-Jamāl b. Mūsā met him and subsequently described him to al-Sakhāwī as an excellent master and *muwaqqit* who had written works on *ʿilm al-mīqāt* and was gifted in those branches that were connected with it from astrology and other (things). He gave *ijāzāt* to many people.[27] Aḥmad b. Ghulām Allāh al-Kawm al-Rishī al-Qāhirī al-Mīqātī (d. 832/1433) knew how to work with a *Zīj* and wrote ephemerides.[28] He was *muwaqqit* at the *madrasa* donated by Sultan al-Mu'ayyad Shaykh (d. 824/1421).[29] The compilation of ephemerides, which included astrological predictions, was not limited to *muwaqqit*s alone, although other entries indicate that it was apparently one of the skills acquired and taught by *muwaqqit*s. Ibrāhīm b. ʿAlī al-Shambārī al-Makkī, known as al-Zamzāmī, studied "with his brother al-Badr Ḥusayn *ʿilm al-farā'iḍ*, arithmetic, algebra, *hay'a*, *handasa*, *ʿilm al-mīqāt*, the derivation of an ephemeris from the (solar, lunar and planetary tables of a) *Zīj* and the (different) eras."[30] The *falakī* (astronomer) Aḥmad b. Ibrāhīm al-Sarmīnī al-Ḥalabī "was an excellent master of *ʿilm al-hay'a*, the work with the *Zīj* and the construction of ephemeris. He was superior in it, unique in Aleppo in his time from where they used to take his ephemerides to the seat of the deputy (of the sultan) who asked for them."[31] This favour brought him only pain since the rest of the Mamluks ruling in Syrian cities bore down on him accusing him of "thinness of religion, disintegration of his creed, the avoidance of prayer and the drinking of wine."[32] As a result he had to leave Aleppo fearing some of the emirs and move to Safad where he died at the age of 80.[33] The *muwaqqit* Ibn Rāzin who had studied timekeeping with al-Nūr b. al-Naqqāsh in Cairo also "excelled in compiling the ephemeris in its perfection, being unique in his precision of the

25 King, *On the Role of the Muezzin*, p. 307.
26 King, *On the Role of the Muezzin*, p. 317.
27 Al-Sakhāwī, *al-Ḍaw'*, vol. I, pp. 5–6.
28 Al-Sakhāwī, *al-Ḍaw'*, vol. II, p. 62.
29 G. P. Matvievskaya, B. A. Rozenfel'd, *Matematiki i astronomy musul'manskogo srednevekov'ja i ich trudy (XIII-XVII vv)*. Moskva: Nauka, vol. 2, p. 479; King, *Survey*, C41.
30 Al-Sakhāwī, *al-Ḍaw'*, vol. I, p. 86.
31 Al-Sakhāwī, *al-Ḍaw'*, vol. I, p. 204.
32 Al-Sakhāwī, *al-Ḍaw'*, vol. I, pp. 204–205.
33 Al-Sakhāwī, *al-Ḍaw'*, vol. I, p. 205.

times and the desired exactitude. Hence many benefited from him."[34] The Alexandrian *ḥāsib* (calculator/astronomer) ʿAlī b. Aḥmad "spent efforts for (learning) *ʿilm al-mīqāt* and was excellent in the knowledge of how to work with a *Zīj* and in the writing of the ephemeris."[35] Then he turned to alchemy or chemistry, something al-Sakhāwī clearly disapproved of saying that he spent his life with works on "what is between evaporation and distillation etc. But he did not achieve a thing in this (respect)."[36] In one case only, namely that of Khalīl b. Ibrāhīm Abū l-Jūd al-Dimyāṭ al-Qāhirī, al-Sakhāwī provided a few more details when talking about the study of *ʿilm al-mīqāt* and the making of ephemerides saying that it included the study of tables, among them the tables (and diagrams) describing the new moon for each month.[37]

As rarely as a *muwaqqit* dealt according to al-Sakhāwī with astrology did practicing Sufis engage with *ʿilm al-mīqāt*. But ʿUmar b. ʿĪsā al-Samnūdī al-Shāfiʿī was knowledgeable in *farāʾiḍ* and *mīqāt* and is remembered for his *karāmāt* (wonders).[38]

2 Studying *ʿilm al-mīqāt*

Al-Sakhāwī confirms what David has argued on the basis of manuscripts and instruments – later generations of *muwaqqit*s acquired their expertise from earlier generations of *muwaqqit*s, i.e., as a rule they did not study *ʿilm al-mīqāt* with teachers who were not themselves *muwaqqit*s. Only a few cases occur where the teacher of *ʿilm al-mīqāt* is not called by al-Sakhāwī somewhere in his dictionary *muwaqqit*. In such cases, al-Sakhāwī explicitly mentioned the previous study of *ʿilm al-mīqāt* by the teacher.[39]

It is in this sense that one can undoubtedly speak of a profession. The evolution of an analogous kind of professional kinship can be observed in al-Sakhāwī's dictionary for teachers of *fiqh*, *uṣūl al-dīn*, *ḥadīth* and perhaps Arabic, although it is less prominent in the last case. The student body of *ʿilm al-mīqāt* was not limited though to these later *muwaqqit*s. A good number of students who took such classes did not become *muwaqqit*s, i.e., they were either not mentioned by al-Sakhāwī as such or do not seem to have written any treatise in the field. They studied treatises on timekeeping as part of their overall education.

> Zakariyyāʾ b. Muḥammad al-Anṣārī al-Sanbakī al-Qāhirī al-Azharī al-Shāfiʿī al-Qāḍī was born in 826 h in Sanbaka. He grew up in his hometown and received his education there. When he was 15 years old, he

34 Al-Sakhāwī, *al-Ḍawʾ*, vol. IV, p. 189.
35 Al-Sakhāwī, *al-Ḍawʾ*, vol. V, p. 169.
36 Al-Sakhāwī, *al-Ḍawʾ*, vol. V, p. 169.
37 Al-Sakhāwī, *al-Ḍawʾ*, vol. III, p. 187.
38 Al-Sakhāwī, *al-Ḍawʾ*, vol. VI, p. 112.
39 See, for example, al-Sakhāwī, *al-Ḍawʾ*, vol. VIII, pp. 75–76.

went to Cairo for further education. He lived for a while at al-Azhar, went home, and then came back for more education . . . He studied *fiqh* with al-Qayatī and al-ʿIlm al-Bulqīnī . . . He took *ʿilm al-hay'a, handasa, mīqāt, farā'iḍ, ḥisāb*, algebra and other things from Ibn al-Majdī. He read with him parts of his works. He also took *farā'iḍ* and *ḥisāb* from al-Shams al-Ḥijāzī and al-Butījī and also on Abū l-Jūd al-Yanabī. He read with him *al-Majmūʿa* and *al-Fuṣūl*. He took *ḥikma* with al-Shirwānī and Jaʿfar, mentioned previously, and medicine with al-Sharaf b. al-Khashshāb and *al-ʿurūḍ* with al-Warūrī and *ʿilm al-ḥarf* with Ibn Qurqmas al-Ḥanafī and *taṣawwuf* with Abū ʿAbdallāh al-Ghumārī et al. . . . And he also heard things from al-ʿIzz b. al-Furāt and from Sarah bint Jamāʿa on *al-Muʿjam al-kabīr* by al-Tibranī in my reading . . .[40]

Muḥammad b. ʿAwad . . . al-Sikandarī . . ., known as Junaybat. He was born in Alexandria in 788 h. He read there the Qur'ān . . . he also studied with al-Muʿizz *ʿilm al-mīqāt* and the beginnings of Uqlīdis . . .[41]

Muḥammad b. Muḥammad al-Shams al-Ḥalabī al-Ḥanafī, . . ., known as Ibn Amīr Ḥajj and Ibn al-Muwaqqit. He was seriously interested in *mīqāt* and led this at the Great Friday Mosque of Aleppo. He lived as a student at the *Halwiyya*. Then he transferred to teach after his father at the *Jaradakiyya* and settled there. Then he led the office of the registrar with the judges in Aleppo. Then he became the tax collector of the markets.[42]

While the students of *ʿilm al-mīqāt* in the first two examples given above studied with one teacher only, several students took such classes with two, three or even more teachers. Judging by the names of the teachers given by al-Sakhāwī, there was evidently a desire to have had classes with a good number of the most prominent *muwaqqit*s and *mīqāt* teachers of a given town and time. The content of the classes may not have mattered as much, since in some cases the same texts were read.[43]

Prominent scholars such as al-Sakhāwī himself, his teacher Ibn Ḥajar al-ʿAsqalānī and his eminent colleague in the field of history Taqī al-Dīn al-Maqrīzī also took classes in *ʿilm al-mīqāt*.[44] The latter studied it under Ibn Khaldūn together with instructions on the astrolabe and the occult sciences *raml* and *zaʿiraja*.[45] Sons of Mamluks apparently were not much drawn to *ʿilm al-mīqāt*. The only two examples given by al-Sakhāwī are ʿAlī b. Sūdūn al-ʿAlā' al-Yushbughāwī

40 Al-Sakhāwī, *al-Ḍaw'*, vol. III, p. 235.

41 Al-Sakhāwī, *al-Ḍaw'*, vol. IX, pp. 272–273.

42 Al-Sakhāwī, *al-Ḍaw'*, vol. IX, pp. 72–73.

43 Al-Sakhāwī, *al-Ḍaw'*, vols. II, pp. 66, 172; III, p. 103; V, pp. 8, 108; VI, pp. 162, 285; VII, pp. 75, 82; IX, p. 254.

44 Al-Sakhāwī, *al-Ḍaw'*, vols. II, p. 24; V, p. 19; VIII, p. 4.

45 Al-Sakhāwī, *al-Ḍaw'*, vol. II, pp. 21–25, in particular p. 24.

(c. 810–86/) and Qāsim b. Qutlūbughā (802–879/1399–1474).[46] Two sons of Mamluks only are known who actually worked as a *muwaqqit* – ʿAlā' al-Dīn ʿAlī b. Ṭaybughā al-Dawādār al-Baklamishī at the end of the fourteenth and beginning of the fifteenth centuries and Ibrāhīm b. Qāyt Bāy in the early sixteenth century discovered by David.[47] A famous *muwaqqit*, i.e., Ibn al-Majdī, was a grandson of a Mamluk. Other sons of Mamluks such as Yūsuf b. Qurqmas al-Ḥamzāwī whose scientific treatises are extant or who are named by al-Sakhāwī as authors of such works focused exclusively on other astronomical disciplines such as *ʿilm al-falak* and *ʿilm al-hay'a* or on astrology and occult topics such as number magic.[48] Ibn Qurqmas' astrological history of events, which happened during the campaign in the region of Adana from 875 to 877, has been identified by David.[49] Mamluks themselves as men of the military are not very likely candidates for studying, teaching or writing treatises on astronomical topics, whether *ʿilm al-mīqāt* or any of the other disciplines. But there was at least one Mamluk who did so. ʿAlā' al-Dīn Ṭaybughā al-Dawādār al-Baklamishī in the fourteenth century wrote on astronomical instruments and compiled a table for the direction of prayer for each degree of latitude and longitude.[50]

3 *Mu'adhdhins*

Based on sources from different regions and periods, David concluded that *mu'adhdhins* mostly used simple shadow-schemes for time-reckoning and thus had little knowledge of mathematical astronomy.[51] Several *mu'adhdhins*, however, were in the same time *muwaqqits*. It can even be speculated that the profession of a *muwaqqit* evolved from that of a *mu'adhdhin*. Several authors of astronomical works, timekeeping and other tables and makers of astronomical instruments of the fourteenth century held positions in both fields. Muḥammad b. Aḥmad al-Mizzī (690–750/1291–1349) studied in Cairo and worked in Damascus. He was a *mu'adhdhin*, but also appointed as *muwaqqit* in a small town near Damascus

46 Al-Sakhāwī, *al-Ḍaw'*, vol. VI, pp. 184–190.

47 King, *On the Role of the Muezzin*, p. 308; King, Survey, C54; David A. King, "Universal solutions to problems of spherical astronomy from Mamluk Egypt and Syria". In F. Kazemi, R. D. McChesney (eds.), *A Way Prepared: Essays on Islamic Culture in Honor of Richard Bayly Winder*. New York: New York University Press, 1988, pp. 153–184, in particular p. 163; reprinted in David A. King, *Astronomy in the Service of Islam*. Aldershot: Variorum, Ashgate, item VII and King, Synchrony, pp. 711–739.

48 King, Survey, C91; Al-Sakhāwī, *al-Ḍaw'*, vols. III, pp. 115, 235; V, p. 217.

49 King, Survey, C91; David A. King, "Mathematical geography in fifteenth-century Egypt: An episode in the decline of Islamic science". In Anna Akasoy, Wim Raven (eds.), *Islamic Thought in the Middle Ages: Studies in Text, Transmission and Translation, in Honour of Hans Daiber* (Islamic Philosophy, Theology and Science. Texts and Studies 75). Leiden: E.J. Brill, pp. 319–344.

50 Matvievskaya, Rozenfel'd, *Matematiki i astronomy*, vol. 2, p. 465; Charette, *Mathematical Instrumentation*, pp. 18–19; King, Survey, C53; King, Universal Solutions, pp. 169–170.

51 King, *On the Role of the Muezzin*, pp. 291–298.

and later at the Umayyad Mosque in Damascus. He constructed astrolabes and quadrants, at least one of which is extant today in St Petersburg. He also wrote several, today extant treatises on such instruments.[52] The famous fourteenth-century *muwaqqits* Ibn al-Shāṭir and Shams al-Dīn al-Khalīlī also were *mu'adhdhins*.[53] Entries in al-Sakhāwī's dictionary confirm too that in the fourteenth century some *muwaqqits* in Egypt and Syria worked as *mu'adhdhins* and some *mu'adhdhins* studied *mīqāt* texts. A certain Muḥammad b. ʿAlī al-Sikandarī al-Shāfiʿī, who lived in Alexandria (731–807/1330–1404), is called by al-Sakhāwī *al-mu'adhdhin al-muwaqqit*.[54] Mūsā b. Muḥammad al-Sharaf al-Muwaqqit (d. 807/1404), a nephew of Shams al-Dīn al-Khalīlī, is described as "the most excellent of those who are left in al-Shām in *ʿilm al-hay'a*. In his hands was the leadership of the *mu'adhdhins* at the *Tūnkūz* Friday Mosque and others."[55] Students who later acted as *mu'adhdhins* for the Mamluks in Cairo also had taken classes in *ʿilm al-mīqāt*. ʿAbd al-Ḥayy b. Mubarakshāh al-Khwārazmī al-Qāhirī al-Ḥanafī, for instance, "was the head of the *mu'adhdhins* at the Friday Mosque of the fortress and others. He benefited in *al-mīqāt* and other (fields) from al-ʿIzz ʿAbd al-ʿAzīz al-Wafā'ī and others."[56] ʿAbd al-Razzāq b. Aḥmad al-Qāhirī al-Ḥanafī was "one of the Sufis of the *Shaykhuniyya*. He read *al-mīqāt* with Ḥasan al-Qaymarī and al-ʿIzz al-Wafā'ī. . . . The sultan made him one of his *mu'adhdhins* after Ibn Khālid and he was inclined towards him until he acted occasionally for him as *imām*."[57] Aḥmad b. Muḥammad al-Dahhān was "the head of the *mu'adhdhins* at the Umayyad Mosque. He had a wailing voice, was knowledgeable in *mīqāt* and lived until he became the senior of the *mu'adhdhins* of the age."[58] The last part of this quote indicates that becoming the leader of any of the groups of scholarly offices was not primarily the result of merit, but of age.

The flow of expert knowledge in astronomical methods for timekeeping apparently did not always and perhaps not necessarily take place from one professional, the *muwaqqit*, to another professional, the *mu'adhdhin*, although this is what the *waqfiyya* of the *Sultan Barqūq Madrasa*, a Sufi convent, stipulated.[59] In one case, al-Sakhāwī tells us that Ibn Khashshāb, one of Jamāl al-Dīn al-Māridānī's students of *ʿilm al-mīqāt*, came usually on Fridays to see Ibn Ḥajar al-ʿAsqalānī for telling him the time to ride out for the *khuṭba*, the Friday sermon.[60] Much more

52 Matvievskaya, Rozenfel'd, *Matematiki i astronomy*, vol. 2, pp. 453–454; King, Survey, C34; François Charette, *Mathematical Instrumentation in Fourteenth-Century Egypt and Syria: The Illustrated Treatise of Najm al-Dīn al-Miṣrī*. Leiden, Boston: Brill, 2003, p. 13.

53 Matvievskaya, Rozenfel'd, *Matematiki i astronomy*, vol. 2, p. 465; King, Survey, C30 and C37; Benno van Dalen, *Preliminary New Zīj Survey*. Unpublished, D11.

54 Al-Sakhāwī, *al-Ḍaw'*, vol. VIII, p. 196.

55 Al-Sakhāwī, *al-Ḍaw'*, vol. X, p. 189.

56 Al-Sakhāwī, *al-Ḍaw'*, *al-Ḍaw'*, vol. IV, p. 40.

57 Al-Sakhāwī, *al-Ḍaw'*, vol. IV, p. 192.

58 Al-Sakhāwī, *al-Ḍaw'*, vol. II, p. 219.

59 King, *On the Role of the Muezzin*, p. 301.

60 Al-Sakhāwī, *al-Ḍaw'*, vol. VI, p. 285.

often the flow of knowledge on timekeeping occurred in al-Sakhāwī's book in the phase of educating people who later would work in the one or the other capacity. It is, however, impossible to find out how many *mu'adhdhins* went through such training in the Mamluk period, because biographical dictionaries are notoriously unsystematic. Their authors did not mean to create a census of education and professional qualifications, but pursued other, often more literary as well as personal goals. It is well known that many of the students of the *madrasa* system later took on a variety of different positions and worked in different fields of occupation.[61] This observation should warn us to assume too stable boundaries between the various *manṣab*s and their responsibilities, including those of a *muwaqqit* and a *mu'adhdhin*. The occasional lack of care in the *waqf* documents David presented in regard to setting *mu'adhdhin*s apart from *muwaqqit*s supports such a reluctance to draw borderlines too rigidly.[62] Accumulating offices was a further important social trend since at the latest the thirteenth century that contributed to increasing the permeability between the various fields of knowledge and their offices.[63] Education itself was not fixed on one or two major disciplines, but tended to be broad and inclusive.

4 *ʿIlm al-mīqāt* in Mecca

When discussing the territories of the Islamic world that either saw or did not see the rise of the *muwaqqit* (or people with similar duties, but other names), the only region of the Arabian Peninsula David discussed is the Yemen.[64] Al-Sakhāwī's dictionary allows extending the geographical scope of the *muwaqqit*. It records that *ʿilm al-mīqāt* was taught in Mecca and Medina at the latest in the late eighth/fourteenth century. The information provided by al-Sakhāwī leaves no doubt that the arrival of knowledge in this discipline and century was due to educational travels by inhabitants of the two Arabian towns to the Mamluk capital Cairo. The important role that Cairo played in this extension of *ʿilm al-mīqāt* towards the Ḥijāz is also pointed out by timekeeping tables for Mecca discovered by David. The extant copies of these tables are all of Egyptian origin.[65]

One Meccan family was particularly influential in teaching this discipline in their hometown in the fourteenth and early fifteenth centuries – the Zamzāmīs, so called because of deputizing the Abbasid caliphs in the distribution of water from the Zamzām well to the pilgrims. Al-Sakhāwī mentions *ʿilm al-mīqāt* in relation to several members of this family, among them Burhān al-Dīn Ibrāhīm b. ʿAlī, Nūr al-Dīn ʿAlī b. Muḥammad, Badr al-Dīn Ḥusayn b. ʿAlī, Muḥammad b. ʿAbd

61 Carl Petry, *The Civilian Elite of Cairo in the Later Middle Ages*. Princeton: Princeton University Press, 1981, chapter IV, pp. 202–272.

62 King, *On the Role of the Muezzin*, p. 302.

63 Berkey, *The Transmission*, pp. 112–119.

64 King, *On the Role of the Muezzin*, pp. 288, 291, 300.

65 King, *Synchrony*, pp. 311–313.

al-ʿAzīz al-Jamāl and Nābit b. Ismāʿīl. Knowledge of the field and other astronomical as well as mathematical disciplines was passed on in the family. Brothers studied with each other, sons with their fathers and nephews with their uncles or uncles of their fathers.[66] Burhān al-Dīn Ibrāhīm b. ʿAlī al-Shambarī al-Zamzāmī I mentioned previously. Nūr al-Dīn ʿAlī b. Muḥammad al-Bayḍāwī al-Makkī al-Zamzāmī al-Shāfiʿī "read with the uncle of his father, our Shaykh al-Burhān al-Zamzāmī. He was (also) educated by his uncle Abū l-Fatḥ. He excelled in *mīqāt*, *farā'iḍ* and other (fields). And he had knowledge in *fiqh* and its *uṣūl* and Arabic. He became the one in which (people) trusted in regard to *mīqāt* and *rūḥānī* and other (things) . . . And there was nobody after him in his arts like him."[67] Muḥammad b. ʿAbd al-ʿAzīz studied *falak* with Nūr al-Dīn and *mīqāt*, *handasa*, arithmetic and *farā'iḍ* in Cairo.[68] Ḥusayn b. ʿAlī al-Bayḍāwī al-Makkī al-Shāfiʿī al-Faraḍī al-Ḥāsib studied a good number of mathematical and astronomical disciplines and topics. While al-Sakhāwī does not mention *ʿilm al-mīqāt* in this list, al-Maqrīzī, as quoted by him, wrote that Ḥusayn was a capacity in this discipline as well as in arithmetic.[69] Ḥusayn b. ʿAlī "was seriously interested in *farā'iḍ* and *ḥisāb*. He took it from al-Shihāb Ibn Zuhayra and al-Burhān al-Burullusī al-Faraḍī . . . He took *ʿilm al-falak* in Cairo from al-Jamāl al-Māridānī and did not stop to add and pay attention until he became the imam, the knowledgeable, the excellent, the perfect whom the people knew in *farā'iḍ*, *hay'a*, arithmetic, *ʿilm al-khaṭa'ayn*, algebra, *handasa*, *falak* and calendars. The leadership in this knowledge came to him finally in the region of the Hijaz, Mecca, Medina and the Yemen."[70] In the biography of his brother Ibrāhīm mentioned previously, al-Sakhāwī listed *ʿilm al-mīqāt* as one of the disciplines Ḥusayn taught.[71] This neglect of *ʿilm al-mīqāt* in Ḥusayn's own biography reinforces my earlier statement that the information given in such entries is neither complete nor was it meant to be so. Hence, when it is said that Nābit b. Ismāʿīl al-Zamzāmī studied with his uncle al-Burhān Ibrāhīm b. ʿAlī, al-Sakhāwī's shaykh, *farā'iḍ*, arithmetic and other (things), it is by no means excluded that he also took *ʿilm al-mīqāt*, since his uncle also gave classes in this discipline.[72] Nābit is the only Zamzāmī whom al-Sakhāwī credits with interest in astrology and an excellent hand in creating horoscopes.[73] Muḥammad b. Abī l-Fatḥ b. Ismāʿīl, a further member of the Zamzāmī family, studied with the uncle of his father Burhān al-Dīn and his cousin Nūr al-Dīn *ʿilm al-falak*. He was an avid religious traveller and led

66 Al-Sakhāwī, *al-Ḍaw'*, vols., I, p. 86; VIII, pp. 61, 278.
67 Al-Sakhāwī, *al-Ḍaw'*, vol. V, p. 291.
68 Al-Sakhāwī, *al-Ḍaw'*, vol. VIII, p. 61.
69 Al-Sakhāwī, *al-Ḍaw'*, vol. III, p. 152.
70 Al-Sakhāwī, *al-Ḍaw'*, vol. III, p. 151.
71 Al-Sakhāwī, *al-Ḍaw'*, vol. I, p. 86.
72 Al-Sakhāwī, *al-Ḍaw'*, vol. X, pp. 194–195.
73 Al-Sakhāwī, *al-Ḍaw'*, vol. X, p. 195.

the prayer call in Mecca.[74] Another cousin of his, Ismāʿīl b. Nābit al-Zamzāmī, also led the *mu'adhdhins* and was the head of those who provided the pilgrims with water.[75] Unsurprisingly, the Zamzāmīs did not restrain their teaching of *ʿilm al-mīqāt* and other astronomical disciplines to family members only. Other Meccans and visitors of the city studied them with Burhān al-Dīn Ibrāhīm b. ʿAlī b. Muḥammad, Badr al-Dīn Ḥusayn b. ʿAlī b. Muḥammad and their great-nephew Nūr al-Dīn ʿAlī b. Muḥammad, among them possibly Ibn Ḥajar al-ʿAsqalānī and al-Sakhāwī himself.[76]

5 *ʿIlm al-mīqāt* and other astronomical disciplines

In addition to the questions raised by David, al-Sakhāwī's entries contain information that supplement the knowledge that he has established about the *muwaqqits*. A surprisingly substantial number of *muwaqqits* and teachers of *mīqāt* also taught *ʿilm al-hay'a*, but did not write about it. Examples are Jamāl al-Dīn al-Māridānī, Ibn al-Majdī, Ḥusayn al-Zamzāmī, Nūr al-Dīn al-Zamzamī, Nāṣir al-Dīn al-Barinbarī, al-Shihāb al-Sijīnī, a former student of Ibn al-Majdī, and others.[77] Unfortunately al-Sakhāwī does not mention which texts exactly were taught. Jamāl al-Dīn al-Māridānī, Ibn al-Majdī, al-Shihāb al-Khawāṣṣ, Ḥusayn al-Zamzāmī, Nūr al-Dīn al-Zamzāmī and al-Shams Muḥammad b. Ayyūb, the head of the ʿUmarī mosque, also taught *ʿilm al-falak*.[78] A few teachers of logic, *tafsīr*, *uṣū al-dīn* taught *hay'a* and *handasa*, but not *mīqāt* or *ʿilm al-falak*. This might have been a result of the discussion of *ʿilm al-hay'a* topics by *uṣūl al-dīn* authors in Iran such as ʿAdud al-Dīn Ījī (d. 757/1355) whose *Mawāqif* was actively studied in fifteenth-century Cairo or al-Sayyid al-Sharīf Jurjānī (741–816/1340–1413) who visited Cairo and studied there for some time.[79] The prominent Anatolian scholar and al-Sakhāwī's teacher Muḥammad b. Sulaymān al-Kāfiyajī (before 790–c. 879/1388–1474), for instance, taught in Cairo several religious and philological disciplines, logic, *ḥikma*, medicine, *ʿilm al-hay'a* and several mathematical disciplines, namely *handasa*, spherics, optics and burning mirrors.[80] Al-Kāfiyajī studied, commented on and possibly taught at least five *hay'a* texts: Athīr al-Dīn al-Abharī's (d. 663/1263) *Mukhtaṣar fī ʿilm al-hay'a*, Jalāl al-Dīn Faḍl Allāh al-Abidī's *Mukhtaṣar fī maʿrifat maqādir al-abʿād wa-l-ajrām*, Maḥmūd Jaghmīnī's (fl. around 620/1223) *al-Mulakhkhaṣ fī l-hay'a*, a *Risāla fī maʿrifat ḥisāb taʿdīl al-kawākib al-khamsa* ascribed to Ibn al-Shāṭir and

74 Al-Sakhāwī, *al-Ḍaw'*, vol. VIII, p. 278.

75 Al-Sakhāwī, *al-Ḍaw'*, vol. II, p. 308.

76 Al-Sakhāwī, *al-Ḍaw'*, vols. II, p. 134; III, pp. 120, 152; X, p. 239.

77 Al-Sakhāwī, *al-Ḍaw'*, vols. I, pp. 300, 310, 317; II, p. 174; III, pp. 115, 235; King, Survey, C 47 (Jamāl al-Dīn al-Māridānī).

78 Al-Sakhāwī, *al-Ḍaw'*, vols. I, pp. 317–318; II, pp. 134, 142; III, p. 151; VIII, p. 278.

79 Al-Sakhāwī, *al-Ḍaw'*, vols. V, pp. 48, 328–330; VII, p. 7.

80 Al-Sakhāwī, *al-Ḍaw'*, vol. VII, p. 261.

an anonymous *Risāla fī ʿilm al-hay'a*.[81] He also studied at least one astrological text, Abū l-Ṣaqr al-Qabīṣī's (d. c. 380/990) *al-Masā'il wa-l-ikhtiyārāt*, which may be the treatise called *Masā'il al-munajjimīn* in Qabīṣī's *Risāla fī imtiḥān al-munajjimīn*.[82] In Mecca, a certain Salām Allāh al-Iṣbahānī taught al-Sayyid al-Sharīf Jurjānī's commentaries on Naṣīr al-Dīn Ṭūsī's *Tadhkira fī ʿilm al-hay'a* and Jaghmīnī's *al-Mulakhkhaṣ* as well as Niẓām al-Dīn al-Nisābūrī's *Shamsiyya fī l-ḥisāb*.[83] Some sons of the Mamluks taught *hay'a, falak, nujūm* and *ḥurūf*.[84]

As indicated knowledge of *hay'a* texts came into Mecca through scholars from Iran being on pilgrimage and occasionally went further to Cairo or Jerusalem and other Northern cities, but *mīqāt* apparently was not taken further east than Mecca or Medina. This however may reflect more al-Sakhāwī's territorial focus than the question as to whether scholars interested in *mīqāt* not only travelled to Eastern cities, but also taught there their knowledge.

6 Conclusions

The most important conclusion from Sakhāwī's dictionary is that the fifteenth century saw a vivacious teaching of a good number of mathematical and astronomical sciences in Egypt, Syria, Palestine and the Arabian Peninsula with close relationships between various groups of the elite. David has made the case for flourishing activities in astronomy on the basis of the body of texts, tables and instruments he has discovered, analysed and interpreted. Al-Sakhāwī confirms that these scientific activities were part and parcel of the educational life in major Mamluk cities. A second important insight is that scholars teaching and writing on these sciences could acquire a solid reputation for their knowledge skills and could become well established among the educated and military elites. This included direct patronage by Mamluk sultans and high-ranking emirs. A third fascinating result is that our categories are too rigid: *muwaqqit*s were – in the representation by al-Sakhāwī – not professionals in the sense of focusing exclusively or even primarily on this field of knowledge, but rather – like others – well educated across a broad range of disciplines, including *fiqh* and *ḥikma*. *Mu'adhdhin*s went in principle through the same educational cycle as *muwaqqit*s. The precise nature of this cycle, i.e., its details, depended on the individuals who taught and studied. Teachers of religious disciplines were the main sources for knowledge about logic and *ḥikma* and

81 Al-Sakhāwī, *al-Ḍaw'*, vol. VII, p. 260. For Jaghmīnī's date see Jamil F. Ragep, "On dating Jaghmīnī and his *Mulakhkhaṣ fī l-Hay'a*". In Mustafa Kaçar, Zeynep Durukal (eds.), *Essays in Honour of Ekmeleddin İhsanoğlu*. Istanbul: IRCICA, 2006, pp. 461–466. *Makhṭūṭāt al-falak wa-l-tanjīm fī maktabat al-mathaf al-ʿirāqī*. Usāma N. al-Naqshbandi, Dhamia M. Abbas (eds.), *Astronomical and Astrological Manuscripts of the Iraq Museum Library*. Dār al-Ḥurriya li-l-Ṭabāʿa, Baghdad, 1982, pp. 100, 212, 215, 219.

82 *Makhṭūṭāt al-falak wa-l-tanjīm*, p. 219, Matvievskaya, Rozenfel'd, *Matematiki*, p. 155.

83 Al-Sakhāwī, *al-Ḍaw'*, vol. V, pp. 12, 78.

84 Al-Sakhāwī, *al-Ḍaw'*, vol. V, p. 217.

occasionally also of theoretical mathematical and astronomical texts. And finally, it is almost impossible to find out what either of these scholars did in scholarly practices other than teaching and occasionally constructing instruments. It is in this sense at least that biographical dictionaries compiled by members of the educated elite define the type of answers we can find to the questions we ask. These authors wrote about themselves and their circles and were not interested in issues that went too far beyond the boundaries set by the category of their work and their social and cultural profile. As repeatedly pointed out their ways of collecting and presenting information differ from our needs in regard to regularity and completeness. Taking their information at face value, both in terms of quantity and quality, will mislead us in our quest for understanding the contexts of the sciences and their practitioners in Islamic societies.

APPENDIX

Affiliations of *muwaqqit*s, *mīqātī*s and teachers of *ʿilm al-mīqāt* according to al-Sakhāwī:

Name	*Affiliation*	*Post*	*Further information*
Aḥmad b. Asad al-Sikandarī al-Qāhirī, known as Ibn Asad (808–872)	–	teacher of children	studied *mīqāt* with Ibn al-Majdī; transformed Ibn al-Majdī's *Risāla fī l-mīqāt* into *Urjūza Ghunyat al-ṭālib fī l-ʿamal bi-l-kawākib*
	al-Ḥakam Friday Mosque, al-Zayniyya Madrasa	notary and imam	
	Barqūqiyya Madrasa	teacher of *qirāʾāt*	
	Fortress	*ḥadīth* classes	
Ibn al-Majdī	al-Jānibākiyya Dawādāriyya Madrasa	head of the teachers	installed by al-Ashraf Barsbay; wrote book about *ḥadīth*; had beautiful handwriting in his *fatwā*s
Aḥmad b. Ṣadaqa al-Makkī al-Qāhirī, known as Ibn Sīrāfī	Ṭaybarṣiyya Madrasa	imam; teacher, *ḥadīth* classes, *fatwā*s	studied with Ibn al-Majdī algebra, arithmetic, *falak*, *muqanṭarāt*, geometry, *hayʾa*, *ḥikma*, wrote *Muqaddima fī l-falak*
	Shaykhuniyya convent Barqūqiyya Madrasa	professor of *fiqh* tutor; *tafsīr*	
Aḥmad b. ʿUbayd Allāh al-Shihāb al-Sijīnī al-Qāhirī al-Azharī al-Faraḍī (810–85)	–	teacher of al-Sharaf al-Jayʿan's children	was adlatus of Ibn al-Majdī in *ḥadīth*, *uṣūl al-fiqh*, Arabic, *farāʾi*, arithmetic, surveying, algebra, geometry, *mīqāt*; and of Shihāb al-Khawāṣṣ in *farāʾiḍ* and *mīqāt;* taught *mīqāt*, including its practical aspects, surveying, *farāʾiḍ*, arithmetic
	al-Azhar Mosque	head of teachers of part called Riwāq Ibn Miʿmār	

Name	*Affiliation*	*Post*	*Further information*
Aḥmad b. ʿUthmān al-Rishī al-Qāhirī, known as al-Kawm Rishī (c. 778–852)	Friday Mosque in Kawm al-Rish; ʿAmru Friday Mosque and others in Cairo	*khāṭib*	excellent chess player studied arithmetic with al-Jamāl al-Māridānī
	–	taught children of al-Tāj b. al-Ẓarīf and Nāṣir al-Dīn b. al-Tansī in Cairo	
Aḥmad b. Muḥammad al-Dahhān	Umayyad Friday Mosque	head of *mu'adhdhin*s	well-versed in *mīqāt*
al-Ḥasan b. ʿAbd al-Raḥmān al-Sharimsāhī al-Muwaqqit (c. 810–93)	Ṭuruqiyya Friday Mosque	head (of *muwaqqit*s?)	studied *mīqāt* with ʿAbd al-Raḥīm b. Rāzin, al-Majdī, al-Badr al-Māridānī
	–	taught children	
	–	notary	
	–	perhaps deputy *khāṭib*	
Ḥasan b. ʿAlī al-Badr al-Qaymarī (d. 885, c. 70 years old)	Qanim bi-l-kabsh (?) Friday Mosque	head	studied with Ibn al-Majdī and Abū l-Jūd, including *mīqāt*. Abū l-Jūd got him the job in al-Ramla; taught *mīqāt*
	Friday mosque of the fortress	head	
	Ḥasaniyya Madrasa	*mu'adhdhin*	
	Jawhar al-Ṣafawī Madrasa, al-Ramla	professor of *farā'iḍ*	
al-Shams al-Tuntada'ī	al-Ẓāhir Friday Mosque	*khāṭib*	taught *mīqāt*
	al-Baybarsiyya Khānqāh	lived there	
Khalīl b. Ibrāhīm Abū l-Jūd al-Dimyāṭī al-Qāhirī, known as Imām Manṣūr	–	was in retinue of Manṣūr b. Ṣafī + his imam; then in that of Jawhar al-Muʿīnī and others; rose to be in entourage of Caliph al-Mutawakkil ʿAlā' Allāh	studied *mīqāt* and other disciplines with several teachers; taught *mīqāt* and other disciplines

Name	*Affiliation*	*Post*	*Further information*
ʿAbd al-Ḥayy b. Mubārakshāh al-Khwārazmī al-Qāhirī (813–80)	Friday Mosque of the fortress	head of the *mu'adhdhin*s	studied *mīqāt* with al-ʿIzz ʿAbd al-ʿAzīz al-Wafāʾī and others
ʿAbd al-Khāliq b. Muḥammad al-Ṣāliḥī, known as Ibn al-ʿUqab (853–after 89)	al-Ḥakam Friday Mosque	head	studied *mīqāt* and other things with al-Badr al-Māridānī; set up sundials and other things
	al-Jānibākiyya Madrasa	head	
ʿAbd al-Raḥmān b. Muḥammad, known as al-Rashīdī (741–803)	several places	head of *mīqāt* and *farā'iḍ*	studied *mīqāt*, arithmetic and *farā'iḍ*
	Amīr Ḥusayn Madrasa	*khāṭib*	
ʿAbd al-Raḥīm b. Muḥammad al-Hamāwī al-Qāhirī al-Muwaqqit (d. 885)	al-Ḥakam Friday Mosque	head deputy of two of his colleagues	studied *mīqāt* with al-Nūr b. al-Naqqāsh
ʿAbd al-Razzāq b. Aḥmad al-Qāhirī	al-Shaykhuniyya	Ṣūfī	studied *mīqāt* with Ḥasan al-Qaymarī and al-ʿIzz al-Wafāʾī
	-	sultan installed him as *mu'adhdhin*	
al-Jamāl al-Māridānī al-Qāhirī al-Ḥāsib (d. 809)	al-Khāṣikī	*shaykh*	was leading scholar of *mīqāt* in his time; Ibn Ḥajar al-ʿAsqalānī and Ibn al-Majdī studied with him; good knowledge in *hay'a* and arithmetic
ʿAbd al-Wahhāb b. Muḥammad al-Shawwā al-Qāhirī (b. 766)	Shaykhuniyya	living there	studied *mīqāt* with al-Shams al-Ghazulī, al-Jamāl al-Māridānī and Ibn al-Majdī; studied ophthalmology; had knowledge on quadrant and astrolabe
	Manṣū Madrasa	determined the time	
	al-Ḥakam Friday Mosque	determined the time	
	hospital	treated eye diseases	
	–	transmitted *hadīth*	
ʿAlī b. ʿAbd al-Qādir al-Naqqāsh al-Mīqātī (d. 886)	*ḥanut* in al-Sagha	engraver	studied *mīqāt* and geometry with Ibn al-Majdī and engraving with his stepfather; taught *mīqāt* to his son and al-ʿIzz al-Wafāʾī
	al-Maqsī Friday Mosque	head (*muwaqqits*?)	
	al-Jamāliyya al-Ṣāḥibiyya Madrasa	head (*muwaqqits*?)	
	tomb of al-Ashraf Īnāl etc.	head (*muwaqqits*?)	
	several places	taught *mīqāt*	

Name	*Affiliation*	*Post*	*Further information*
ʿAlī b. ʿUmar al-Qāhirī al-Maqsī (d. after 893)	Mu'ayyadiyya Madrasa	served Qujmas in *mīqāt mu'adhdhin* of the sultan Ṣūfī	
ʿAlī b. Muḥammad al-Haythamī al-Tibnawī al-Qāhirī (d. 888)	al-Ashrafiyya Barsbay Madrasa	Ṣūfī	studied *mīqāt* with al-Shams Muḥammad b. Ḥusayn al-Sharinbabilī; wrote two introductions on the *Sinus Quadrant* and one on *muqanṭarāt*, but did not teach them
	–	in entourage of Amīr al-Jamāl	
ʿAlī b. Muḥammad al-Nūr al-Qāhirī (d. after 896)	al-Zayn al-Ustādār Friday Mosque	*muwaqqit*	
ʿUmar b. Aḥmad al-Siraj al-Hilālī al-Ḥamawī al-ʿAnbarī, known as Ibn al-Khadar (b. 816)	Great Friday Mosque Hama	head	studied *mīqāt*
Muḥammad b. Aḥmad al-Makhzūmī al-Qāhirī, known as Ibn al-Khashshāb (793–873)	al-Ashrafiyya Barsbay Madrasa	determined time	studied medicine; studied *ʿilm al-waqt* with al-Jamāl al-Māridānī, al-Shihāb al-Saṭḥī, al-Baridānī, Ibn al-Majdī
	al-Ṣāliḥ Friday Mosque	determined time	
	al-Manṣūriyya Madrasa	determined time	
	–	informed Ibn Ḥajar every Friday of time for sermon	
	al-Ẓāhiriyya Madrasa (the old one)	librarian and lived there	
Muḥammad b. ʿAbd al-Laṭīf al-Āqsarāy'ī, known as al-Mahallī (d. c. 860)	al-Sharabshiyya Madrasa near al-Aqmar Friday Mosque	lived there and led the teachers	studied *mīqāt* with al-Shihāb al-Khawāṣṣ, Ibn al-Majdī, al-Nūr al-Naqqāsh; taught *mīqāt* to al-Muẓaffar al-Amshatī, ʿAbd al-ʿAzīz al-Mīqātī (al-Wafā'ī?) and Nāṣir al-Dīn al-Akhmīmī

Name	*Affiliation*	*Post*	*Further information*
al-Badr (Sibṭ) al-Māridānī (826–912)	al-Ḥājibiyya Madrasa	teacher	studied *mīqāt* with Ibn al-Majdī; al-Nūr al-Naqqāsh gave him his post at the Ibn Ṭūlūn Friday Mosque; wrote more than 200 treatises, among them text on his work at the Manṣūriyya Madrasa
	Ibn Ṭūlūn Friday Mosque + other places	head (*muwaqqits*/ teachers?)	
	al-Manṣūriyya Mosque	worked there (*muwaqqit*?)	
Muḥammad b. Muḥammad al-Shams al-Ḥalabī, known as Ibn Amīr al-Ḥajj and Ibn al-Muwaqqit (791–868)	Great Friday Mosque in Ḥalab	led *mīqāt* led the signing of appointment certificates with the office of the judges	
Muḥammad b. Muḥammad al-Gharraqī (795–858)	Nabulsiyya Madrasa	lived there and taught there, among other things *mīqāt*, *fiqh*, *uṣūl al-fiqh*, Arabic, arithmetic, prosody and *rūḥānī* (automata?)	studied *mīqāt* with Nāṣir al-Dīn al-Barinbarī, al-Shams al-Gharraqī and Ibn al-Majdī
Muḥammad b. Yūsuf al-Fāraskūrī al-Ḥarīrī (d. c. 870)	al-ʿAtiq Friday Mosque in Fāraskūr	imam + *muwaqqit*	knowledgeable in *ʿilm al-waqt* and *farāʾiḍ*
Mūsā b. Muḥammad al-Sharaf al-Muwaqqit, nephew of al-Shams al-Khalīlī (d. 807)	Tūnkūz Friday Mosque and others	head of the *muʾadhdhin*s	the most excellent in *hayʾa* in Syria

3

AYYUBID PRINCES AND THEIR SCHOLARLY CLIENTS FROM THE ANCIENT SCIENCES

The Ayyubids (1171–1260 CE) are often seen as a dynasty that patronized generously physicians, but contributed to ruining the philosophical disciplines in Syria and Egypt with their execution of Shihāb al-Dīn Suhrawardī in 586/1191 and their removal of Sayf al-Dīn al-Āmidī (d. 631/1233) from his chair at the *ʿAzīziyya Madrasa* after 1229. In this paper I discuss what it meant to be a scholar patronized by an Ayyubid prince. I will survey the relationships of scholars to the ruler or the prince they served and to the world outside the princely sphere. I will ask where they lived, how they earned their livelihood, with whom they interacted and which fields of knowledge they considered important. I will discuss types of patronage and their features. I will point to components that challenge the traditional understanding of patronage as a hierarchical relationship between two people – a protector and a dependent. I will suggest that this deviation from standards is caused by two main factors – the specifics of the profession of the dependent and the specifics of the power set up of the family of princely patrons. I will highlight the position of the mathematical sciences through the particular lens of the physicians' interest in them. Finally, I will argue that the philosophical sciences were not persecuted by the Ayyubid dynasty, but supported through their patronage of physicians and other scholars.

The richest source for a study of Ayyubid patronage for medicine, astrology, astronomy, philosophy and other ancient sciences is Ibn Abī Uṣaybīʿa's (d. 668/1270) *ʿUyūn al-anbāʾ fī ṭabaqāt al-aṭibbāʾ*. Some information can also be found in the historical chronicles of the period, while almost none can be gleaned from Ibn al-Qifṭī's (d. 646/1248) *Taʾrīkh al-ḥukamāʾ*. Thus, at the moment there is almost no other source available that can be used to corroborate the basic features of patronage described by Ibn Abī Uṣaybiʿa, let alone the detailed and at occasions very colourful stories he interspersed his biographical entries with. Thus, I will not assume that Ibn Abī Uṣaybiʿa delivers facts, reports neutrally on real events or surveys fair and square the main features of patronage for the ancient

 DOI: 10.4324/9781003372455-4

sciences. I assume, though, that the picture of the lives of his immediate predecessors, friends of his family and contemporaries he draws will not be fiction either, if only because he shared it with several of these people, including patrons, and hence must have met their expectations to some degree at least. A systematic analysis of Ibn Abī Uṣaybiʿa's presuppositions, assumptions and goals is beyond the scope of this paper. I rather try to survey the author's terminology, types and content of information and narrational foci. Given the paucity of source material available at the moment, my interpretations of my condensing description of Ibn Abī Uṣaybiʿa's reports are more speculative and provide more questions than answers.

Physicians at Ayyubid courts

Ibn Abī Uṣaybiʿa had a particular interest in the educational and professional features of the lives of the physicians since this is the foremost subject matter of his dictionary. As a result, the entries he compiled have a certain regular structure that agrees with other biographical dictionaries of the period. They list names, relatives, locations, educational activities, professional activities including the oeuvre, additional activities and skills, dates of birth and death, if known, and burial sites plus accompanying events. Ibn Abī Uṣaybiʿa moved beyond this standard pattern by telling stories about the doings of a person and the events that involved him. The content of these stories demonstrates that Ibn Abī Uṣaybiʿa was eager to report social and cultural aspects of the physicians' conditions in society and to gossip or comment on their personality marks on a fairly broad scale. An analysis of the chapters on physicians in Egypt and Syria in the late twelfth and during the first half of the thirteenth centuries CE shows that Ibn Abī Uṣaybiʿa had two focal points in his narration – patronage and excellence. Patronage appears as the uniting factor of the group of physicians Ibn Abī Uṣaybiʿa described. Excellence defines the paragons among the patrons and the clients. Excellence is defined by position or rank in the respective group plus quantitative and qualitative features separating a member presented as excellent from other members of the group. In the group of the physicians, exemplary status is defined by knowledge, experience, success and behaviour. In the group of the patrons, exemplary status is defined by the position in the ruling family and in regard to the family, the number of clients, the clients' status and the patron's behaviour.

These focal points are made visible by the fact that many, if not most of the physicians in Ibn Abī Uṣaybiʿa's biographical entries regarding the Ayyubid dynasty and its immediate predecessors were linked in patronage relationships with one of the warlords, city rulers or heads of dynasties of the region at some or the other point in their life. Physicians with no patronage relationships are rare exceptions. Although princely patronage was not the only possibility to earn a living as a physician, the other options, that is, engaging with lower-ranking patrons such as *umarāʾ*, working in a hospital, working for the fortress or the troops, working for a number of patients in town or working as an itinerant doctor, appear less often, with the exception of hospital service, in Ibn Abī Uṣaybiʿa's entries.[1] It is, however, not very likely that almost all physicians of the Ayyubid period earned their living in the service of an Ayyubid prince, while almost none worked for an *amīr* or the notables of Damascus, Cairo or any of the other Ayyubid cities and towns. Hence, this lesser visibility of alternative sources of income must reflect a conscious choice by the author. This focus on physicians in patronage relationships with Ayyubid princes corresponds with Ibn Abī Uṣaybiʿa's own position and that of his family in the hierarchical world of Ayyubid doctors.[2] It reflects in all likelihood the social values held by the author and the group of patronized physicians. The focus indicates that the position of a physician was only partially defined by his knowledge, skills and achievements in healing. It depended secondly on his success in acquiring patrons among the members of the Ayyubid family and in assuring the continuity of patronage. A third part that went into building the position of a physician was his behaviour towards patrons, patients and other members of the medical community. Patients were of least importance in Ibn Abī Uṣaybiʿa's reports and comments on behavioural aspects. Patrons and their social equals could be occasionally rejected, overruled or ridiculed. What mattered most in Ibn Abī Uṣaybiʿa's stories was exercising support and help to younger and less well-established members of the professional community and to be courteous and linked in friendship with colleagues of equal or higher ranking. Physicians who managed to achieve high marks in all three domains were paragons of knowledge, virtue, sociability and success.

The focal points of patronage and excellence also structure the presence of the Ayyubid princes and their *umarāʾ* in Ibn Abī Uṣaybiʿa's biographical entries. Most of their activities take place in the framework of patronage relationships indicating their intent of establishing such relationships, their contributions to and behaviour in the relationships and the criteria they applied for evaluating clients and their deeds. The group of Ayyubid princes is explicitly ranked according to the historiographical identification of the dynasty with its founder and first head. Frequency of naming in relationship to cities and fortresses shows that in addition Ibn Abī Uṣaybiʿa implicitly ranked the princes according to localities and thus status within the family.

It is thus not surprising that the most coveted patron, in the view of Ibn Abī Uṣaybiʿa, was the founder of the dynasty, Ṣalāḥ al-Dīn ibn Ayyūb (r. 1171–93 CE). Many of the physicians of the Ayyubid period described by Ibn Abī Uṣaybiʿa were his clients. Such a distribution is not likely to reflect the realities of 90 years of medical practice in Ayyubid Egypt and Syria. Nor is it reasonable to assume

1 Examples are found in Ibn Abī Uṣaybiʿa (1965), 630, 673, 682, 696, 706 (here Ibn Abī Uṣaybiʿa reports about his move to the fortress Sarkhad where he entered the service of its lord, the *amīr* ʿIzz al-Dīn al-Muʿaẓẓamī, in the year 634/1236), 728.

2 His grandfather, father, uncle and he himself served several Ayyubid princes beginning with Ṣalāḥ al-Dīn ibn Ayyūb.

that Ibn Abī Uṣaybiʿa disposed of more precise information for the early period of Ayyubid patronage of the medical arts than for that of his own lifetime. The pre-eminence of Ṣalāḥ al-Dīn as a patron in Ibn Abī Uṣaybiʿa's report about the Ayyubid dynasty is rather the author's version of the story of the extraordinary qualities of this warlord become ruler. His pre-eminence is also made visible by his appearance in stories about physicians that give space for describing the components of his exemplary behaviour and for highlighting that at times he even had an inclination to do too much of the good. In addition, in relative terms the number of physicians who are reported to have served Ṣalāḥ al-Dīn as clients is indeed fairly high (twenty-nine). The other members of the family appear less often as a patron in Ibn Abī Uṣaybiʿa's entries: al-ʿĀdil (r. 1199–1218 CE) – eleven clients, al-Kāmil (r. 1218–38 CE) – ten clients, al-Muʿaẓẓam ʿĪsā (r. 1218–27 CE) – eight and al-Ashraf Mūsā (r. 1229–37 CE in Damascus) – seven. All the many other Ayyubid princes appear as patrons not more than five times, but often even only once or twice.[3] These numbers are not fully reliable, since in some cases Ibn Abī Uṣaybiʿa does not say explicitly that a patronage relationship existed, while in other cases he mentions the same patron-client combination in more than one entry.[4] Occasional remarks by Ibn Abī Uṣaybiʿa as well as tradition suggest, however, that all Ayyubid princes extended their patronage to more than one or two physicians. As a rule, they made several physicians look after their health and consult about the most commendable therapy in cases of sickness. Thus, the quantitative differences in the group of the patrons are part of Ibn Abī Uṣaybiʿa narrative strategies in regard to his two focal points.

Although Ṣalāḥ al-Dīn's pre-eminence in the group of the patrons is the result of the author's selection and presentation, it is not reasonable to assume that the number of physicians he had surrounded himself with is an exaggeration. Ṣalāḥ al-Dīn obviously bound many of the senior physicians in Damascus and Cairo as clients. Ibn Abī Uṣaybiʿa even seems to suggest that the Ayyubid ruler tried to establish exclusive relationships with them, that is, invited them to serve him alone, when he remarked in one of the entries that Ṣalāḥ al-Dīn paid and treated those who agreed to such a relationship excessively well.[5] This desire for exclusive relationships

3 Ibn Abī Uṣaybiʿa mentions in addition to Ṣalāḥ al-Dīn the following Ayyubid princes as patrons of physicians: al-ʿĀdil (Aleppo, Mayyāfāriqīn, Damascus, Cairo), al-ʿAzīz ʿUthmān ibn Salāḥ al-Dīn (Cairo), al-Afḍal ʿAlī ibn Ṣalāḥ al-Dīn and his sons (Damascus, Sumaysāṭ), al-Ẓāhir (Aleppo), al-Muʾayyad Najm al-Dīn Masʿūd ibn Ṣalāḥ al-Dīn, al-Fāʾiz ibn al-ʿĀdil (Cairo), al-Muʿaẓẓam ʿĪsā (Damascus), al-Ashraf Mūsā (Karak, Ruhāʾ, Sinjar Mayyāfāriqīn, Damascus), al-Ḥāfiẓ Arslanshāh ibn Abī Bakr (al-ʿĀdil) (fortress Jaʿbar), Awḥad Najm al-Dīn Ayyūb ibn al-ʿĀdil (Khilāṭ, Mayyāfāriqīn), al-Ṣāliḥ Ismaʿīl ibn al-ʿĀdil (Damascus), Taqī al-Dīn ʿUmar (Ḥamāh), al-Manṣūr ibn Taqī al-Dīn (Ḥamāh), Ibrāhīm ibn Majāhid ibn Asad al-Dīn Shīrkūh (Ḥimṣ), Najm al-Dīn Ayyūb ibn al-Kāmil (Cairo), ʿIzz al-Dīn Farrukhshāh (Baʿlbak), al-Amjad Majd al-Dīn Bahrām Shāh (Baʿlbak), al-Masʿūd Aqsīs [sic] ibn al-Kāmil (al-Yaman), al-Muʿaẓẓam Turanshāh [sic] (Cairo, Damascus), al-Nāṣir Yūsuf ibn Muḥammad (Aleppo, Damascus).

4 For instance, see Ibn Abī Uṣaybiʿa (1965), 737f.

5 See ibid. 652.

and the high number of clients suggest that other than purely medical reasons will have motivated Ṣalāḥ al-Dīn's approach to patronage. Building prestige through acquisition of reputation was one such additional motive. Another motive hinted at by Ibn Abī Uṣaybiʿa through the aspects he emphasized and highlighted was the display of royal qualities in an Islamic context that combined urban with tribal groups and mentalities – generosity, justice, loyalty and lenience. In special cases, though, as in the case of Moshe ben Maimon, or Maimonides (d. 1204 CE), professional reputation was the most important motivation to establish such a relationship.[6]

General elements

Beyond the two focal points, Ibn Abī Uṣaybiʿa's biographical entries display a number or other general features. They suggest first that Ayyubid princes did not introduce substantial novelties into patronage patterns for the ancient sciences, but rather continued the practices of their predecessors. The Ayyubids provided physicians who had served the last Fatimid caliph al-ʿĀḍīd Abū Muḥammad ʿAbdallāh (r. 1160–71 CE), the Zangid ruler Nūr al-Dīn Zangī (r. 1146–73 CL) or the Artuqid prince Nūr al-Dīn ibn Jamāl al-Dīn ibn Artuq (r. 1174–85 CE) with positions in their own entourage and supported or imitated the institutions of healthcare they had installed, that is, hospitals, heads of the main medical branches (physicians, ophthalmologists, surgeons) per region (Egypt, Syria) and administrators of hospital *awqāf*.[7] They granted various kinds of monetary and nonmonetary remuneration to the physicians whom they patronized – monthly or yearly stipends (*jāmakiyya; jirāya*) mostly paid from the patron's private treasury, tax farms (*iqṭāʿ*) and robes of honour (*khilʿa*).[8] The only indirectly emphasized difference between the Ayyubids and their predecessors is in terms of quantity caused by the size of the family and their holdings. According to Ibn Abī Uṣaybiʿa's entries, the system of power distribution among the Ayyubid princes created more outlets for physicians looking for patronage than those of their predecessors. In terms of quality as summarized above, Ibn Abī Uṣaybiʿa's picture is one of continuity and uniformity of patterns from the predecessors of the Ayyubids, among the members of the dynasty and to their Mamluk successors.[9] This impression of continuity and uniformity is strengthened by Ibn Abī Uṣaybiʿa's usage of a stable vocabulary

6 See ibid. 582.

7 Examples of physicians taken over by Ṣalāḥ al-Dīn are Muhadhdhab al-Dīn ibn al-Naqqāsh (d. 574/1178; served previously Nūr al-Dīn Zangī), Raḍī al-Dīn al-Raḥbī (d. 63 I/I 233; served previously at the fortress and the hospital of Nūr al-Dīn Zangī in Damascus), Sadīd al-Dīn ibn Raqīqa (d. 635/1237; served previously Nūr al-Dīn ibn Jamāl al-Dīn ibn Artuq), Abū Sulaymān Dāwūd ibn Abī l-Munan (d. unknown; served previously the Fatimid caliphs). Examples for the continuation of the posts of head physician of Egypt or Syria can be found in: ibid. 585f, 672.

8 See ibid. 589, 599, 600, 637, 646, 660, 671f, 718 et passim.

9 See ibid. 599.

when talking about patronage relationships. When compared with other sources, the rhetoric of patronage between these three periods appears to have been stable insofar as it concerned the direct relationships between the patron and his client and the region ruled by the Ayyubids. Outside this region and in indirect or mediated relationships the rhetoric of patronage was much more variable than Ibn Abī Uṣaybiʿa's work allows for. At the centre of his patronage terminology are words like service (*khidma*), benefit (*niʿma*), honour (*ikrām*), respect (*iḥtirām*) and favour (*ḥuẓwa*).[10] A client served and a patron provided benefits, favours and graces. A client was knowledgeable and skilled in his service and the patron respected and honoured him in return. Occasionally, the relationship between a patron and a client is designated as companionship (*ṣuḥba*), a term that described in addition to the hierarchy of positions emotional ties close to friendship as some remarks of Ibn Abī Uṣaybiʿa about friends of his grandfather or father demonstrate.[11] Nonetheless, some physicians refused to enter relationships termed companionship.[12] Ibn Abī Uṣaybiʿa wrote emphatically about ʿImrān al-Isrāʾīlī (d. 637/1240) that he did not serve any of the princes in companionship and did not go with any of them on a journey, but every single one of them asked for him in cases of disease.[13] ʿImrān al-Isrāʾīlī's behaviour suggests that there was a difference between a companionship and a service. The latter term covered most likely a broad variety of relationships, while the companionship seems to have implied a more exclusive, personal bond. The physician refused for instance al-ʿĀdil's ardent wish to enter in a companionship with him, but accepted to come for treatments whenever he was called by one of the princes or notables. He also agreed to travel to far away smaller Ayyubid holdings when he was mature and stay there for some time until the patient had recovered. This kind of relationship, called by Ibn Abī Uṣaybiʿa *khidma*, could be remunerated with a monthly stipend.[14] When the payments continued after one and a half years the physician refused them.[15] Thus ʿImrān seems to have preferred relationships of a more focused kind in which he indeed treated an acute disease over a long-term, general kind of relationship. Although physicians regularly treated female members of Ayyubid princely families, it is only in rare cases that Ibn Abī Uṣaybiʿa reports that an Ayyubid woman had a patronage relationship of the *khidma* type with a physician.[16] One such rare case is the patronage that Sitt al-Shām, the sister of al-ʿĀdil, exercised in regard to Ibn Abī Uṣaybiʿa's uncle Rashīd al-Dīn ibn Khalīfa (d. 616/1219). The second case is that of the sister of Aḥwad Najm al-Dīn ibn al-ʿĀdil (d. 607/1210 in Mayyāfāriqīn) who was married to Ṣalāḥ al-Dīn ibn Yāghīsān, a Saljuq *amīr*

10 See ibid. 581f, 589, 591, 600, 630, 635, 637, 652,661, 718 et passim.
11 See ibid. 639, 646, 683f, 698.
12 See ibid. 673, 696.
13 Ibid.
14 See ibid. 697.
15 Ibid.
16 For a story about the treatment of Sitt al-Shām, the sister of al-ʿĀdil, see ibid. 721.

who ruled Antioch.[17] Loyalty, apparently expected on both sides, was expressed through duration of service, promotion and continuation of existing relationships after the death of either party.[18] Thus it is not surprising that most physicians in patronage relationships with Ayyubid princes stayed for years in their service until either of the two died. In several cases the sons continued the relationship either as the patron or the client.[19] Due to the Ayyubid family network shifts in patronage could also involve other male relatives such as brothers, uncles or cousins.[20] In some cases, a client outlived more than two successive patrons.[21] The transfer from one Ayyubid prince to another involved issues of maintenance of status, remuneration and other allocations and possessions. Ibn Abī Uṣaybiʿa does not speak in great detail of such material aspects of patronage transfer, but refers to them repeatedly.[22] The case of Raḍī al-Dīn al-Raḥbī (d. 631/1233), who worked at the fortress and in the Nūrī hospital, is particularly interesting since he refused to accept the patronage offer by the new head of the family. Despite this refusal al-ʿĀdil accepted to continue paying the physician all his entitlements from his relationship with Ṣalāḥ al-Dīn. Raḍī al-Dīn continued to remain on the pay list of al-ʿĀdil until the ruler's death. Then his son al-Muʿaẓẓam took over the responsibility reducing, however, the monthly payments.[23] In a few cases, a transfer of patronage relationships could not be achieved.[24]

Loyalty also included demands for mobility, although it was not a necessary condition for the continuation of a patronage relationship. When an Ayyubid prince moved to another position some of his clients were asked to accompany him and some indeed accepted the offer.[25] Other clients, however, left a particular patronage position or refused to move to a new locality.[26] Raḍī al-Dīn al-Raḥbī refused to follow orders of two of his Ayyubid patrons, Ṣalāḥ al-Dīn and al-ʿĀdil, neither of whom insisted on their demands.[27] In some of the cases when a client rejected leaving the place of his living, in particular if it was Damascus, the payments continued when the client agreed or continued to work at the hospital or the

17 See ibid. 706.
18 See ibid. 585, 599, 601, 659–661, 671, 718 et passim.
19 See ibid. 5821: 585, 599, 601, 661.
20 See ibid. 585, 659.
21 See ibid. 673, 700.
22 Three examples are the confirmations of the rights and possessions of Raḍī al-Dīn al-Raḥbī by al-ʿĀdil after Ṣalāḥ al-Dīn's death, of Muwaffaq al-Dīn Yaʿqūb b. Saqlāb (d. 626/1229) by al-Nāṣir Dāwūd (r. 1227–9 CE in Damascus; 1229–49 CE at Karak) after the death of his father al-Muʿaẓẓam and of Saʿd al-Dīn ibn ʿAbd al-ʿAzīz al-Sulamī (d. 644/1246) by al-Kāmil after the death of his brother al-Ashraf Mūsā. See ibid. 672f, 699.
23 See ibid. 673.
24 See, for instance, ibid. 689.
25 See ibid. 585.
26 See ibid. 637, 673, 696, 700.
27 See ibid. 673.

fortress or, perhaps, when there was an understanding that the physician would serve his patron whenever he visited Damascus.

A further necessary element of a patronage relationship was the exchange of gifts. Physicians mostly wrote or dedicated medical treatises to their patrons.[28] Occasionally they also offered works on other subject matters, including panegyric poetry.[29] Patrons presented gold, silver, jewels, precious ceramics or porcelain and other luxury objects to their physicians for particularly successful services.[30] When the head of the family recovered from a severe disease the physician who led the treatment was not only generously rewarded by the patient and patron, but also by his sons and other male relatives. In addition to al-ʿĀdil's 7,000 *dīnār* Muhadhdhab al-Dīn ʿAbd al-Raḥīm ibn ʿAlī (d. 628/1230) received gold from his relatives, robes of honour, mules with gold bridles and other items.[31] When al-Kāmil and many of his *khawāṣṣ* recovered from an illness, they sent Muhadhdhab al-Dīn gold amounting to 12,000 *dīnār*, fourteen mules with gold bridles, robes of honour made from atlas and other items.[32]

Entering patronage relationships with princes was in the view of Ibn Abī Uṣaybiʿa undoubtedly the best professional position a physician could acquire. A second, very important position was the service at the major hospitals in Damascus and Cairo. While Ibn Abī Uṣaybiʿa states that Nūr al-Dīn Zangī hired a physician to work at his hospital, he only rarely says so about the Ayyubid princes.[33] One of the cases where he imparts a little more information about the place of the hospitals in Ayyubid princely patronage is the case of al-Kāmil after the death of his father al-ʿĀdil. One of the obligations al-Kāmil had to face was to decide which of his father's physicians he would take over as his own clients. Among those who received an extension of patronage was Ibn Abī Uṣaybiʿa's father and Muhadhdhab al-Dīn ʿAbd al-Raḥīm ibn ʿAlī. While the former apparently became a member of the group of physicians who took care of the new ruler, al-Kāmil assigned a *jāmakiyya* and a *jirāya* to the latter and ordered him to reside in Damascus and to work at the Nūrī hospital.[34] Another case is that of al-Ashraf after he had conquered Damascus in 1229 CE. In this case Ibn Abī Uṣaybiʿa not only says that al-Ashraf ordered a physician to take care of the princely houses in the fortress, but also reports that al-Ashraf told him to practice regularly healing the sick in the great hospital.[35] In addition, Ibn Abī Uṣaybiʿa's occasional remarks that it was the princes who paid the physicians for their hospital services suggests seeing these activities as a second mode of patronage relationships

28 See ibid. 583, 599–602, 630, 638, 647, 651, 6581: 703, 717, 750 et passim.
29 See ibid. 630, 635.
30 See ibid. 592.
31 See ibid. 730.
32 See ibid. 731.
33 See ibid. 628.
34 See ibid. 731.
35 See ibid. 706.

where the client did not deliver personal services to the patron.[36] The fact that the leading physicians of Damascus served at the Nūrī hospital for long years and several of them even most of their lives and that they treated there *umarā'*, *a'yān*, and ordinary people speaks of the importance of this hospital in the career of an Ayyubid physician and of the social reputation that this institution possessed in the first half of the thirteenth century CE.[37] Although Ibn Abī Uṣaybi'a's rhetoric and narrative emphasis clearly single out personal service for a patron as the most desirable professional career goal the life-long affiliation of such leading physicians to the hospital and its being chosen by some of them rather than a personal patronage relationship suggests considering it in this period as a very competitive career option. In addition to the high quality of its personnel, the variety of its patients and the vivacity of its scholarly atmosphere, it provided the more conscientious physicians like 'Imrān al-Isrā'īlī, who apparently preferred to stay away from worldly power, an excellent alternative to personal service engagements. A second position of such a general type was that of a fortress physician. As the patronage relationships of Ibn Abī Uṣaybi'a's grandfather, his father, his uncle and himself indicate, the position of a fortress physician who was responsible for all the houses in the fortress brought high reputation.[38] A third position of a general type an Ayyubid prince could offer to a physician was that of a military doctor.[39]

Two other options for patronage relationships a physician could enter were those with Ayyubid *umarā'* and those with successful physicians. Young physicians, in particular when they travelled and arrived at a new location, often turned to physicians with high reputation and a broad network of relationships for supervision, introductions, recommendations and advice. Ayyubid *umarā'* and the lesser princes of the family served as entrance opportunities into the circle of patronage. Introductions to them, if they lived in the same city or town, were done in person. The local physician accompanied his young client to a sick *amīr* for an official visit of inspection and installed him as an executor of the therapy and personal caretaker. If the sick and the physician happened to live at different locations, a *ghulām* and letters of introduction served the same purpose. Letters of recommendation were an important instrument for a client who wished to enter into the service of a patron. When brought from scholars in other cities letters of recommendation eased the access to local dignitaries. When solicited from an influential local client, letters of recommendation were a crucial asset when approaching an Ayyubid patron for the first time. The literary quality of the letters mattered as much in a successful application as did their calligraphy. Advice included medical treatment, behaviour and dress codes.

36 See ibid. 697, 706.

37 See ibid. 731t: 737, 739.

38 See ibid. 738f.

39 See ibid. 673, 697, 706, 729.

Particular elements

In addition to these general features of patronage between Ayyubid princes and physicians, Ibn Abī Uṣaybiʿa also reports details, mostly as part of a story, that allow us to establish a denser web of cultural features characterizing these relationships. This web includes Ayyubid interfamilial relationships, particularities in relationships between patrons and clients, living arrangements, mobility, social advancement and values. The more prominent and the more successful a physician was the more stories Ibn Abī Uṣaybiʿa included in his biography that give testimony to the events that marked the physician's life and characterize the physician's position within the medical community and in regard to the princes. Several of such details challenge the understanding of patronage relationships between Ayyubid princes and their physicians. They suggest that patronage was not a closed system between two people. They point to the relative flexibility of boundaries between the military and physicians as distinct professions, carriers of particular values and recipients of specific kinds of benefits or favours. Some of the details show that physicians could not only rise to positions of trust with a patron due to their professional skills, but move beyond into administrative and advisory posts. The terminology used in such cases seems to imply that this meant a transformation of the patronage relationship not merely to a higher, but also to a more abstract, less personal level of meaning. Other details indicate that power was not solely concentrated in the hands of the patron. A client disposed of various ways to determine whether a patronage relationship was established and what its conditions were. Although Ibn Abī Uṣaybiʿa is very opaque about the contractual aspects of the relationship there are some whispers in his entries about negotiations and the conclusion of agreements before a decision about a patronage relationship was reached. He describes more often and more clearly the fact that patrons as well as clients sought such relationships. They asked for them, invited them, ordered them, applied for them, modified them by posing conditions or declined them.

Client-patron relationships are often understood as relationships of service and obedience on the side of the client, and power and control on the side of the patron. This is indeed the image produced by Ibn Abī Uṣaybiʿa through his terminology of service, benefits, favours and honours. It portrays a model of spikes radiating from one centre. This model is, however, not sustained by the behaviours displayed by the physicians in the stories Ibn Abī Uṣaybiʿa collected. The image produced through the stories rather suggests a model of a web of spider nets of different sizes connected with each other. The language of order and service is contradicted by the behaviour of invitation or proposal and refusal of negotiation. An important indicator of this web-like quality of the patron-client relationship in Ibn Abī Uṣaybiʿa's narrative is the fact that physicians did not only serve a succession of patrons, but could have several patrons at the same time. Another important reflection of this quality is the repetition of statements that an Ayyubid prince desired the companionship or service of a particular physician or ordered him into his service, wishes the coveted client declined and orders he refused. Two motives given by Ibn Abī Uṣaybiʿa for these reactions were the wish to remain where one was, mostly in Damascus, and the affiliation to a hospital, mostly the Nūrī hospital in Damascus. A fourth feature that led to the web-like quality of the patron-client relationship was the spread of the Ayyubid family over some major urban centres and numerous lesser holdings. Thus, clients had alternatives in terms of patrons and professional opportunities, while patrons desired to be served by particular clients. At the heart of the different model suggested by the various reactions of physicians to patronage offers by Ayyubid princes was, however, the profession of the clients and the needs of the patrons. It was the particular skills and type of knowledge involved in patronage relationships between a physician and his patient that modified the spike model of a unilateral direction of the patronage relationship from top to bottom, patron to client. The dependence of the patron in case of sickness on his client shifted the power potential between the involved parties. As a result, clients had a certain freedom to choose and to determine some of the conditions of the relationship. Professional expertise was an important factor in this bargaining power and was acknowledged as such by both sides.

Professional excellence appears in Ibn Abī Uṣaybiʿa's narration as an important criterion for establishing and regulating a patronage relationship. It was applied by both parties at various occasions. Merit served as an argument in salary and status negotiations when a new offer for patronage was made and a particular position was at stake. Muhadhdhab al-Dīn ʿAbd al-Raḥīm ibn ʿAlī who became one of the most prominent physicians of the Ayyubid period was at the beginning of his career in the service of Ṣafī al-Dīn ibn Shukr, the vizier of al-ʿĀdil. Ṣafī al-Dīn paid him a comfortable income (*jāmakiyya*) since 'he knew . . . his rank in the medical arts.'[40] In 1207–8 CE, al-ʿĀdil informed his vizier that he wished to add a second physician to the medical service of the troops who would cooperate with Muwaffaq al-Dīn ʿAbd al-ʿAzīz (d. 604/1208). The vizier recommended his own client for the job and told Muhadhdhab al-Dīn of the promotion: 'I praised you in front of the sultan and these thirty Nāṣirī *dīnār* are for you every month in the service.'[41] Muhadhdhab al-Dīn refused answering: 'O my Lord, the physician Muwaffaq al-Dīn ʿAbd al-ʿAzīz has every month one-hundred *dīnār* and a [further] payment equivalent to them. I know my rank in the science and I do not serve below his settlement.'[42]

Ayyubid princes used the criterion of professional excellence in addition to determining rank and payment for gauging the behaviour of the clients and deciding how much aberration from the norm could and should be tolerated. Arrogance, excessive pride, violations of social norms and eccentric behaviour could be overlooked by the patron if the client was seen as possessing above-average theoretical

40 Ibid.
41 Ibid.
42 Ibid.

knowledge and a successful record in exercising his craft. At the same time it was expected that clients who had successfully amassed goodwill on the side of their patrons, power and material means applied themselves to mediating conflicts, in particular issues or remuneration, for clients in less fortunate positions and to support the profession at large. Arrogance and pride, while occasionally tolerable when applied to *umarāʾ* and other people of high rank in the Ayyubid dynasty, were not behaviours well looked upon where the professional community was concerned. The exemplary case of both ways of behaviour was Muwaffaq al-Dīn ibn al-Muṭrān (d. 587/1191). He served Ṣalāḥ al-Dīn for many years and rose to a position of great wealth and power. His status was so high that Ibn Abī Uṣaybiʿa described him as "concealed" by the ruler and as having stopped treating people.[43] When he began treating even the princes with arrogance and excessive pride, Ṣalāḥ al-Dīn accepted it because he respected and revered him due to his professional excellence.[44] Ibn al-Muṭrān sulked and refused to serve for days when his patron graced another of his clients with excessive generosity or curbed some of Ibn al-Muṭrān's excesses of haughtiness such as setting up during a campaign against the crusaders not only a tent of the same red colour as that of Ṣalāḥ al-Dīn, but one that had more luxury than that of the ruler. In order to appease his main physician, the ruler sent him each time a small fortune.[45] Ibn al-Muṭrān's behaviour was of a very different kind, claims Ibn Abī Uṣaybiʿa on the authority of Ibn al-Qifṭī and others, in regard to the scholarly community at large and the medical community in particular. When he went to study at the mosque he walked, although many of the *mamālik* accompanied him riding. He respected the *shaykh* who held the class, sat among the students and participated like them speaking in a soft voice and with great courtesy. He lent his authority to all those physicians who needed help for settling payment disputes, introduction to wealthy clients or an opportunity for earning a living. Those who succeeded came to deliver the often substantial benefits they had received. Paragon that Ibn al-Muṭrān was, he declined magnanimously the fees and assured his clients of his continued good will and support.[46]

The most unusual marker among the qualifiers of good behaviour that sets Ibn Abī Uṣaybiʿa's rhetoric apart from other biographical dictionaries of scholars in the Ayyubid and Mamluk period is the term *muruʾa*. He used it to describe Ibn al-Muṭrān's behaviour towards young professional newcomers to Damascus and a young man fallen on hard times with no knowledge of the medical profession whom he secured nonetheless entrance to patronage with an Ayyubid prince. It is a term that appears in the entries of the leading Ayyubid physicians, often only as the short claim that the person 'possessed the complete ideal of manhood

43 See ibid. 652.
44 Ibid.
45 See ibid. 652f.
46 See ibid. 653–655.

(*dhū l-murū'a al-kāmila*).[47] Often the word seems to refer primarily to acts or generosity and support for the less fortunate as in the case of Ibn al-Muṭrān whom Ibn Abī Uṣaybiʿa described as having much *murū'a* and a noble soul explaining this by adding that he gave his students books and robes of honour when they healed a patient.[48] A story about one of Ibn Abī Uṣaybiʿa's teachers and most prominent Damascene physicians, Muhadhdhab al-Dīn ʿAbd al-Raḥīm ibn ʿAlī, makes clear, however, that the concept means more than mere generosity and charity. In Ibn Abī Uṣaybiʿa's view it is closely related to "tribal or group solidarity" (*ʿaṣabiyya*), which a man of excellence should possess in abundance.[49] The system of Ayyubid patronage apparently was sustained on the side of its clients by a strong sense of belonging to a social group and by a code of honour that included the obligations to exercise *murū'a* and *ʿaṣabiyya*. The social group often centred on the practitioners of the medical arts, but could also include scholars of other disciplines and those who served in the judiciary and possibly other public positions.

The web-like quality of the patron-client relationship between Ayyubid princes and their physicians reflects the Ayyubid practice of power sharing within the family on the one hand and of contesting power by military means on the other. This practice made mobility a key component of Ayyubid rule, including their patronage relationships. Ayyubid mobility encompassed campaigns, moves to a new location due to an advancement or downfall in the family hierarchy, family visits, pilgrimage and a few leisure activities. Some of the strongest terms of appreciation Ibn Abī Uṣaybiʿa used when talking about a patron's perceptions of his client express the desire of the patron to be with his client. This desire could be described by saying that the prince spent most of his time with the physician. It also could be expressed by saying that he rarely was separated from him (*wa-lā yufāriquhu fī akthar awqātihi*) or that he wished to have him with him whether in residence or on a voyage (*fī l-ḥaḍar wa-l-safar*).[50] While described as one of the greatest honours an Ayyubid prince could bestow upon his client, this mobility did not, however, please all physicians. Not only did physicians resist the demand to relocate with their patrons, some apparently also found travelling around a nuisance. When Rashīd al-Dīn, Ibn Abī Uṣaybiʿa's uncle, received a patronage offer from al-Muʿaẓẓam due to his skills as an orator it was specifically agreed that his service was of a kind that he was exempted from the sultan's movements.[51] Many physicians though accepted travelling with their Ayyubid patrons, be it to a new location or on a campaign against the Crusaders, members of the Ayyubid family and other Muslim rulers.[52] The language used by Ibn Abī Uṣaybiʿa for describing the participation of physicians in military campaigns only rarely reveals the

47 See ibid. 703. Also, see ibid. 590.
48 See ibid. 655.
49 See ibid. 629.
50 See ibid. 635, 663, 739.
51 See ibid. 737.
52 See ibid. 585f, 700.

reasons why physicians accompanied their patrons into an encampment or even a battle. Some provided medical service to the soldiers. Others accompanied their patron as part of their overall duties as a client of high standing as was the case of Ibn al-Muṭrān.[53] A few participated voluntarily in such campaigns as fighting men as was the case of ʿAbd al-Munʿim al-Jilyānī (d. after 587/1191) and perhaps also Rashīd al-Dīn ibn al-Ṣūrī (d. 639/1242).[54] Some of the physicians who accompanied their patrons on a campaign also continued teaching medical texts to their students as happened to Ibn Abī Uṣaybiʿa himself at the beginning of his studies with Muwaffaq al-Dīn Yaʿqūb, a Christian physician of al-Muʿaẓẓam ʿĪsā.[55]

The web-like quality of patronage relationships between Ayyubid princes and physicians was supported by the practice that heads of Ayyubid households did not only establish themselves as patrons for their own health and that of their wives, concubines and under-age children. They assigned at least occasionally physicians who were bound to them as clients to adult male relatives, mostly sons, who ruled their own sub-domains of the Ayyubid territory. Such an assignment could impose on the respective client additional obligations, among them either to travel or to relocate from an Ayyubid centre to an Ayyubid periphery.[56] Another form of assigning physicians as clients to male relatives was the fulfilment of an obligation offered to a client as payment for a special service. When the Christian physician Abū Sulaymān Dāwūd ibn Abī l-Munan (d. *c.* 583/1187) had given voluntary astrological advice to Ṣalāḥ al-Dīn that convinced him to campaign against the Crusader kingdom of Jerusalem ending with its conquest, the Ayyubid prince offered to rely on him. As a response the physician asked for his protection of his five sons. The ruler agreed to look after them and attached them to his brother al-ʿĀdil.[57] Al-ʿĀdil in his turn assigned three of the brothers to his son al-Kāmil and two to his other son al-Muʿaẓẓam ʿĪsā.[58] Elder male relatives occasionally invited physicians of the head of the dynasty to serve them in a particular kind of disease without addressing their request to their family head and patron of the physician.[59] Ayyubid princes also asked physicians to travel on their behalf to their relatives for a check-up and treatment.[60] While in cases of princely assignments it remains unclear whether clients could have declined accepting them, in the case of invitations and requests, the physician apparently could make his own decisions, even if it meant several days of travel and longer absences from the patron or position he served. In 1219 CE, when Rashīd al-Dīn ʿAlī ibn Khalīfa was in the service of al-Muʿaẓẓam, his patron's brother al-Ṣāliḥ Ismāʿīl (r. 1237–8 and 1239–45 CE

53 See ibid. 656.
54 See ibid. 630, 700.
55 See ibid. 698.
56 See ibid. 590.
57 See ibid. 588.
58 See ibid. 589–591.
59 See ibid. 652.
60 See ibid. 741.

in Damascus) sent him a letter in his own hand, as Ibn Abī Uṣaybiʿa proudly notes, asking him to come to Basra since his mother and other people in his retinue had fallen ill. Rashīd al-Dīn, although well advanced in years, accepted the invitation, went to Basra and healed the princess, a service which they remunerated by paying him gold and giving him robes of honour.[61]

Promotion within a patronage relationship reflected the dependence of the patrons on their clients. Clients were promoted within the domain of their professional competence for their knowledge of medical theory, their skills in healing, their capability of producing powerful theriacs and their familiarity with a broad range of simple and compound drugs. A patron chose to rely on a particular physician in these three domains of theory, therapy and pharmacy, which Ibn Abī Uṣaybiʿa expressed by describing the physician as a pillar on which the prince leaned.[62] When Muwaffaq al-Dīn Yaʿqūb ibn Saqlāb served al-Muʿaẓẓam ʿĪsā and was with him in a companionship, the prince was filled with good, firm belief in him until he relied on him in many medical and other views. He profited from them and commended their effects.[63] Sadīd al-Dīn Abū Manṣūr (fl. thirteenth century CE) served al-Nāṣir Ṣalāḥ al-Dīn Dāwūd in a companionship in al-Karak. The prince raised him to a high position and relied on him in the art of medicine.[64] In addition to this singling out for particular professional expertise as a physician, clients were also tapped for their additional skills in the mathematical sciences, as a companion in sessions of conviviality and as political or financial advisors and managers. Ibn Abī Uṣaybiʿa's uncle Rashīd al-Dīn served al-Amjad on the prince's request as an author of a manual on arithmetic.[65] Al-Muʿaẓẓam implored him to become the head of the administration of the troops.[66] Several physicians of Ayyubid princes rose to become their viziers, namely Muhadhdhab al-Dīn Yūsuf ibn Abī Saʿīd (d. 624/1227), his nephew Amīn al-Dawla Abū l-Ḥasan ibn Ghazāl ibn Abī Saʿīd (d. 648/1250), Fakhr al-Dīn Riḍwān ibn Rustam al-Sāʿātī (d. *c.* 629/1230), Najm al-Dīn al-Lubūdī (d. *c.* 65/1265) and Najm al-Dīn ibn Minfākh (d. 652/1254). Muhadhdhab al-Dīn Yūsuf ibn Abī Saʿīd was highly appreciated for healing Sitt al-Shām when she suffered from dysentery. He served ʿIzz al-Dīn Farrukhshāh (d. 587/1191) and his son al-Amjad in Baʿlbak as a physician. The latter turned more and more to him for advice in his affairs (*umūr*) and relied upon him in his powers (*aḥwāl*). Intelligence and virtue made the physician's advice sound and approved of by the prince who finally made him his vizier. Muhadhdhab al-Dīn Yūsuf ibn Abī Saʿīd rose to a high rank until he was the manager of all of his realm (*dawla*) and powers (*aḥwāl*). Nobody equalled

61 Ibid.
62 See ibid. 738 et passim.
63 See ibid. 698.
64 See ibid. 699.
65 See ibid. 738.
66 See ibid. 740.

him in authority and command.[67] In a few cases, the head of the family honoured a client's success with his relatives by appointing him to high-ranking positions in the administration of Egypt or Syria.[68] However, not all clients accepted this promotion above and beyond the medical profession, underscoring the possibilities of choice for a client in the Ayyubid patronage system. Al-Muʿaẓẓam ʿĪsā wished Muwaffaq al-Dīn Yaʿqūb ibn Saqlāb to chair some part of the management of his realm (*dawla*) and to administer it, but the physician refused, limiting himself exclusively to his medical work.[69]

The fact that physicians did not live, as a rule, in the households of their patrons, but had their own houses in town, may have enforced the flexible character of the patronage system since it gave clients spaces of their own and distance from their patrons. While physicians did not live in the houses of their patrons, they met there regularly for their service. The expression by which Ibn Abī Uṣaybiʿa describes time and again the presence of the physicians in the houses of their patrons is being at the *bāb al-dār*. One day early in his career, Muhadhdhab al-Dīn ʿAbd al-Raḥīm ibn ʿAlī, for instance, was 'at the *bāb dār al-sulṭān* and with him was a group of the physicians of the houses,'[70] that is, the houses at the fortress, when a eunuch came with a slave girl to consult the doctors about something that ailed her. Many times Muhadhdhab al-Dīn met with Muwaffaq al-Dīn Yaʿqūb at this very place 'where the physicians sat at the *dār al-sulṭān* and discussed medical research matters.'[71] Since it is very unlikely that a medical consultation between a eunuch and a physician would have taken place in the open air, in front of the princely residence, not to mention the presence of a slave girl in such an environment, and since it is not plausible to assume that the two physicians always met in a courtyard for discussing their professional opinions, the *bāb al-dār* does not signify the entrance of the house or the fortress, but rather a space in the house that preceded the rooms of the princely patron. It is there where the physicians of an Ayyubid prince worked and spent their service time as doctors. Unfortunately, Ibn Abī Uṣaybīʿa is silent about the equipment of this space, that is, whether it contained furniture for medical instruments and remedies as well as shelves for manuscripts.

If the clients were prominent and well-established physicians, they were able to accrue much wealth. They expressed their economic fortunes through large and richly endowed houses, the quality of their attire, and the beauty and education of their person. Most of their wealth consisted of cash and in some cases of landed property in villages. The separation of the living spaces of patrons and clients reflects the lack of a formalized court, the flexibility of Ayyubid lifestyles

67 See ibid. 722.
68 See ibid. 663.
69 See ibid. 698f.
70 See ibid. 729.
71 See ibid. 698.

including their accommodations while in town and the fact that not all Ayyubid princes engaged in exclusive patronage relationships with their physicians, that is, not all physicians served only a single princely patron. Abū Saʿīd ibn Abī Sulaymān (d. 613/1216–7), that is, one of the sons of the Christian physician Abū Sulaymān whom Ṣalāḥ al-Dīn had promised to protect, is named by Ibn Abī Uṣaybiʿa as the client of three Ayyubid princes – Ṣalāḥ al-Dīn, al-ʿĀdil and al-Kāmil – with the implication that he served them simultaneously, not subsequently.[72] Ibn Abī Uṣaybiʿa's relatives also served several Ayyubid princes at the same time – Ṣalāḥ al-Dīn and his children or al-ʿĀdil, al-Muʿaẓẓam ʿĪsā and al-Amjad.[73] Even in cases where Ibn Abī Uṣaybiʿa seems to imply that a physician had entered an exclusive service with one patron only, such a status did not prevent him from exercising his profession among other patients in town.

Many, but not all, Ayyubid princes lived in their fortresses when in town.[74] Al-ʿĀdil for instance lived in Cairo in a residence called the "House of the Viziers".[75] Physicians lived mostly outside of the fortresses, but could occasionally also live in a house inside a fortress. Another of Abū Sulaymān's sons and a client of al-ʿĀdil, Abū Shākir (d. 613/1216–7), had a house in the citadel of Cairo in order to live with and take care of al-Kāmil according to al-ʿĀdil's wish.[76] In other cases, the house of a physician could be linked through a door or a hallway to the house of the Ayyubid prince for the delivery of special services. When al-Kāmil ruled the city of al-Ruhāʾ, his residence was connected with the house called the "House of Ibn al-Zaʿfrān" where the physician al-Fāris Abū l-Khayr ibn Abī Sulaymān (fl. twelfth and thirteenth centuries CE) lived. Whenever al-Kāmil went for a bath, al-Fāris was supposed to deliver fruits and rosewater.[77]

The group of physicians that served Ayyubid princes as clients was composed of members of various Christian and Jewish communities and of Muslims. Religious affiliations beyond the generalities were not unimportant for Ibn Abī Uṣaybiʿa as his occasional hints at specifics in cases of Christians and Jews indicate. But they appear to have been of less importance in the case of physicians than in other fields of intellectual enterprise. The Ayyubids also were willing to disregard stipulations that discriminated against the *ahl al-dhimma* when they wished to include non-Muslim physicians among their clients, something they did regularly and in relatively great numbers. A particular favour and right that Ayyubid princes granted their Muslim, Christian or Jewish physicians was the permission to ride in town, to ride into their fortresses and up to the door of their residences and to remain on

72 See ibid. 589.
73 See ibid. 736–738.
74 For instance, see ibid. 707.
75 See ibid. 589.
76 Ibid.
77 See ibid. 590. Ibn al-Zaʿfrān was a former ruler of al-Ruhāʾ who had been disposed by the Ayyubids.

horseback even in the presence of their patrons.[78] The Christian physician Abū Shākir ibn Abī Sulaymān, for instance, was allowed to ride into the fortresses of Damascus, Karak, al-Ruhā᾿ and Ja῾bar. When al-῾Ādil became head of the family residing in Cairo, he extended this permission to the citadel of Cairo. One day he visited his son in the citadel, went to the physician's house, ordered him to mount a mule that al-῾Ādil had brought along with him from his own residence and ride on it from the citadel to the "House of the Viziers", while the *umarā᾿* and his son walked there.[79] Not all Muslim inhabitants of the Ayyubid territories acquiesced to such revocations of Muslim legal stipulations. An inhabitant of the famous Sufi *Khānaqāh al-Sumaysāṭ* took it upon himself to punish in Cairo a physician whom he perceived of unlawful behaviour. Ibn Abī Uṣaybi῾a described him as being acquainted with Najm al-Dīn Ayyūb who ruled in al-Jazīra and his brother Asad al-Dīn, having 'a heavy spirit', living an ascetic life, being hard in religion and 'eating the world by law and order.'[80]

> When Asad al-Dīn moved to Cairo, he followed him and settled in the mosque close to the House of the Viziers. He began slandering the people of the fortress and cursed them when praising God, becoming very influential. Whenever he saw a *dhimmī* riding he wished to kill him. One day, he saw the Jew al-Muwaffaq ibn Shū῾a [d. 579/1183–4], one of the senior physicians of Ṣalāḥ al-Dīn, riding. He threw a stone at him and hit his eye in a way that it could not be saved.[81]

Ibn Abī Uṣaybi῾a's choice of words indicates that he did not condone such behaviour against a colleague. But he did not provide further comments on this event. Hence it remains unclear whether Ayyubid patronage relationships included the obligation to extend legal and material protection in such cases of bodily harm.

When particularly successful Christian or Jewish physicians serving an Ayyubid prince agreed to convert to the religion of their patrons, elaborate and lavish ceremonies took place to celebrate the event. One such case described by Ibn Abī Uṣaybi῾a on the authority of Ibn al-Qifṭī was the conversion of Ibn al-Muṭrān. He received panegyric poems and an extraordinary amount of other gifts.[82] The *umarā᾿ al-dawla* competed with each other for giving him the best of the best. Ṣalāḥ al-Dīn married him to Jawza, one of the preferred servants of his wife Khawand Khāṭūn. Jawza was a very rich woman in her own right and improved her husband's position and affairs, 'putting them in order, rectifying his conditions, taking care of his attire and beautifying his interior and exterior.'[83] His

78 See ibid. 589.
79 Ibid.
80 See ibid. 581.
81 Ibid.
82 See ibid. 657.
83 See ibid. 656.

conversion was celebrated with a royal announcement (*wa-ṣāra lahū dhikr sāmin fī l-dawla*).[84]

Physicians were highly appreciated for their medical skills. This alone did not bring, however, necessarily an invitation to an evening session of conviviality, although some of the leading physicians seem to have been included in the leisure activities of their patrons, due to their intellectual status and ranking. Those physicians who were well-versed in poetry, played *ʿūd* or had a good singing voice received such invitations, independent of their medical skills, as well as patronage offers that emphasized their capabilities as artists and delightful companions. Fakhr al-Dīn ibn al-Sāʿātī excelled in calligraphy and poetry. In addition to being a physician and a vizier of two Ayyubid princes, he also became a boon companion to one of them, namely al-Muʿaẓẓam, and played the *ʿūd* for him.[85] Rashīd al-Dīn, Ibn Abī Uṣaybiʿa's uncle, was an exemplary client being highly talented as a musician and capable of reciting poetry in Arabic, Persian and Turkish in addition to his medical skills.[86] On 15 Ramaḍān 605/23 March 1209, al-Muʿaẓẓam ʿĪsā 'invited him for listening to his speech. He was very pleased by his performance, benefited him [and] ordered him into his service.'[87] Another time, when al-Kāmil visited his brother al-Muʿaẓẓam the two princes spent a night of conviviality together. Rashīd al-Dīn participated in the event and received for his contributions a perfect robe of honour and five hundred Egyptian *dīnār*.[88] Other festive events (*ḥafla*) were organized as scholarly sessions. ʿAbd al-Laṭīf al-Baghdādī (d. 629/1231) reported that one evening in Jerusalem after Ṣalāḥ al-Dīn had concluded a truce with the Christian troops in Akko he felt it necessary to pay him a visit. He collected from his books of the ancients what was possible and made his way to Jerusalem. He found the ruler completely pleased, with a happy heart and all of his companions joyous like him. The first night of his presence in Jerusalem a *ḥafla* session with the *ahl al-ʿilm* (i.e., scholars, primarily of the religious disciplines, but also of other fields) took place. They presented talks and readings about the various kinds of sciences, making it a thoroughly pleasant and enjoyable event.[89]

Ayyubid patronage relationships existed between individuals, not between groups or institutions, even in the case of the more general types of patronage referred to above. This personal character is reflected in the numerous particularities that separate one relationship from the other. Patrons could override both law and custom if they wished to honour a client in a special way. The transfer of property rights was one of the most sensitive issues, in particular when land was involved. Such a transfer of land figures only occasionally in Ibn Abī Uṣaybiʿa's accounts, but occurred nonetheless with a certain regularity insofar as it is reported

84 See ibid. 654.
85 See ibid. 662.
86 See ibid. 737.
87 Ibid.
88 See ibid. 739.
89 See ibid. 688.

for the Zangids, the Ayyubids and the Mamluks.[90] Patrons could also interfere in the private affairs of their physicians if they wished to. Ibn Fāris loved to dress in military dress and to fight although he served as a physician. When he had a little son, he also dressed him in uniform. One day al-Kāmil was served in his bath in al-Ruhā' fruits and rosewater by this little boy. Looking at him he recognized him as a relative of his physician. He called the father and said to him:

> This your son is intelligent. Do not teach him to be a soldier. We have many soldiers. You are a fortunate family. You are blessed with your medicine. Send him to the physician Abū Saʿīd in Damascus that he will read with him medicine.[91]

The 'father had no option but to follow the order'[92] and the son, Rashīd al-Dīn ibn Ḥalīqa, became one of the most prominent Ayyubid physicians who at the end of his life served the Mamluk sultan al-Ẓāhir Rukn al-Dīn Baybars (r. 1260–77 CE).[93] Clients could approach a patron for special favours such as financial help in family affairs. The Christian physician Abū l-Faraj served Ṣalāḥ al-Dīn and his household.

> One day he said to the sultan that he had daughters and that he needed to endow them. He demanded that he grant him what he needed in this respect. Ṣalāḥ al-Dīn said to him: write down on a piece of paper all that is necessary for endowing them and bring the paper. Abū l-Faraj departed and wrote on the paper jewellery, fabric, utensils etc. of about 3,000 *dirham*. When Ṣalāḥ al-Dīn read the paper, he ordered the treasurer to buy for Abū l-Faraj all that it contained, leaving nothing out.[94]

Scholars of the mathematical sciences at Ayyubid courts

Ibn Abī Uṣaybiʿa's work is only in one respect a rich source for studying the places that the mathematical sciences occupied under Ayyubid rule, that is, in regard to the education and preferences of physicians. This general aspect includes as a sub-theme the professions in which people had earned their living, before they switched to medicine. A second sub-theme is the skills that a physician possessed in addition to his medical knowledge and proficiency in healing. A third subtheme is the kind of people with whom the physicians met, related to in friendship and visited for learning from them. In all three cases, patronage relationships

90 See ibid. 591, 638.
91 See ibid. 591.
92 Ibid.
93 Ibid.
94 See ibid. 652.

appear as one component of the story. The three possible situations when Ibn Abī Uṣaybiʿa talks about the mathematical sciences indicate that while only a minority of physicians seems to have studied one of them these sciences had a very high reputation. Those that were of interest to patrons were astrology, the compilation of astronomical handbooks, the construction of instruments, arithmetic, engineering and music; the latter in particular when combined with musical practice. Other disciplines such as geometry or *ʿilm al-hayʾa* (mathematical cosmography) were considered as theoretically challenging and caused amazement and wonder if mastered fast and profoundly.

The mathematically related professions that were exercised by several physicians before they acquired their medical training were architecture and instrument making, in particular the construction and repair of sundials. These professions when exercised on public buildings like the Nūrī hospital or the Umayyad mosque were financed by stipends from the princes, an approach already practised by Nūr al-Dīn Zangī.[95] Muhadhdhab al-Dīn ibn al-Ḥājib (d. after 583/1193), praised by Ibn Abī Uṣaybiʿa as strong in theoretical thinking (*naẓar*) in geometry, was one of those physicians who began their career with serving the clocks at the Umayyad mosque.[96] Another was Fakhr al-Dīn al-Sāʿātī who had learned the craft and astrology from his father, a migrant from Khurasan.[97] Some physicians also built other astronomical and geometrical instruments. Abū Zakariyyāʾ Yaḥya al-Bayāsī (fl. twelfth and thirteenth centuries CE) came from the Maghrib and excelled in the mathematical sciences. He made geometrical instruments for his teacher in medicine, Ibn al-Naqqāsh al-Baghdādī.[98] He also played *ʿūd* and the organ.

Rashīd al-Dīn ibn ʿAlī ibn Khalīfa, Ibn Abī Uṣaybiʿa's uncle, studied not only medicine, but also arithmetic, music, geometry, *ʿilm al-hayʾa* and astrology. He was very gifted in all these fields.[99] Rashīd al-Dīn's knowledge of the theory and practice of these various mathematical sciences improved his chances of receiving honourable and lucrative patronage offers. When al-Muʿaẓẓam asked him to chair the administration of the troops he felt that he had to comply with the demand. After some time of sitting on the council he discovered that all of his time went into running the chancellery and the accountancy, while nothing remained for his personal life or his study of the rational sciences. Although his position brought him much power and connections, he decided it was not worth it. He turned to the

95 See ibid. 661, 670.

96 See ibid. 659.

97 See ibid. 661.

98 See ibid. 637.

99 He learned arithmetic from Abū l-Taqī Ṣāliḥ ibn Aḥmad al-Maqdisī, astrology from Abū Muḥammad ibn al-Jaʿdī, the son of a Fatimid *amīr* who had served the Fatimid caliphs and was considered one of their *khawāṣṣ*, mathematical cosmography from ʿAlam al-Dīn Qayṣar ibn Abī l-Qāsim (d. 649/1251), 'the leading scholar of the mathematical sciences in his time' (ibid. 736) in the Ayyubid realm and music from practitioners as well as writers about the field according to Arab and Persian teachings. See ibid. 736f, 740.

sultan and demanded his release. But the sultan kept him as one of his *khawāṣṣ* until he finally discharged him.[100]

Several of Ibn Abī Uṣaybiʿa's teachers also studied various mathematical sciences and excelled in them. Muhadhdhab al-Dīn ʿAbd Raḥīm ibn ʿAlī studied *ʿilm al-hayʾa* and astrology. He took classes in these two disciplines with the astrologer Abū l-Faḍl al-Isrāʾīlī. He purchased the brass instruments that were needed in this art and possessed 16 treatises on the astrolabe written by various authors.[101] Muwaffaq al-Dīn Yaʿqūb ibn Saqlāb was born and grew up in Jerusalem. There he learned Greek and studied in the monastery Dayr al-Siyaq with a monk who was a virtuous man, a philosopher and very knowledgeable in natural philosophy, geometry and astrology.[102]

The interest in the mathematical sciences motivated a few physicians to undertake long-distance travel to study with a leading expert. When Sharaf al-Dīn Ṭūsī (d. c. 607/1210) was in Mosul, Muhadhdhab al-Dīn ibn al-Ḥājib and Muwaffaq al-Dīn ʿAbd al-ʿAzīz travelled there to meet with him and study with him. When Sharaf al-Dīn returned to his native town in north-eastern Iran, the two Damascene physicians accompanied him and stayed there with him for some time. On his way back to Damascus, Muhadhdhab al-Dīn ibn al-Ḥājib passed through Irbil where he met another gifted scholar of the mathematical sciences, the astrologer Ibn al-Dahhān al-Thuʿaylab (d. 590/1194). The two scholars then travelled together to Damascus. There, the astrologer was very well received by Ṣalāḥ al-Dīn and all the leading men. The Ayyubid ruler honoured the astrologer greatly and settled a monthly stipend (*ujra*) of 30 *dīnār* on him. In the Kalāsa quarter an unnamed patron built a lodge (*maqṣūra*) for the astrologer.[103]

A field studied only in exceptional cases, if one is to trust Ibn Abī Uṣaybiʿa, was the construction of automata where the work of the Banū Mūsā (fl. ninth century CE) was used for instruction and guidance to build some such machines.[104] The fact that ʿAlam al-Dīn Qayṣar is also known for his activities in this domain, having constructed on al-Ashraf's order many beautiful things, among them an octagonal palace in Raʾs al-ʿAyn, and that there existed a certain, albeit not well-documented interest in the work of ʿAbd al-Razzāq al-Jazarī (fl. 1200–22 CE) in the Ayyubid period, suggest that this subject matter was not of major relevance to Ibn Abī Uṣaybīʿa and his work and hence received only little attention.[105]

In addition to such information linked to scholars who practised medicine Ibn Abī Uṣaybiʿa has little to say of Ayyubid patronage for the mathematical sciences. He even does not apply the language of patronage to the few scholars of these sciences whom he mentions and who did not belong to the medical community.

100 See ibid. 740.
101 See ibid. 733.
102 See ibid. 698.
103 See ibid. 659.
104 See ibid. 704.
105 See Ibn Wāṣil (1977), V:135f.

Other sources such as Ayyubid historical chronicles and biographical dictionaries written outside the Ayyubid realm confirm, however, that the few leading scholars of the mathematical sciences who were linked to Ayyubid princes and are known to us through their work, that is, Mu'ayyad al-Dīn al-'Urḍī (d. 660/1260), Muḥyī l-Dīn ibn Abī Shukr al-Maghribī (d. 681/1282), and 'Alam al-Dīn Qayṣar, talked of themselves as serving Ayyubid princes and were perceived by others as their clients. Mu'ayyad al-Dīn al-'Urḍī described his relationship with the last Ayyubid ruler al-Nāṣir Ṣalāḥ al-Dīn Yūsuf as one of service when he told his new friends and colleagues in Maragha how he had escaped through his own courage the bloodbath in which all members of the court had died except for himself and two sons of the ruler.[106] Jamāl al-Dīn ibn Wāṣil (d. 697/1298), a historian of the Ayyubid dynasty and well educated in various disciplines including mathematics and philosophy, characterized his relationship to the patron of his father, Nāṣir al-Dīn Dāwūd, as well as the relationship of his older colleague in the mathematical sciences 'Alam al-Dīn Qayṣar to al-Ashraf exclusively by one term: service (*khidma*).[107] Historical chronicles, while not overly rich in information about Ayyubid patronage of the mathematical sciences, also suggest that the impression that Ayyubid princes had only limited interest in astrological counselling and hence only supported very few practitioners of this discipline may not reflect the historical conditions in their complexity. Sibṭ ibn al-Jawzī (d. 654/1257) described in his chronicle *Mir'āt al-zamīn* how al-Ashraf went about killing his relative Bahrām Shāh ibn Farrukhshāh (r. 1191–1230 CE) in Ba'lbak against whom he carried an old grudge in his heart. Part of the intrigue was to make him leave his own territory and come to Damascus. Bahrām Shāh agreed, but planned to take the city. In preparation for the campaign his Jewish astrologer Ibn Fahīd determined the auspicious hour. The astrologer used an astrolabe for this purpose and undertook the operation in the presence of his patron who was playing a game with his secretary 'Abbās ibn Akhī al-Sharīf al-Bahā' according to their habit.[108] Unfortunately, the hour proved the astrologer wrong and his patron was killed. The intimacy of the story as well as the religious affiliation of the astrologer paint Ayyubid patronage of the mathematical sciences in colours not known from other sources. Astrologers like physicians were not merely professional experts who delivered services. They were persons of trust who operated in the immediate and often private environment of the patron.

Philosophers, princes, and their predilections

Ibn Abī Uṣaybi'a included in his biographical entries 75 scholars who lived and worked in Ayyubid Egypt and Syria. Twenty-four of them are presented in chapter

106 Ibn al-'Ibrī (1958), 280.
107 Ibn Wāṣil (1977), V:145f.
108 Sibṭ ibn al-Jawzī (1371/1952), 667.

XIV about Egypt and fifty-five in chapter XV about Syria. Although most of them were practising physicians, some of them had at best a literary knowledge of the field. Scholars like Shihāb al-Dīn Suhrawardī, Sayf al-Dīn al-Āmidī, Rafīʿ al-Dīn al-Jīlī (executed between 637–43/1240–5), Shams al-Dīn Khusrawshāhī (d. 652/1251–2), Shams al-Dīn al-Khūyʾī (d. 637/1240) and Afḍal al-Dīn Khūnjī (d. 640/1243) are better known for their contributions to the rational sciences than for their medical expertise. Their presence in Ibn Abī Uṣaybiʿa's biographical dictionary of physicians is one of the indicators for the presence of and interest in the philosophical disciplines and their relatives among the religious disciplines, in particular *uṣūl al-dīn*, in the Ayyubid period. These five scholars apparently were the leading representatives of this intellectual domain in Ayyubid Damascus and Cairo. To ignore them was obviously not a choice Ibn Abī Uṣaybiʿa wished to make. The relevance of this decision becomes clearer when we consider that the author did not treat leading astronomers and mathematicians of the period in the same manner.[109] Ibn Abī Uṣaybiʿa's inclusion of the five leading and often controversial figures of the rational sciences in Ayyubid Syria and Egypt into his dictionary was neither an accident nor a decision based merely on a wish for comprehensiveness. In some cases, he obviously wished to report the events in which the scholars had been involved and suggest his view of what had happened to them. But critiquing his contemporaries and their deeds was not a consistent approach that Ibn Abī Uṣaybiʿa followed in his book.[110] The five mentioned scholars and their entries are rather linked by the fact that they were famous representatives in their respective fields, that some of them were the author's teachers and that most of them were friends of his family and some of his teachers. In 1239–40 CE, Ibn Abī Uṣaybiʿa read in Cairo with Afḍal al-Dīn Khūnjī parts of the *kulliyyāt* of Ibn Sīnā's medical *Qānūn*.[111] In Damascus, he studied with Shams al-Dīn Khusrawshāhī and Rafīʿ al-Dīn al-Jīlī a philosophical text of Ibn Sīnā and other works.[112]

Among the 70 physicians presented by Ibn Abī Uṣaybiʿa, 26 are mentioned as students of either *ḥikma* or *falsafa*. This number is most likely an inferior boundary since there are cases like the one of Sadīd al-Dīn ibn Abī l-Bayyān (d. after 638/1240) where Ibn Abī Uṣaybiʿa does not mention their training in these disciplines in their own entry, while referring to it in other biographies.[113] In most cases

109 Despite the fact that Sharaf al-Dīn Ṭūsī spent some time in Damascus and was acknowledged by Ibn Abī Uṣaybiʿa as the leading exponent of the mathematical sciences and a major figure in the philosophical disciplines, he did not devote an entry to him. The same applies to Muʾayyad al-Dīn al-ʿUrḍī and Muḥyī l-Dīn ibn Abī Shukr al-Maghribī, both leading scholars of the mathematical sciences of their time and clients of Ayyubid princes, the former at least having been known and appreciated by Ibn Abī Uṣaybiʿa. See Ibn Abī Uṣaybiʿa (1965), 768.

110 He is, for instance, conspicuously silent about Sayf al-Dīn al-Āmidī's loss of position and its causes.

111 See ibid. 586.

112 See ibid. 649, 729.

113 See ibid. 583, 737.

Ibn Abī Uṣaybiʿa merely enumerates the disciplines in an additive form. He obviously did not feel a need for justifying the presence of the philosophical sciences in his dictionary and saw no reason for hiding a scholar's interest in them. A certain hint towards undercurrents of change may be contained in juxtapositions between the philosophical and the legal disciplines when Ibn Abī Uṣaybiʿa emphasizes that a physician in addition to his acquisition of philosophical knowledge also was versed in *sharīʿa*.[114] The shift from philosopher-physician to jurist-physician has been marked by Sabra as one important component of the transformation in the cultural role of the ancient sciences in the late Abbasid period.[115] The developments in the Ayyubid realm contributed substantively to this process.

When we turn to the question how Ayyubid patrons looked upon the philosophical sciences the information delivered by Ibn Abī Uṣaybiʿa tells us first and foremost that the princes sponsored almost all of the physicians with training in philosophy. They did not question their future clients about it nor did they reprimand them for having sessions on such topics. Some of them accepted treatises on philosophical topics by their clients as gifts.[116] Other indicators of the overwhelmingly either positive or neutral position that Ayyubid princes took towards the philosophical studies of their clients are to be found in the fact that all viziers who came from the medical profession except for one either taught or wrote about philosophical subjects, that several head judges appointed by Ayyubid princes were deeply involved with the philosophical sciences and that the four Ayyubid princes who were involved either in the execution of Shihāb al-Dīn Suhrawardī or in the problems of Sayf al-Dīn al-Āmidī, that is, Ṣalāḥ al-Dīn, al-Ẓāhir, al-Kāmil and al-Ashraf, all patronized other scholars with strong and well-known philosophical interests.[117] Those princes who cared at all for these intellectual debates and the tensions resulting at times from them may have well believed like Rashīd al-Dīn, Ibn Abī Usaybiʿa's uncle, that *ḥikma* meant 'following the example of God, the Most High'.[118]

Some of the scholars with only a literary or theoretical competence in medicine also were clients of Ayyubid princes or of an Ayyubid vizier, who honoured them, assigned teaching positions in the newly built *madāris* to them or appointed them as head judges. Several of them had studied with Fakhr al-Dīn al-Rāzī (d. 606/1209) or one of his students. The most influential student of Fakhr al-Dīn al-Rāzī in Ayyubid Damascus was Shams al-Dān Khusrawshāhī. He was sponsored by al-Muʿaẓẓam and appointed as tutor of his son al-Nāṣir Dāwūd. He taught philosophy in both terms, that is, as *ḥikma* and *falsafa*, to several of the physicians mentioned by Ibn Abī Uṣaybiʿa, including Ibn Abī Uṣaybiʿa himself.[119]

114 See ibid. 600.
115 Sabra (1987), 236f.
116 Ibn Abī Uṣaybiʿa 1965), 648.
117 See ibid. 718, 721, 751.
118 See ibid. 745 (*al-ḥikma al-iqtidāʿ biʾllāh taʿāla*).
119 See ibid. 683, 649, 768.

Shams al-Dīn al-Khūy'ī was another one of them. Al-Muʿaẓẓam allocated stipends to him, appointed him head judge of Damascus and gave him a professorship in the *ʿĀdiliyya Madrasa* where Shams al-Dīn took his residence.[120] Shams al-Dīn, on his side, linked himself with the prince in a companionship.[121]

Scholars trained in the philosophical sciences who worked in Ayyubid Egypt and Syria had often acquired their respective knowledge outside the Ayyubid realm in al-Andalus, Iran, Baghdad and al-Raqqa.[122] They taught their knowledge in Cairo, Damascus, al-Karak and Ḥamāh.[123]

The most prominent scholar studied in such classes was Ibn Sīnā. The work that according to Ibn Abī Uṣaybiʿa attracted most attention was his *Kitāb al-ishārāt wa-l-tanbihāt*. Najm al-Dīn al-Lubūdī wrote an epitome of this treatise.[124] Ibn Abī Uṣaybīʿa studied the original with Rafīʿ al-Dīn al-Jīlī in Damascus.[125] Muhadhdhab al-Dīn ʿAbd al-Raḥīm ibn ʿAlī studied with Sayf al-Dīn al-Āmidī several of al-Āmidī's books on the philosophical sciences, among them his commentary on Ibn Sīnā's work.[126] Rafīʿ al-Dīn al-Jīlī composed a commentary on Ibn Sīnā's treatise for the Ayyubid ruler of Baʿlbak, Taqī al-Dīn ʿUmar.[127] A second type of philosophical work of Ibn Sīnā studied by the scholars in Ayyubid cities were his encyclopaedias. Shams al-Dīn Khusrawshāhī wrote an epitome of the great encyclopaedia, the *Kitāb al-Shifāʾ*.[128] Muwaffaq al-Dīn Yaʿqūb al-Sāmarrī (d. 681/1283) imitated it and composed an introduction to the sciences of logic, natural philosophy and metaphysics.[129] The reason for this focus was Fakhr al-Dīn al-Rāzī's critique of this treatise and the debates about Ibn Sīnā's philosophy at large that Fakhr al-Dīn's students brought from the East to Damascus and Cairo. Other philosophers whose oeuvre was taught and discussed in the medical communities under Ayyubid rule were Aristotle and Fakhr al-Dīn al-Rāzī. Ibn Abī Uṣaybiʿa's uncle studied Aristotelian texts with ʿAbd al-Laṭīf al-Baghdādī, a well-known scholar with a broad philosophical education and friend of Ibn Abī Uṣaybiʿa's father from their shared studies at the *Niẓāmiyya Madrasa* in Baghdad. Teacher and student discussed difficult passages found in these works.[130] Rashīd al-Dīn also went to philosophy classes held by another teacher who had a high reputation in these disciplines in his time, Sadīd al-Dīn ibn Abī l-Bayyān.[131] In comparison to ʿAbd al-Laṭīf al-Baghdādī's exposure not only to Aristotle, Ibn

120 See ibid. 646.
121 Ibid.
122 See ibid. 662.
123 See ibid. 683, 699.
124 See ibid. 668.
125 See ibid. 725.
126 See ibid. 733.
127 See ibid. 648.
128 See ibid. 650.
129 See ibid. 767.
130 See ibid. 737.
131 Ibid.

Sīnā and Fakhr al-Dīn al-Rāzī, but also to Plato, the great commentators Alexander of Aphrodisias (fl. second and third centuries CE) and Themistios (d. 388 CE), Abū Naṣr al-Fārābī (d. 339/950) and other ancient and medieval scholars as well as his own philosophical oeuvre the philosophical studies in Ayyubid Damascus and Cairo appear to have been of a fairly limited scope and depth. But this was not the result of any kind of interference by the Ayyubid patrons. It rather reflects the then dominant perception, as ʿAbd al-Laṭīf reported about his own education, that philosophy was the study of Ibn Sīnā's *œuvre* and the commentaries available on it.[132]

In this kind of atmosphere and activities, why did Sayf al-Dīn al-Āmidī lose his teaching post and why did he stay confined to his house? Was his confinement part of the punishment or a reaction to the loss of his teaching position? Can it be corroborated that he suffered this exclusion from the public sphere for having taught philosophy and related sciences? Why has his case been taken as a symptom of an overall anti-philosophical attitude of the Ayyubid dynasty? Or, if other reasons caused the confinement, why were philosophical activities accepted as the major culprit, both by medieval and modern scholars? These questions have puzzled me for a long time.[133] Not all of them can be answered with the available sources. The biographical dictionaries and historical chronicles written in the thirteenth century CE offer, however, a much richer array of rhetorical strategies and narrative plots than allowed for so far by modern commentators on the event.[134] The medieval writers do not offer variations of one, but of several perspectives on the events. Their differences on the factual as well as narrative level are so profound that it seems if not altogether impossible, but at least very difficult to harmonize them into a single, coherent story. The thirteenth-century authors constructed their own representations of what happened, when and why. They either highlighted intrigues in the scholarly communities of Baghdad and Cairo, but kept their silence about the collegial climate in Damascus. Or they pointed to Sayf al-Dīn's pride and conviction of being the leading scholar of the rational sciences in his time, a position that brought him into conflict with the faction of the "Persians" as well as with one of the Ayyubid rulers of Damascus. Others emphasized al-Āmidī's true or alleged breach of etiquette in his relationship with the Ayyubid princes. The fourth type of narrative singled out Sayf al-Dīn's study and teaching of philosophy as the main cause for his downfall. A fifth approach is a strategy of silence or, in a softer variant, of imprecision. As a rule, more than one of these narrative lines make up an author's biographical record. Sayf al-Dīn's "biographers" of the thirteenth century CE were Ibn al-Qifṭī, Ibn Abī Uṣaybiʿa, Ibn Khallikān (d. 681/1282), Sibṭ ibn al-Jawzī, and Ibn Wāṣil. I will focus here exclusively on the latter's story because of the author's love for detail.

132 See ibid, 688.

133 For an earlier version presenting a number of stories told by different authors see Brentjes (1997).

134 See, for instance, Goldziher (1915), 393f; Sourdel (1986), 434; Chamberlain (1995), 83f.

Ibn Wāṣil took up elements of Ibn al-Qifṭī's, Ibn Khallikān's and Sibṭ ibn al-Jawzī's reports in non-trivial ways. He shared with them a certain taste for the dramatic. In contrast to the previous writers, Ibn Wāṣil though had a penchant for the particular and the precise. He tells more individual stories than any of the other biographers and presents himself as possessing numerous specific details. Although it is by no means clear whether this is more than a narrative technique, Ibn Wāṣil may well have had access to previously untapped information because he and his father served several Ayyubid and Mamluk patrons and were members of the scholarly elite of al-Shām. The nucleus of Ibn Wāṣil's report on how and why Sayf al-Dīn lost his professorship agrees with Ibn al-Qifṭī's story, based like it on unnamed informants:

> Al-Malik al-Ashraf turned away from al-Āmidī because he loathed him. Then al-Malik al-Masʿūd, the lord of Āmida, sent for him, demanding him [to come]. Thereafter happened what we mentioned about the taking away of Āmida from her lord. After she was taken from him, al-Malik al-Kāmil said to the lord [of Āmida] according to what I was told: 'In your city, you have no one of excellence.' But he erred and the lord answered: 'I had sent for the Shaykh Sayf al-Dīn demanding him [to come to me] and I was promised that he would come to me.' This was so painful for al-Malik al-Ashraf and al-Malik al-Kāmil that the two grew very angry against Sayf al-Dīn. As a consequence, al-Malik al-Ashraf dismissed him from the teaching at the *ʿAzīziyya Madrasa*. He withdrew to his garden and stayed there as someone who had been treated unjustly until he died in this year being half a year older than eighty years.[135]

It is, however, more specific and detailed than that of Ibn al-Qifṭī, which enabled Ibn Wāṣil to construct a meaningful story about the loss of face of the two Ayyubid princes in front of a defeated Artuqid ruler. The historian set up this culmination of the drama by informing his readers at the beginning of the biography that the ruler of Āmid had already invited Sayf al-Dīn to this very same position when the scholar was serving the Ayyubid prince of Ḥamāh. But the prince did not want him to depart and so the scholar had declined the invitation. After the death of al-Manṣūr ibn Taqī al-Dīn (d. 617/1220), Ibn Wāṣil writes, al-Muʿaẓẓam invited Sayf al-Dīn to come to him to Damascus promising him beautiful rewards. Sayf al-Dīn accepted this offer and al-Muʿaẓẓam honoured him with the teaching post at the *ʿAzīziyya Madrasa* and a wonderful house in a beautiful street. After al-Muʿaẓẓam's death, his son al-Nāṣir Dāwūd honoured the scholar greatly and gave him the lavish sum of 8,000 *dīnār* for buying a palace and a garden. He joined Sayf al-Dīn as a disciple, although he was already affiliated with Shams al-Dīn Khusrawshāhī, and asked him to write a book on the rational sciences for

135 Ibn Wāṣil (1977), V:37–40.

him, the *Farāʾiḍ al-qalāʾid*.[136] Thus while al-Āmidī served three Ayyubid princes as a much-honoured client and rejected the offers of the lord of his native town in their favour, at the end of his life he is portrayed as someone who had changed sides. This alone might have been of no consequence had it happened under "normal" circumstances. Changing patrons was not at all uncommon. Moving from one principality to another was part of the patron-client relationships of the period. But the fact that this switch of patrons had become an element in a public taunt on the battlefield may have indeed created the atmosphere where it was felt obligatory to retaliate not merely against the military enemy but also against the scholar who was thrown as the gauntlet into the face of the victors causing them a loss of face.

A second major point of difference between Ibn Wāṣil's biography of Sayf al-Dīn and those told by his predecessors is the non-trivial reformulation of the involvement of the philosophical sciences in the conflict presented in similarly dramatic tones. Ibn Wāṣil spoke colourfully about tensions between Sayf al-Dīn and the faction of the Persians and about why the Ayyubid princes disliked the scholar. Already in Ḥamāh when writing many books, Sayf al-Dīn was obsessed with refuting, criticizing and maligning Fakhr al-Dīn al-Rāzī.[137] Ibn Wāṣil found this behaviour so reprehensible that he offered bluntly his own opinion: Sayf al-Dīn violated good tone and taste because he envied Fakhr al-Dīn and because he was convinced that he was superior or at least equal to him in *ʿilm*.[138] To his dismay, however, Sayf al-Dīn had to see that people preferred Fakhr al-Dīn and his works. Moreover, Ibn Wāṣil claims, Fakhr al-Dīn had found a much more generous patron than Sayf al-Dīn and more students pooled around the Iranian scholar. This difference in success and public respect was the main reason for Sayf al-Dīn's incessant battles, Ibn Wāṣil opined. A second, equally important component for this state of affairs in Ibn Wāṣil's view was the cultural difference between the *ʿarab* (Sayf al-Dīn) and the *ʿajam* (Fakhr al-Dīn).[139]

Sayf al-Dīn's opinion of and attitude towards Fakhr al-Dīn was at the heart of the displeasure that al-Muʿaẓam felt for him. One day namely, Sharaf al-Dīn (d. 630/1229–30), known as Ibn ʿUnayn, visited Sayf al-Dīn and listened to him deriding Fakhr al-Dīn. But this Sharaf al-Dīn, being one of the zealots of Fakhr al-Dīn, became infuriated with what he heard and ran to al-Muʿaẓẓam to complain. His report diminished Sayf al-Dīn's standing in the eyes of the Ayyubid ruler.[140] Despite this, al-Muʿaẓẓam continued to visit the Friday night sessions of the *ʿulamāʾ* where Sayf al-Dīn shone with eloquence and intelligence so that nobody could live up to him in a disputation or could surpass him in figurative

136 See ibid. V:40.
137 See ibid. V:36.
138 Ibid.
139 See ibid. V:36f.
140 See ibid. V:38. I chose to use the reading *majāzāt* given in footnote 10 as an alternative to the text's *majārāt*.

speech as stressed by al-Muʿaẓẓam's son al-Nāṣir Dāwūd in an encounter with Ibn Wāṣil:

> The Sultan al-Malik al-Nāṣir Dāwūd – may God have mercy with him – told me, and we were in his service at al-Karak, saying: 'When Shaykh Sayf al-Dīn was with my father – may God have mercy with him – I went to the meeting in order to hear his speech. I marvelled at his rhetoric and his excellent Arabic, the beauty of his arguments and his towering over the others in the disputation.' I answered al-Malik al-Nāṣir: 'Who of the two men is more excellent in the view of my Lord Sultan – Shams al-Dīn Khusrawshāhī or Sayf al-Dīn [al-Āmidlī]?' He said: 'Subḥāna llāh, how can you say that! All of those were chicken to slaughter for Sayf al-Dīn! Sayf al-Dīn saw himself as more excellent than their teacher Fakhr al-Dīn. Hence, he considered them as insignificant.[141]

Such an outspoken judgement by an Ayyubid prince may have caused the medieval reader of the biography to chuckle or to sneer, depending on his position towards the involved scholars. It also subtly counterbalanced the unfavourable comments Ibn Wāṣil himself had made earlier in his story about Sayf al-Dīn. He now let the reader know that yes, the scholar was arrogant and excessive in his critique towards Fakhr al-Dīn, but he also was in many respects superior to his peers and a true star in the courtly sphere of Ayyubid Damascus, a fact that father and son obviously appreciated.

Ibn Wāṣil's biography of Sayf al-Dīn al-Āmidī is undoubtedly the one richest in detail and narrative power. It elucidates the edges of the scholar's personality, the intellectual fights that took place under Ayyubid patronage about philosophical and related issues and the tensions that flared up occasionally because of the overpowering standing of Sayf al-Dīn in comparison to Fakhr al-Dīn's students and his own feelings of deprivation. In Ibn Wāṣil's view none of these elements, however, led to Sayf al-Dīn's dismissal from his teaching post at the *Azīziyya Madrasa* by al-Ashraf. They rather seem to have dwindled away since several scholars who had participated in these sessions chaired by al-Nāṣir Dāwūd left Damascus with him after al-Ashraf's victory or followed him later to al-Karak on his invitation.[142] Thus, the last, like the first, narrator of the story about how and why Sayf al-Dīn al-Āmidī lost his professorship agreed that it was a case of confrontation on the battlefield over an issue of patronage that gave the defeated enemy a chance to retaliate and cause a public loss of face to two Ayyubid princes, an affair of public shame for which the scholar as a publicly claimed client of an inimical patron had to be punished. Hence, the benefits given to him by his former patrons were removed from him. Sayf al-Dīn's withdrawal into his house and garden appears

141 See ibid, V:39.
142 Ibid.

then less as an exclusion from the public domain of Damascene scholars than as a deliberate refusal to communicate with his vengeful former patrons.

The material, its arrangement and presentation, the judgements and evaluations of the thirteenth-century historians served writers between the fourteenth and seventeenth centuries in Mamluk and Ottoman Egypt and Syria for their transmissions of Sayf al-Dīn's biography as well as modern scholars of the twentieth century for their evaluations of the relationship between the Ayyubids and the ancient sciences. The most often used source was Ibn Khallikān's report. Some writers also drew on Sibṭ ibn al-Jawzī and Ibn al-Qifṭī. The story about the patronage conflict receded into the background, while the claim that the ancient sciences were at the heart of all the conflicts in which Sayf al-Dīn was involved during his life, including his loss of the *ʿAzīziyya Madrasa*, became not merely predominant, but also got sharper and more rigidly framed. Examples of these transforming interpretations can be found in the works of Shams al-Dīn al-Dhahabī (d. 748/1348) or ʿAbd al-Qādir al-Nuʿaymī (d. 927/1520).[143] A similarly narrow focus on Ibn Khallikān and Ibn al-Qifṭī, occasionally bolstered by some of the 'testimonies' of the later writers, can be found among the modern scholars who wrote about Sayf al-Dīn's life and his loss of position. They too considered the ancient sciences and their opponents among the *ʿulamaʾ* as the reason that motivated Ayyubid activities against Sayf al-Dīn in particular and adherents of the ancient sciences in general.[144] The belief in a widespread hostility of rulers and scholars of Islamic societies against the ancient sciences characterized the views of many historians of Islamic societies throughout the entire twentieth century. The sources they used and the methods they applied for their interpretation seemed to confirm that their expectations were justified. Ideological and historiographical presuppositions as well as a paucity of method consciousness among modern scholars cooperated with ideological and disciplinary presuppositions among medieval scholars writing about Sayf al-Dīn al-Āmidī and led them to overlook the other elements that filled the stories about the conflicts in his life.

Conclusions

The culture of patrons and clients as talked about by the author who provided me with most of my material is surprisingly rich, amazingly flexible and astonishingly happy. Physicians as clients wielded power due to their expertise, acted as patrons and mediators and were free to choose. They were sought after and patrons competed with each other for their favours. They could amass riches as any of the princes or *umarāʾ*. While they were pious believers they did not have to belong to the dominant religion, nor were they puritans. They loved poetry, music, luxury, beautiful women and men as well as a good debate about all sorts of topics, from

143 See Brentjes (1997), 28–33.
144 See Goldziher (1915), 394; Sourdel (1986), 434.

religion to philosophy, from diseases and their critical days to the stars and their omina. They were broadly educated in most of the ancient sciences and appear as their main guardians and customers. The Ayyubid princes and their households are portrayed as the main patrons of the physicians, a perspective most likely set by Ibn Abī Uṣaybiʿa's social position and value system. They were generous, lenient, supportive, and intelligent enough not to drive their most capable clients into the arms of other patrons. They understood that professional expertise in hours of sickness weighed more than a particular creed or character trait. In a sense, the world of the princes and their physicians as portrayed by Ibn Abī Uṣaybiʿa is perfect. If the content of the profession was not disease and death the picture would almost be too perfect. There are no severe conflicts between patrons and clients, except if one of them acts like a tyrant oppressing the people. Then draconian punishment can be meted out, against the misbehaving client of course or an inferior prince. There are no severe intrigues between the clients either. Nobody tries to triumph over a rival. All are nice, polite, linked in friendship, striving to live up to the ideal of manliness. Such an ideal state of affairs cannot be but a construct. Unfortunately, other sources than Ibn Abī Uṣaybiʿa's biographical dictionary do not focus in a comparable manner on issues of patronage and hence do not provide us with material to pierce through the peaceful, happy appearance of the princely physician and his enjoyable life with his colleagues and patrons. If we consider the artificial harmony and conviviality of Ibn Abī Uṣaybiʿa's picture as a kind of panegyric laudatio of both the patron and the colleague, and focus our attention on the various activities of the many who were participants of the system we are well served by Ibn Abī Uṣaybiʿa's rich terminology of patronage, his references to material aspects of patronage, his description of the particularities that honours, graces and benefits comprised and the singular moments celebrated in the many stories that he collected, embellished and perhaps occasionally even made up. While there are no terms for court or patronage in any of the Ayyubid sources I am familiar with, Ibn Abī Uṣaybiʿa had a particular liking for three terms – *dawla*, *murūʾa* and *bāb dār al-sulṭān*. I have argued that the second term meant more than the ideal of an individual manliness prescribing generosity, courtesy and honour. I suggested that the third term represents, in contrast to its literal meaning, a space inside the residence of a prince close to his living quarters or bedroom although where an Ayyubid prince resided was subject to change. I did not discuss the first term as extensively as it would have been possible on the basis of the available material. The variations between the different usages and their individual contexts, even if only taken within Ibn Abī Uṣaybīʿa's dictionary, seem to be too large to be covered in this paper. My impression is that it can cover such different terms as time period, dynasty, realm, power, administration, and perhaps even household and thus may be the word that could come closest to the term court found in Christian cultures. Finally, Ibn Abī Uṣaybiʿa's work leaves no doubt that he, his relatives and his colleagues felt free to study, teach and practise any of the ancient sciences they liked, be it for professional purposes or for their pleasure and cultured lifestyle. Their Ayyubid patrons profited from this broad educational ideal and practice in their sessions on intellectual topics as well as in the nights filled with poetry and music. They enjoyed speeches for their linguistic and rhetorical brilliance as well as for their intellectual bite and fire. Although they may have despised opinions on intellectual matters that they did not share they did not interfere yet as often into the affairs of the *ʿulamāʾ* as became the norm under their successors. If they deposed or executed them it was mostly about issues of power and public offence.

Bibliography

Brentjes, S. (1997) '"Orthodoxy", Ancient Sciences, Power, and the Madrasa ("College") in Ayyubid and Early Mamluk Damascus', *International Workshop: Experience and Knowledge Structures in Arabic and Latin Sciences*, Berlin: Max-Planck-Institut für Wissenschaftsgeschichte.

Chamberlain, M. (1995) *Knowledge and Social Practice in Medieval Damascus, I/90–1350*, Cambridge: CUP.

Goldziher, I. (1915) *Die Stellung der alten islamischen Orthodoxie zu den antiken Wissenschaften*, Berlin: Reimer.

Ibn Abī Uṣaybiʿa (1965) *ʿUyūn al-anbāʾ fī ṭabaqāt al-aṭibbāʾ*, ed. N. Riḍā, Bayrūt: Dār Maktabat al-ḥayāt.

Ibn al-ʿIbrī, Ghrighūriyūs al-Malatī (1958) *Taʾrīkh mukhtaṣar al-duwal*, Bayūt: Maṭbaʿat al-kāthūlikiyya.

Ibn Khallikān, Aḥmad ibn Muḥammad (1968–72) *Wafayāt al-āʿyān wa-abnāʾ al-zamān*, ed. I. ʿAbbās, 8 vols., Bayrūt: Dār al-thaqāfa.

Ibn al-Qifṭī (1903) *Taʾrīkh al-ḥukamāʾ*, 1st Edition, Lippert, Leipzig: Dieterich.

Ibn Wāṣil, Jamāl al-Dīn Muḥammad ibn Sālim (1977) *Mufarrij al-kurūb fī akhbār Banī Ayyūb*, ed. Ḥ.M. Rabīʿ and S.'A. ʿĀshūr, al-Qāhira: Dār al-kutub.

Sabra, A.I. (1987) 'The Appropriation and Subsequent Naturalization of Greek Science in Medieval Islam: A Preliminary Statement', *History of Science* 25, pp. 223–243.

Sibṭ ibn al-Jawzī, Yūsuf ibn Qizughlū (1371/1952) *Mirʾāt al-zamān fī taʾrīkh al-āʿyān*, vol. 8/2, Ḥaydarābād: Maṭbaʿat majlis dāʾirat al-maʿārif al-ʿuthmāniyya.

Sourdel, D. (1986) 'al-Āmidī, Sayf al-Dīn', *Encyclopaedia of Islam*, New Edition, vol. 1, Leiden: Brill, p. 434.

4

PATRONAGE OF THE MATHEMATICAL SCIENCES IN ISLAMIC SOCIETIES

Structure and rhetoric, identities, and outcomes

Patronage is a term that has no direct equivalent in any of the languages in which mathematical sciences were practised in Islamic societies.[1] If we take one of the ancient Roman meanings of patronage, namely the existence of a clientele relationship between a protector and a protégé, as the starting point for exploring social relationships involving the mathematical sciences in Islamic societies, we will see that practitioners of these sciences could indeed enter relationships with people of higher social status and greater wealth, which provided them with relatively stable positions and various benefits. The complexity of the terminology used to talk about such relationships, and the diversity of forms of exchange involved in them, suggest that there was not one single type of patronage relationship in every Islamic society through the centuries. Indeed, whether all such relationships actually qualify as patronage demands further study.

Patronage and other forms of support for the mathematical sciences constituted part of a much wider phenomenon of social and cultural dependence between men and women of different social status, upbringing, and access to resources. These fields of knowledge shared basic rituals that configured clientele relationships with sectors such as medicine, the arts, administration, and even the military. At the same time, there were certain differences within the mathematical sciences and their practitioners' access to patrons and their resources. Studies of these manifold aspects of patronage for the sciences, particularly in the mathematical disciplines, are relatively rare compared to ancient Rome or early modern Catholic and Protestant Europe. In particular, theoretically grounded investigations are lacking (Elger 2007). There is no clarity about the types of patronage and other clientele

1 The names and dates of all the dynasties, scholars, and patrons discussed in this chapter are given in Tables 4.1 and 4.2.

DOI: 10.4324/9781003372455-5

relationships that existed in different Islamic societies. Only one study has tried to survey the rhetoric, rituals, and, to some extent, the efficacy of patronage relationships and their specific forms, in ninth to eleventh-century Iran and Iraq (Mottahedeh 1980). Being limited to the military and the administration, it pays no attention to the practitioners of the ancient sciences. A second study focuses on patronage in medieval Cairo relating to scholars of Islamic legal theories (*fiqh*), the transmission of Muḥammad's tradition (*ḥadīth*), the reading and interpretation of the Qur'ān (*qirā'a*; *tafsīr*), and the fundaments of religion and law (*uṣūl al-dīn*; *uṣūl al-fiqh*) (Berkey 1992). Berkey's observations and insights agree with most of what I have read in primary sources about patronage relationships for the non-religious sciences, including the mathematical disciplines. As the only available studies of patronage in Islamic societies, these two works gave me guidelines for examining the terminology and social practices of patronage in other Islamic societies, applying their questions whenever possible to the mathematical sciences. In order to enhance the conceptual basis of the paper, I also sought inspiration from studies of patronage in early modern Catholic and Protestant societies, particularly their theoretical and methodological approaches (Kettering 1986, 2002; Asch and Birke 1991; Danneskiold-Samsøe 2004).

In this chapter I use the term 'mathematical sciences' in accordance with its definitions and descriptions in primary sources from Islamic societies. Until the classical system of knowledge was replaced during the nineteenth and early twentieth centuries, the notion of the mathematical sciences, their values, and practices were based on Platonic, Aristotelian, and neo-Pythagorean classifications of philosophy enriched by Stoic and sceptical elements on the one hand and perspectives developed within the religious disciplines of Islam, in particular law and theology, on the other. The mathematical sciences comprised the four main theoretical disciplines of Antiquity, namely number theory, geometry, astronomy, and musical theory (theory of proportions). They also included a variety of applications, such as optics, burning mirrors, mental or written calculation, algebra, magic squares, business arithmetic, surveying, timekeeping, architecture, and water lifting. The twelfth-century scholar Bahā' al-Dīn Kharaqī summarized this view as follows:

> The mathematical (sciences) are called the four teaching (disciplines). They are four because their subject matter is quantity. Quantity is either that which is continuous or that which is discrete. The continuous is either in movement or in rest. The moving is *hay'a* (mathematical cosmography) and the non-moving is *handasa* (theoretical geometry). The discrete is either that which has a compound ratio, and this is music, or that which does not have it, and this is the numbers.[2]
>
> (Bayhaqī 1996, 173)

2 Inna l-riyāḍiyyāt tusammā l-ta'līm al-arba'a, wa-innamā kānat arba'a li-anna mawḍū'uhā al-kamiyya, wa-hiya immā an takūna muttaṣila aw munfaṣila, wa-l-muttaṣila mutaḥarrika aw ghayr

Table 4.1 Islamic dynasties

Abbasid caliphs (750–1256), capital Baghdad
Umayyad *emirs* and caliphs in al-Andalus (756–1031), capital Cordoba
Samanid *emirs* (819–1005), capital Bukhara
Fatimid caliphs (910–1171), first capital in Mahdiyya, second capital in Cairo
Buyid *emirs* (945–1055), capitals Baghdad and Shiraz, with semi-independent courts in Rayy, Hamadan, Isfahan, Kirman: governed Iran and southern parts of Iraq under the Abbasid caliphs
Khwarazmshahs, capitals Kath, Gurganiye, Urgench (Kunya-Urgench): pre-Islamic title of Afrighids ruling since the early centuries CE, converted to Islam in the early ninth century; destroyed in 995 by their commercial rivals and successors, the Ma'munids (995–1017) who were overtaken by the Ghaznavids; the last dynasty with this title existed from 1077 to 1231. They began as Saljuq slave governors; after gaining independence their title was first *emir* and later *shah*; they conquered large parts of the Saljuq empire, were finally destroyed by the Mongols
Ghaznavid sultans (975–1187), capital Ghazna
Great Saljuq sultans (1037–1157), capitals Baghdad, Isfahan, Nishapur, governed Iran, Iraq, Anatolia, Syria, and Palestine under the Abbasid caliphs
Zangid *atabegs* and *emirs* (1127–83), capitals Mosul, Aleppo, Damascus, Sinjar, Jazira: recognized Abbasid caliphs as overlords
Ayyubid *maliks* (1169–1260), capital Cairo with semi-independent courts in Damascus, Aleppo, Hama, Hims and other cities in Syria, Palestine and north Iraq: recognized Abbasid caliph as overlord
Mamluk sultans (1250–1516), capital Cairo
Ilkhanid *khans* (1258–1336), capitals Tabriz, Maragha, Sultaniyye: recognized Mongol rulers of China as overlords
Ottoman sultans (1299–1922), capitals Bursa, Edirne, Istanbul
Timurids (1370–1506), capitals Samarkand, Herat
ʻĀdil Shāhis (1518–1686), capital Bijapur
Mughal shahs (1526–1857), capitals Delhi, Agra, Lahore

Astrology was rarely included explicitly in the classifications of the mathematical sciences. But since it was often seen as the ultimate goal of astronomical theory, and since most practitioners of astronomy also exercised the craft of astrology either as textbook writers or as composers of horoscopes, astrology will be included in the spectrum of disciplines discussed in this chapter.

The multiple cultural backgrounds that defined the mathematical sciences and their status explain some of the specifics of patronage for the various disciplines united under this heading. Linked to philosophy, cosmography, *fiqh*, business,

mutaḥarrika, wa-l-mutaḥarrika hiya al-hay'a, wa-ghayr al-mutaḥarrika hiya al-handasa, wa-l-munfaṣila immā an takūna lahā nisba ta'līfiyya wa-hiya al-mūsīqā, aw lā takūna wa-hiya al-aʻdād.

administration, and licit magic, the mathematical sciences commanded a rich but contradictory social prestige. Books by Euclid, Archimedes, Ptolemy, and other ancient authors were translated into Arabic, Persian, and occasionally other languages such as Ottoman Turkish. These translations, and works based on them, were studied by scholars involved in mathematical research and by students preparing for religiously sanctioned positions such as that of muezzin, the caller for prayer, and *muwaqqit*, the person responsible for determining prayer times, the directions of prayer towards Mecca, and the beginning of the new month. Names of Greek scholars were given as epithets to people seen as experts in their fields. Copies of, and commentaries on, their texts were prepared for individual reading, for classes at schools and colleges, and lavishly illuminated if destined for the library of a collector or wealthy connoisseur.

In the twelfth century, under the impact of Abū Ḥāmid Ghazālī's writings on the relationship between the religious and non-religious sciences, however, theoretical geometry became to be seen by some as a threat to the faith of the believer because of its claim to absolute truth. By contrast, calculation, algebra, surveying, and astronomy were appreciated as useful knowledge for determining inheritance shares and taxes, measuring fields, and predicting eclipses. Excessive practice, however, was also seen as a threat to the mental and psychic balance of the believer (Rebstock 1992, 19–22).

Astronomy, astrology, and music equally found themselves both cherished and challenged. Astrology was heavily disputed and eloquently defended by philosophers, astronomers, theologians, and transmitters of *ḥadīth*. In the view of some practitioners, their competitors did not know enough philosophy to understand the workings of the universe and the impact of its different spheres on the climate, affairs of the state, and the lives of individuals. Others accused their colleagues of not having mastered the technicalities demanded by these disciplines (Burnett 2002, 206–211). A number of specific issues of theory and reliability of results were also at stake, for instance, whether astrology could deliver forecasts only for universal events in nature and politics or was also applicable to the life of the individual (Burnett 2002, 207). The unreliability of astrological predictions could lead to ridicule and occasionally to imprisonment or even execution. Music, when practiced, not so much when dealt with as a mathematical theory, was regarded by various Sunni and Shiʿi legal authorities as compromising the true path of a believer, while Sufis and courtiers appreciated it as support for spirituality and the enhancement of courtly life. As a result, scholars of various legal orientation wrote *fatwā*s and other kinds of texts either commending or condemning these fields of mathematical knowledge and their practitioners.

In a sweeping simplification, in the classical Islamic world one can differentiate two major periods for patronage of scientific knowledge, including the mathematical sciences. The first period spans the eighth to the later twelfth century, while the second period stretches over six hundred years to the nineteenth century. The grounds for the earlier period were laid during the two first centuries of the Abbasid dynasty, when the so-called Translation Movement, centred on Baghdad, made Middle Persian, Greek, Syriac, and Sanskrit scientific works accessible, in Syriac and Arabic. Interested readers included physicians, theologians, astrologers, courtiers, princes, and rulers. Scientific enterprises resulting from these translations were first undertaken in the Abbasid capital and other cities of the empire. With the emergence of powerful governors, local rulers, and independent Muslim dynasties such as the Fatimids, the Umayyads, and the Ghaznavids a richer spectrum of courtly patronage became available. The second period began in the twelfth century, when endowed teaching institutions became a regular feature of scholarly practice in most of the lands from Central Asia to the Atlantic.

Patronage from the eighth to the twelfth centuries

Until the twelfth century, patronage relationships, including those in the mathematical sciences, were mostly located in courts and among wealthy urban groups such as viziers and their families, tutors of princes, and other court officials. The relationships were primarily forged between individuals and between families. Mottahedeh's (1980) study of loyalty and leadership, which also deals with forms of patronage, concentrates on the period of Buyid rule in Iran and Iraq under the formal suzerainty of the Abbasid caliphs. In his view, loyalties were created through three main features: benefits, various forms of formal commitment (oath, vow, guarantee of safe conduct), and gratitude. Benefits (*ni'ma*) given by a ruler to his subjects served to create stable, reciprocal ties. Generosity (*samāḥa*) was expected from the ruler when handing out benefits, but when combined with indulgence (*musāmaḥa*) could diminish the ruler's reputation (Mottahedeh 1980, 90–91). Formal commitments regulated 'duties and obligations that could be enforced without coercion' (Mottahedeh 1980, 43). A personal oath of allegiance (*bay'a*) followed by a payment (*rasm al-bay'a*) bound rulers and (male) members of their families to their soldiers and high-ranking officials. Such personal oaths were also exchanged between leading members of the administration, but were often concealed from the rulers, to maintain the façade of a unilateral top-to-bottom hierarchy of power (Mottahedeh 1980, 70–79).

It is unclear whether such personal oaths were also exchanged between rulers or viziers and practitioners of the mathematical sciences. Such practitioners did benefit, however, from the second type of formal commitment, the vow. The vow was made from man to God and contained the promise to do some good work on the basis of a good intention. Repairs of irrigation channels and bridges, or the building of mosques and later madrasas, could be publicly announced in the form of such a vow and contributed, in addition to the actual work of repair, to elevating the symbolic and real power of the ruler. Practitioners of the mathematical sciences, in particular surveyors and *muhandisūn* (geometers, engineers, architects), were often involved in the repair and replacement work. The recipient of benefits was obliged to show gratitude (*shukr al-ni'ma*), since masculine honour was based on acknowledging received benefits, reinforced by the Qur'an's repeated reminder that the (true) believer is grateful for God's continuous benefits (Mottahedeh 1980, 72–76).

Mottahedeh (1980, 42) argued that patronage during Buyid rule resulted from this triangular net of benefits, formal commitments, and gratitude. A central form of the Buyid patronage system was a relationship called *iṣṭināʿ*, which saw the patron as a parent or protector (*muṣanaʿ*) and the client as a child or protégé (*ṣāniʿ*). This relationship was introduced in the middle of the ninth century, together with slavery, by the Abbasid caliphs to rebuild their army, complementing and eventually replacing the older form of clientship (*walāʾ*). In the texts of the tenth and eleventh centuries, *iṣṭināʿ* dominates, while *walāʾ* has disappeared almost completely. Its verb, *iṣṭanaʿa*, often meant that someone's career was fostered. In exchange for this promotion, the patron expected his protégé to serve him in various ways as long as either of the two lived (Mottahedeh 1980, 82–84). Indeed, many of the scholars who worked as astrologers or physicians at the Buyid court, as well as other courts of the period, stayed in courtly service until they died, often through the reign of more than one ruler. It is no exaggeration to claim that many of the most of productive and innovative mathematical scholars of this period, such as Thābit b. Qurra, Abū l-Wafāʾ Būzjānī, Abū Sahl Kūhī, Abū Rayḥān Bīrūnī, Ibn al-Haytham, and ʿUmar Khayyām, spent long years in courtly service. This means that the impressive results of nearly five centuries of mathematical research and teaching were produced through the institution of patronage at its major centres – central and local dynastic courts.

The patronage relationships that rulers, viziers, and other courtiers extended to physicians, astrologers, and other scholars were more often discussed in a variety of other terms. A patron extended honour (*ikrām*) and benefits, often in a generous manner, to the men who healed him, cast his horoscope, or observed the stars. The physicians and occasionally also the astrologers are said to have served (*khadama*) their patrons, often until their patron's death. The content of honouring included precious robes, monetary gifts, privileged ranking in seating at court during official occasions, and the inclusion into caliphs', sultans', or viziers' groups of boon companions (*nadīm*). Patrons also shared other, more personal favours with their clients. Ibn Abī Uṣaybiʿa and Shams al-Dīn Shahrazūrī report, for instance, how Caliph al-Muʿtaḍid greatly honoured his astrologer and boon companion Thābit b. Qurra by apologizing for a mistake in etiquette the caliph had committed while walking with Thābit in one of the palace gardens (al-Shahrazūrī 1976, I 4–5; Ibn Abī Uṣaybiʿa 1965, 295). Thābit was one of the leading translators of Greek mathematical works, himself an excellent mathematician and astronomer, but also a philosopher, physician, and writer on the beliefs of the Sabeans, his religious community. He also took his obligations as a caliph's boon companion very seriously. When al-Muʿtaḍid's uncle and official predecessor as caliph, al-Muʿtamid, was imprisoned on the instigation of his brother al-Muwaffaq in the house of Ismāʿīl b. Bulbul, the caliph's vizier, Thābit visited the prisoner three times a day at Ibn Bulbul's invitation to keep the caliph company. He entertained him with topics from philosophy, geometry, and astrology. The caliph hungered for Thābit's company and felt pleased in his presence. When freed, al-Mu'tamid described Thābit as the man he valued most after his preferred military slave al-Badr. He made him sit close to him in audiences for the military and civil elite (*al-khāṣṣ*) and the general public (*al-'āmm*) while leaving al-Badr and the vizier to stand (al-Shahrazūrī 1976, I 5–6).[3]

The scholars engaged in patronage relationships were expected to offer expertise in areas such as healing, observing the planets and stars, casting horoscopes, constructing instruments, writing books, making automata, and repairing clocks, water wheels, channels, and other infrastructural components. Some of the best-known services of mathematical scholars and instrument makers carried out for courtly patrons are the expeditions for measuring the length of 1° of a meridian, patronized by Caliph al-Ma'mūn in the early ninth century (King 2000). Later he was both praised and cursed for this patronage of the 'foreign' (in particular ancient Greek) sciences (Gutas 1998). His support and encouragement of the geodetic measurements became standard narrative fare in geographical, astronomical, and also some historical writings. Patronage of the sciences was thus transformed into a powerful discursive instrument, either initiating the reader into its heritage and shaping his scientific identity or admonishing him to abstain from this heritage due to its danger to one's religious beliefs.

An important professional service of physicians and astrologers was to accompany their patrons on military campaigns, pilgrimages, and other travels. Abū Ma'shar, a favourite of the Caliph al-Muwaffaq, was in his retinue as an astrologer when al-Muwaffaq besieged the Zanj, the mutinous slaves of the great estates in Southern Iraq (Ibn al-'Ibrī 1958, 149). Teaching sons of rulers, their relatives, and other courtiers was another service scholars of the mathematical sciences provided for their patrons. Scholars of all disciplines dedicated books and instruments to patrons from the early Abbasid period onwards It is, of course, not always clear whether the dedication was an expression of gratitude for benefits received or a proposal to enter into a patronage relationship The vocabulary used in histories of scholars and bibliographies includes translating, writing, composing, or doing a work for a ruler, a vizier or a scholar. Besides texts, instruments too were named after patrons and designed specifically for them (Charette 2006, 133 Tables 2a, 2b). Dedications could take various forms, such as joining a part of the patron's name to the title of the book or including the name and titles of the patron as well as wishes for him in the introduction of the work. Dedications indicate another important service that scholars were supposed to deliver: appealing to God for their patrons' worldly and otherworldly well-being. The belief in such a possibility of intermediation was fundamental to all forms of patronage, including that of endowed teaching institutes through which the donors, in addition to protecting their wealth for their children, aspired to secure their salvation

3 Al-Shahrazūrī reports this story for al-Mu'taḍid. But al-Muwaffaq was the latter's father. Al-Muwaffaq usurped the power of Caliph al-Mu'tamid, who was his brother.

through the continuous recitation of Qur'anic verses and other religious texts. Writing eulogies for rulers was as a rule the obligation of poets and historians, but in dedications mathematical scholars too waxed eloquently about their patron's eminence, goodness, care for his subjects, and religious steadfastness. Occasionally practitioners of the mathematical sciences are even remembered in the biographical literature as writers of panegyric verses.

Other ways of talking about patronage relationships in the mathematical sciences included terms like linking oneself with someone (*ittaṣala bi*), being a *ghulām* (slave/apprentice?), carrying someone away (*ḥamala*), elevating someone or making him great or powerful (*a'ẓama*; *'azzaza*), binding someone by an obligation to someone else or being bound by an obligation to someone (*irtabaṭa*), engaging someone's service (*istakhdama*), summoning someone (*istaḥḍara*), ardently desiring someone (*ishtāqa ilā*), being in someone's favour or good graces (*ḥaẓiya 'inda*), and inviting someone (*da'ā*). Some of these expressions, such as *irtabaṭa*, *istakhdama*, and *da'ā*, were applied to both rulers and viziers, while others such, as *ḥamala*, were used only of rulers. *Irtabaṭa*, *istakhdama*, and *da'ā* describe relationships in which power, although asymmetric, is shared because while scholars looked for patrons, rulers and viziers tried to attract clients to their courts, occasionally going to great lengths to do so. But the scholars did not always accept the prospective patron's offers, invitations, or gifts. The Saljuq sultan Sanjar sent the enormous sum of 1000 dinars to ʿAbd al-Raḥmān Khāzinī, a freed slave of Byzantine origin. Khāzinī refused to take the money, saying that he possessed ten dinars, three of which sufficed to finance him for a year, plus a cat. He also rejected the same sizable monetary gift when offered by the wife of one of the Saljuq emirs (Bayhaqī 1996, 181).

The term *ḥamala* 'to carry someone away' reflects the often violent nature of medieval society and the role of scholars as pawns in conflicts between rulers, invaders, rebels, and other claimants for power. When Sebügtegīn and his son Maḥmūd conquered parts of Central Asia and Iran in the late tenth century, they forced numerous scholars, such as the eminent physician and philosopher Ibn Sīnā, to flee westwards into the protection of the Buyid dynasty or else carried them away to Ghazna (in modern Afghanistan). Among those who had to relocate after the final campaign in 1017 were the scientist Bīrūnī and the physician Abū l-Khayr al-Ḥasan (Bayhaqī 1996, 36; Matvievskaya and Rozenfel'd 1985, II 264). The Ghaznavids were not the only dynasty that forced scholars into their patronage. Various Khwarazmshahs brought scholars to their court in this manner, as did the Mongols in the thirteenth century and Timur and his family in the late fourteenth to fifteenth centuries (Bayhaqī 1996, 36, 173).

The historical sources paint a complex and rich picture of different forms and formats of patronage. Successful relationships that lasted a lifetime are found, as are those of a more fluid nature. Cases of stability over two or even three successive rulers occurred at several courts during this period. Cases of conflict and failed continuation of patronage were, however, unexceptional. Important factors in the fluidity of courtly patronage were enmity among members of the

Table 4.2 Scholars and patrons of the mathematical sciences

Abū l-Jūd, Khalīl b. Ibrāhīm (fifteenth century)
Abū l-Khayr al-Ḥasan (tenth/eleventh century)
Abū Ma'shar (c. 787–886)
Abū l-Wafā' Būzjānī (940–998)
'Aḍud al-Dawla (r. 949–982), Buyid emir
Aḥmad b. Khalaf (ninth century)
Akbar (r. 1556–1605), Mughal padishah
'Alam al-Dīn Qayṣar (d. 1251)
'Alī b. 'Īsā (ninth century)
al-Ashraf (r. 1229–1237), Ayyubid ruler of Damascus
Awrangzīb 'Ālamgīr (r. 1659–1707), Mughal padishah
Bīrūnī, Abū Rayḥān (973 – after 1053)
Fanarī, Shams al-Dīn (d. 1435)
Fātiḥ 'Alī Tippu (1750–1799), second and last sultan of Mysore
Ghāzālī, Abū Ḥāmid Muḥammad b. Muḥammad (1058–1111)
Ghāzzān (r. 1295–1304), Ilkhanid ruler
Ḥasan b. 'Alī al-Qaymarī (d. 1480)
Hülägü (r. 1258–1265), founder of the Ilkhanid dynasty
Ibn Abī Uṣaybi'a (1194–1270)
Ibn al-Akfānī (d. 1348)
Ibn al-A'lam (tenth/eleventh century)
Ibn al-'Amīd (d. 970), Buyid vizier
Ibn al-Fuwaṭī (1244–1323)
Ibn al-Haytham (d. after 1042)
Ibn al-Lubūdī = Najm al-Dīn Yaḥyā b. Muḥammad (1210 – c. 1265), Ayyubid vizier
Ibn al-Majdī (1365–1447)
Ibn al-Nadīm (d. 995)
Ibn Naqqāsh, Nūr al-Dīn (d. 1475)
Ibn al-Shāṭir (1306–1375)
Ibn Sīnā (d. 1037)
Ibn Wāṣil, Jamāl al-Dīn (1207–1298)
Ismā'īl b. Bulbul (879–890), Abbasid vizier
al-'Izz Abū Naṣr Beg Arslān, Saljuq emir
'Izz al-Dīn al-Ḥasan (fl. c 1314)
Karajī, Abū Bakr Muḥammad b. al-Ḥasan (d. c. 1030)
Kāshī, Ghiyāth al-Dīn (d. 1429)
Khafrī, Shams al-Dīn (d. 1550)
Kharaqī, Bahā' al-Dīn Abū Bakr Muḥammad (twelfth century)
Khayyām, 'Umar (d. 1131)
al-Khāzin, Abū Ja'far (tenth century)
Khāzinī, 'Abd al-Raḥmān (twelfth century)
Kūhī, Abū Sahl (tenth century)
Maḥmūd b. Sebügtegīn (r. 998–1030), founder of the Ghaznavid dynasty
al-Ma'mūn (r. 813–833), Abbasid caliph
al-Māridānī, Jamāl al-Dīn (d. 1406)
Ma 'ūd b. Ma ḥmūd (r. 1030–1041), Ghaznavid sultan
Mehmet Fātiḥ (r. 1451–1481), Ottoman sultan
Mu'ayyad al-Dīn al-'Urḍī (d 1266)
Mu ḥammad 'Ādil Shāh (r. 1626–1660),

(*Continued*)

Table 4.2 (Continued)

ʿĀdil Shāh sultan Muḥammad b. Armaghān (fifteenth century)
Muḥammad b. Khalaf (ninth century)
al-Mustan ṣir (r 1226–1242), Abbasid caliph
al-Muʿtaḍid (r. 892–902), Abbasid caliph
al-Muʿtamid (r. 870–892), Abbasid caliph
al-Muwaffaq (r 875–891), Abbasid regent
Nāṣir al-Dīn Dā'ūd (r. 1227–1229), Ayyubid ruler of Damascus
Nāṣir al-Dīn Yūsuf (r. 1250–1260), Ayyubid ruler of Damascus
Rāzī, Abū Yūsuf Yaʿqūb b. Muḥammad (tenth century)
Rukn al-Dawla (r. 932–976), one of the three founders of the Buyid dynasty
Ṣadr al-Sharīʿa (fourteenth century)
al-Samaw'al, Abū Naṣr (d. 1175)
Ṣamṣām al-Dawla (r. 989–998), Buyid emir
Sanjar (r. 1117–1157), Saljuq sultan
Sebügtegīn (976–997), Samanid governor
Shāh Jahān (r. 1628–1659), Mughal padishah
Shahrazūrī, Shams al-Dīn (d. after 1288)
Shīrāzī, Quṭb al-Dīn (1236–1311)
Sibṭ al-Māridānī, Muḥammad b. Muḥammad (1423–c. 1495)
Sijistānī, Abū Sulaymān (c. 912–c. 985)
Ṣūfī, ʿAbd al-Raḥmān (903–986)
Thābit b. Qurra (d. 901)
Timur (r. c. 1369–1404), founder of the Timurid dynasty
Ṭūsī, Aṣīl al-Dīn (d. 1316)
Ṭūsī, Naṣīr al-Dīn (1201–1274)
Ulugh Beg (r. 1447–1449), Timurid ruler
al-ʿUrḍī, Mu'ayyad al-Dīn (d. 1266)

patron's family and towards his clients as well as military, economic, and political instability. Competition among different factions within a ruling family and among families of administrators, and the growing influence of the Turkish military slaves upon the choice of the next caliph, made patronage an unreliable field of social relations. As a result, the figure of the itinerant, often impoverished, scholar emerged. After the Buyid ruler ʿAḍud al-Dawla's death, one of his two teachers of astronomy, al-Sharīf Ibn al-Aʿlam, fell into poverty when ʿAḍud's son, Ṣamṣām al-Dawla, did not take over his father's patronage obligations (Ibn al-ʿIbrī 1958, 174). The asymmetry of power and the instability inscribed in it are reflected in the various acts of punishment that a patron could heap on a client, either on his own behalf or on behalf of another client. Extortion, loss of office and property, exile, imprisonment, and execution were widespread means of regulating access to power and funds among the civil elites, in particular at the level of viziers and their immediate entourage. Scholars of the mathematical sciences, even those who were powerful patrons in their own right, also suffered under the arbitrariness of the patronage system. Several were exiled, incarcerated, or killed.

Boundaries that separated patronage from other kinds of relationships, for instance those that existed in the realm of craftsmanship, are often blurred due to a lack of precise information in the sources. The tenth-century bookseller Ibn al-Nadīm, a member of the circle of philosophers and literati surrounding the philosopher Abū Sulaymān Sijistānī and well connected with courtly life under the Buyids, wrote a work about the intellectual life in the Abbasid caliphate. He used the term *ghulām* (plural *ghilmān*), which in the period primarily meant a military slave, to describe relationships among scholars of the mathematical sciences and instrument makers. In most cases, the owner or master of the *ghulām* was either an astrologer or an instrument maker. Instrument makers like Aḥmad and Muḥammad, the sons of Khalaf, often started their career as *ghilmān* and later had *ghilmān* themselves. Both brothers had been *ghilmān* of the astrolabe maker ʿAlī b. ʿĪsā and then trained six other men as their *ghilmān*, all of whom became known as instrument makers (Ibn al-Nadīm 1971, 343). Hence it is perhaps reasonable to interpret this relationship as one between a master and an apprentice rather than one of patronage.

As scholars rarely focused on one discipline alone but were often well educated in several branches of knowledge, neither did patronage focus only on one type of knowledge. Scholars who excelled in the mathematical sciences can be found heading the administration of a hospital founded by their patron or serving as head of courtly protocol, as in the case of Abū l-Wafāʾ (Kraemer 1993, 182, 216). Abū l-Wafāʾ's eminence in courtly affairs was such that he also participated, together with elders and representatives of the religious sciences, in political and financial negotiations between his Buyid patron and the city of Baghdad about protecting the populace against the threat from the Byzantine army's invasion of northern Iraq (Kraemer 1993, 100). Another tenth-century mathematical scholar, Abū Jaʿfar al-Khāzin, acted as ambassador for his patron, the head of the Samanid dynasty in Samarqand, in a war between the Samanids and the Buyids. After he had successfully concluded negotiations between the defeated Samanid general and his Buyid opponent, Abū Jaʿfar was patronized by the Buyid overlord Rukn al-Dawla, who esteemed him, made his vizier Ibn al-ʿAmīd employ him, and strongly recommended that he emulate the scholar (Kraemer 1993, 252).

Patronage at courts and endowed teaching institutions from the late twelfth century

The proliferation of richly endowed teaching institutions from the twelfth century changed the framework of patronage of the mathematical sciences. The dynasties that contributed most to this new development were the Saljuqs, the Zangids, the Ayyubids, and the Mamluks. The success of the new system of endowed teaching institutes was such that after the twelfth century many other dynasties also donated funds for building madrasas, Sufi *khānqāh*s (lodges), tombs, mosques, houses for teaching the Qur'an and *ḥadīth*, as well as hospitals. The *waqf* (religious endowment) typically included the costs of new and regularly funded posts as well as stipends for students. The new system extended the scope and complexity of patronage relationships between scholars and courts on the one hand and among scholars themselves on the other, by opening new possibilities for participating in the distribution and redistribution of the donated funds and endowed offices (Berkey 1992, 96). Socially, it was characterized by four major features. First, the donors selected and appointed administrators of the endowment, the ruling military aristocracy, in particular the rulers themselves and the highest-ranking officials of the court hierarchy, the scholars, the head judges of the different legal schools, and other leading religious officials shared responsibilities and opportunities for making appointments to professorships and other posts. Second, professors amassed positions in more than one discipline and more than one endowed institute. Third, it became possible to buy, sell, and outsource professorships and other offices. Fourth, it became feasible to choose one's own successor, preferably from among immediate family members (including minors) or one's own students, which created the phenomenon of hereditary teaching posts (Berkey 1992, 96–97, 102–119, 121–125).

Most of the available teaching posts were within law, *ḥadīth*, Qur'anic studies and languages. But there were also positions for medicine at *madrasas* and mosques in Anatolia, Iraq, Syria, and Egypt; for timekeeping (*'ilm al-mīqāt*) in Mamluk Egypt and Syria as well as in regions of the Ottoman Empire; and for a special mathematical discipline linked to law and dealing with the determination of inheritance shares and legations (*'ilm al-farā'iḍ*) in different regions of the Islamic world (Petry 1981, 65, 428, n90; Berkey 1992, 69; al-Sakhāwī, III 119). In the fifteenth century posts for the mathematical sciences in general were created, when the Ottoman sultan Mehmet Fātiḥ began a tradition of endowing new *madrasas*, while the Mughal padishah Akbar initiated a similar practice in some of the teaching institutes.

Mathematical sciences were also taught in *madrasas* in north Africa, Iran, and Central Asia. This is verified by colophons and ownership marks in mathematical manuscripts as well as the donation of such books as *waqf* to *madrasa* and mosque libraries. But there is little to no information about whether the teachers held positions specifically linked to the one or the other of the mathematical disciplines in these regions. Occasionally writers of biographical dictionaries report a scholar of the mathematical sciences settling in a *madrasa*, such as al-ʿUrḍī who lived in the ʿIzziyya Madrasa in Maragha, northwest Iran, founded by the Saljuq emir al-ʿIzz Abū Naṣr Beg Arslān (Ibn al-Fuwaṭī 1962–1965, IV/1, 387–388). The fluidity of the teaching system in regard to where and what was taught, and its character as a scholarly network, opened wide doors to teachers and students of non-religious disciplines living in *madrasas* and Sufi *khānqāh*s, or simply visiting them for classes.

During the Ayyubid period, the rhetoric of patronage lost some of its previous richness. As main terms there remained service (*khidma*), honour or grace (*ikrām*), and benefit (*ni'ma*). Most of the patronage relationships between a ruler or prince and a physician were described by this terminology (Ibn Abī Uṣaybiʿa

1965, 584–586, 589–591, 598–601, 635–357 *et passim*). Occasionally a ruler is reported to have esteemed (*iḥtarama*) one of his clients excelling in the mathematical sciences. Although several Ayyubids sponsored astrologers at their courts, the authors of biographical dictionaries seldom applied to them this standard vocabulary of patronage. Writers of historical chronicles and mathematical scholars, however, talked about patronage of the mathematical sciences in such terms. In the mid-thirteenth century, al-ʿUrḍī described his relationship with the last Ayyubid ruler Nāṣir al-Dīn Yūsuf as one of service. He told his new friends and colleagues in Maragha how he had escaped through his own courage, announcing his professional expertise to the enemy soldiers and his future usefulness to their lord during the bloodbath in which all members of the court died except for himself and two sons of the ruler (Ibn al-ʿIbrī 1958, 280). On the day of the attack the scholar had just had a session with the unfortunate Ayyubid prince to cast his horoscope. Al-ʿUrḍī remained silent, though, about whether he had predicted his patron's imminent demise. Ibn Wāṣil, court historian of the Ayyubid dynasty and well educated in various disciplines including mathematics, characterized his relationship to his father's patron, the Ayyubid prince Nāṣir al-Dīn Dā'ūd, exclusively by one term – service – which he also used for the relationship between his older colleague ʿAlam al-Dīn Qayṣar and al-Ashraf, an earlier Ayyubid ruler of Damascus (Ibn Wāṣil 1977, 145–146).

While the Ayyubids continued to exercise their patronage to a substantial degree at court, which for this dynasty was usually the city's fortress, male and some female members of the family put a huge investment into building *madrasas* and other endowed teaching institutions and in maintaining those built by their predecessors in Syria, northern Iraq, and Egypt. This meant that an increasing part of their patronage relationships included endowing such institutions for a particular client, or providing a client with a professorship or another office in an already created institution. Whether practitioners of the mathematical sciences received such dynastic support within the framework of endowed teaching institutions needs to be further explored.

In other regions, the rhetoric of patronage also increasingly privileged the vocabulary of service. Ibn al-Fuwaṭī was a student of Naṣīr al-Dīn Ṭūsī and head of the library of the Mustanṣiriyya Madrasa in Baghdad, founded by the caliph al-Mustanṣir. He spoke of service when talking about relationships between a variety of scholars, including practitioners of the mathematical sciences, and Ilkhanid rulers, their viziers, and governors of cities (Ibn al-Fuwaṭī 1962–1965, IV/1, 40, 392, 458, 620; 3, 103). Ibn al-Fuwaṭī frequently also used an earlier term that emphasized the character of the patronage as a connection (*ittiṣāl bi*), occasionally adding new ones of a similar vein by speaking of establishing a relationship or link, and being devoted to or depending on (*nasaba, taʿallaqa*) (Ibn al-Fuwaṭī 1962–1965, IV/1 40, 132, 149, IV/2 513, 754, IV/3 513).

The successors of the Ayyubids, the Mamluks, removed patronage of the mathematical sciences from their courts and linked it almost exclusively to endowed teaching institutions. Sultans and high-ranking court officials promoted scholars of the mathematical sciences to professors at prestigious *madrasas*, donated professorships and stipends for students at their *madrasas*, and appointed to the fortress mosque religious functionaries such as *muezzins* who had taken classes in timekeeping, arithmetic, and other mathematical disciplines (Brentjes 2008). The rhetoric of patronage under the Mamluks continued to revolve around the same terms as in the Ayyubid period. Clients served, linked themselves with patrons, and patrons honoured or benefited them (al-Sakhāwī, *al-Ḍaw'*, III, 296; VI, 112, 235; VIII, 191; IX, 15; X, 48). New terms are to settle or establish someone finally or firmly (*istaqarra bihi*), to appoint or establish someone (*qarrara hu* or *fī*), and to favour, distinguish, or confer distinction (*ikhtaṣṣaṣa bi*). All reflect the demands and opportunities created by the system of endowed teaching institutions with their secure posts.

The mathematical sciences became part of the general education provided by the endowed teaching institutions, although only some madrasas had a professor capable or willing to teach them. Students and teachers of the mathematical sciences were fully integrated into this new framework. They could acquire paid positions and offices as muezzins, *muwaqqit*s, teachers, preachers, imams, or physicians. Patronage proved useful for acquiring more than one position and keeping them, since competition was strong and interference in the distribution of posts widespread. Offices could also be lost frequently, either due to envy and greed among patrons and scholars or because the office holder was sick or meddled in dynastic policy.

As in the previous period, scholars acted as ambassadors and mediators in moments of war, conflict, or other needs. Naṣīr al-Dīn Ṭūsī, one of the leading philosophers, mathematicians, and astronomers of the thirteenth century, negotiated for his Isma'ili patron the peaceful surrender of the fortress of Alamut in northwestern Iran with the attacking Mongols, who, however, did not honour the negotiated contract. Ṭūsī changed sides and later wrote diplomatic letters for his new Mongol patron Hülägü to the last Ayyubid ruler Nāṣir al-Dīn Yūsuf. His student Quṭb al-Dīn Shīrāzī, a Sufi, physician, philosopher, and gifted practitioner of the mathematical sciences, served the Ilkhanid ruler Ghāzzān as ambassador in Mamluk Cairo. Authors of mathematical treatises, such as Ibn al-Lubūdī, were appointed as viziers or took other administrative positions. Ṭūsī and later his son Aṣīl al-Dīn, in addition to directing the observatory in Maragha, headed the *dīwān* for religious donations of the Ilkhanid dynasty and were thus powerful patrons in their own right who were served by scholars as well as princes.

Dedications of manuscripts and instruments continued during this period. Due to the impact of courtly patronage on the arts of manuscript production, scientific books grew ever more lavish in decoration and design. Several of the most beautifully decorated scientific manuscripts went from one princely library to the next, acquired as gifts, bought for a high price on the market, or taken as booty in a war between neighbours. A few examples are the copy of 'Abd al-Raḥmān Ṣūfī's *Star Catalogue* allegedly made from a copy owned by Naṣīr al-Dīn Ṭūsī for the library of Ulugh Beg; 'Alī Qushjī's work on mathematical cosmography written for Ulugh Beg in Persian and translated into Arabic for Mehmet Fātiḥ; a partial copy of Bīrūnī's astronomical magnum opus dedicated to the Ghaznavid sultan Maṣʿūd b. Maḥmūd, acquired by a courtier of Shāh Jahān's court in 1649; a copy of Ṭūsī's edition of Euclid's Elements, which in 1659 was part of the library of Sultan Muḥammad ʿĀdil Shāh in Bijapur and later came into the library of the Mughal ruler Awrangzīb ʿĀlamgīr; and a copy of Ghiyāth al-Dīn Kāshī's work on arithmetic dedicated to Ulugh Beg, which was bound for Sultan Fātiḥ ʿAlī Tippu's library (Blochet 1900, 87; Schöler 1990, 172; Loth 1877, 215–216, 220).

Patronage exercised by scholars, including those in mathematical disciplines, broadened too. Visiting, travelling together, deputizing, finding a position for a former student, marrying one's daughters to one's students, choosing a family member or a former student as one's successor: all were often-used means to secure one's own and one's family's position, influence, and possibly fields of knowledge. Ḥasan b. ʿAlī al-Qaymarī was excellent in arithmetic, algebra, timekeeping, the determination of inheritance shares and legations, and prosody, also possessing good knowledge in law and grammar, and was a student of Ibn al-Majdī and Khalīl b. Ibrāhīm Abū l-Jūd from Damietta in Egypt. He received a chair in the field of inheritance shares and legations at the Jawhar al-Ṣafawī Madrasa in al-Ramla after his teacher Abū l-Jūd had talked to its donor (al-Sakhāwī, *al-Ḍawʾ*, III 119).

The vocabulary of courtly patronage found its way into describing relationships between scholars, teachers, and students. Having travelled widely in the East, including China, the son of an educated family from Wasit in Iraq, ʿIzz al-Dīn al-Ḥasan turned to Syria where he 'joined the service (*khidma*) of the judge [Muḥammad b. Wāṣil al-Ḥamawī], the judge of Hama. He was well versed in the *Almagest* and mathematics. I [read] with him for some time' (Ibn al-Fuwaṭī 1962–1965, IV/1 101–102).[4] Sons were in the service (*khidma*) of their fathers when studying with them, were launched into the service of princes through serving their fathers, and professors served the books of their intellectual fathers by writing commentaries and glosses on them (Ibn al-Fuwaṭī 1962–1965, IV/3 318, 338; al-Sakhāwī, *al-Ḍawʾ*, X 48). Students formed connections (*ittiṣāl*) with teachers and married their daughters (Ibn al-Fuwaṭī 1962–1965, IV/3 418–419). The obligation embedded in offers of marriage to a professor's student could lead to severe friction if rejected. In the fifteenth century, Muḥammad b. Armaghān, a student of Shams al-Dīn Fanarī and later a prominent scholar, fell out with his teacher's sons because he refused their father's offer to marry their sister. Armaghān justified his decision by pointing to an earlier promise of marriage to another scholar's daughter (Tasköprüzade 1978, 45–46). Occasionally, a holder of a chair preferred to appoint one of his students rather than a relative as his successor. The *muwaqqit* Sibṭ al-Māridānī was the grandson of Jamāl al-Dīn al Māridānī,

4 wa'jtamiʿtu bi-khidmat al-qāḍī [Muḥammad ibn Wāṣil al-Ḥamawī] qāḍī Ḥamā wa-huwa ʿārif bi-l-Majisṭī wa-l-riyāḍī [qara'<tu>] ʿalayhi mudda.

himself a *muwaqqit* and successful teacher of the mathematical sciences in Cairo who left at least six treatises on timekeeping and astronomical instruments. Sibṭ al-Māridānī, a prolific writer on arithmetic, algebra, timekeeping, astronomical handbooks, and instruments, received his teaching position at the Ibn Ṭūlūn Friday Mosque due to desire of the previous professor and *muwaqqit* Nūr al-Dīn b. Naqqāsh that he should take the office (al-Sakhāwī, *al-Ḍaw'*, IX 36). The increasing transformation of the positions at endowed teaching institutions into family holdings was not limited to the professorships. Other posts could also be handed down in the family, such as that of the librarian of a madrasa or mosque library (al-Sakhāwī, *al-Ḍaw'*, II 154).

Professional identities and remuneration

In the eighth to twelfth centuries, the first period of patronage, courtly patrons paid their clients in two forms, although the information given in the sources regarding them is so irregular and sparse that it is difficult to get a clear picture. The form mentioned most often are gifts such as robes of honour and one-time monetary payments made from the patron's personal treasury (al-Ṭabarī 1989, 313; Ibn al-ʿIbrī 1958, 131, 137, 182). The second form was regular monthly and yearly payment (*rizq*; *ujra*) allocated mostly from the general treasury of the court or the private treasury of the patron, but also from the *dīwān* for religious donations (King 2000, 211; Ibn Abī Uṣaybīʿa 1965, 198–200; Halm 1997, 75).

By the twelfth century, regular stipends, paid annually or monthly, seem to have been the more widespread format. It is not always clear whether they came exclusively from the personal treasury of the patron. In addition, other forms of payment were also used, in particular the turning over of an *iqṭāʿ*, a taxable region, although occasionally already in the first period a village had been given as a gift to a physician or other scholar. The stipends (*jāmakiyya*; *jirāya*) and the *iqṭā'* had previously been standard forms of paying the military. In the twelfth and thirteenth centuries, the Ayyubid dynasty applied such forms to reconcile members of the civil elite of a city with the conquest and destruction wrought by Ayyubid troops, to express their highest appreciation for a local notable or a courtly client, and to pay for services rendered (Eddé 1999, 280). Some Ayyubid court physicians received such remuneration and gratitude but it is unclear whether astrologers or engineers were also recompensed in this manner. With the proliferation of endowed teaching institutes, regular salaries in conjunction with non-monetary components such as food and lodging for professors, teaching assistants of different kinds, *muwaqqit*s, muezzins, and other positions linked to these institutes became the norm. The highest salaries often, but not always, went to the professors. The amount depended on the endowment and thus on the wealth and status of the donor. Even among the members of the Mamluk elite in Cairo, the endowments and hence the stipulated salaries varied considerably (Berkey 1992, 77–78). *Muwaqqit*s at Mamluk mosques and *madrasa*s got substantially less, but a little bit more than the muezzins. A *muwaqqit* in a major mosque or *madrasa* in Cairo earned at the highest 60 percent of what a Sufi shaykh and 40 percent of what a professor of law received. More often, however, this post was supported by a salary of less than a third of a professor or an imam (King 1996, 302). The further bureaucratization of the system under the Ottomans fixed salaries for positions according to their distance from the capital, their status within their city, and the access it could open to positions in the religious and civil administration of the empire. Salaries were paid on a yearly, monthly, or daily basis. The money for the salaries came from profits gained from agricultural production in donated villages, fruit gardens, and shops or were taken from the dynasty's waqf treasury, from the *dīwān* for taxes, or from economic enterprises monopolized by a dynasty, such as salt mining. In addition to the salary fixed by the donor or prescribed by the administration, patrons could add substantial monetary and other gifts to honour the person they had appointed and keep him from looking for other lucrative income (Tasköprüzade 1978, 59).

Earning a living through teaching was a lifestyle that was heavily contested over several centuries in a number of Islamic societies. Many remarks and discussions about which kinds of knowledge it was permissible to be remunerated for, and which kinds of offices a devout believer should accept or refuse, fill the pages of manuscripts. Algebra, for instance, could be safely taught for money, while knowledge of *ḥadīth* should be shared for free (Berkey 1992, 95–97; Tasköprüzade 1978, 26, 57). Taking the position of a judge or other, less prestigious positions for discharging the law for a salary was seen as an acceptable form of paid employment after long years of learning. The rise of endowed teaching institutions altered the framework of access to remuneration for knowledge, bringing with it new employment opportunities (Berkey 1992, 96). Regularly remunerated positions increased substantially in the centres and spread to the provinces and even villages. Most opportunities were, of course, found in capitals and major administrative centres. But the often fluid forms of power, and the distribution of governance of provinces and smaller cities among male members of a dynasty or members of the military, offered men of the lower ranks their own fields of patronage. In addition, everyone with sufficient wealth could donate an endowed teaching institute, even if it consisted of one position only. Hence families who could afford it were able to create posts for the scholars of their own family and stipulate the kind of disciplines that would be taught. A number of medical madrasas were created in this manner, but no case of a scholar of the mathematical sciences is known who founded his own madrasa chair specifically dedicated to these disciplines. Most teachers of mathematical disciplines at endowed institutes in Mamluk Cairo, for instance, were appointed by courtly patrons to teach a broad range of religious and other disciplines, or succeeded their former teachers or relatives in their positions. The biographical literature shows clearly that most of them focused on teaching a particular branch of law, the determination of inheritance shares and legates. This focus allowed them to carve out a substantial space for teaching mathematical knowledge covering arithmetic, algebra, practical and theoretical geometry, three astronomical disciplines (timekeeping, planetary theory, compiling astronomical handbooks and ephemerides), and the construction of scientific instruments. In this sense, the mathematical sciences underwent a process of stabilization and professionalization once they became integrated into the new framework of endowed teaching institutions.

This substantial and important gain of territory, opportunity, and stability is well documented through the thousands of texts introducing these disciplines to many generations of students, extant today in manuscript libraries across the globe. It was accompanied, though, by a focus on elementary content and a submission to the teaching methods and values that governed the religious and philological disciplines. Learning by heart was seen as the highest expression of scholarship and reason for fame. Scholars were praised and admired for the speed and quantity of their memorizing. Studying the texts of one's teachers and of their teachers in turn – that is, chains of texts determined by chains of scholars – became the norm not only for transmitting *ḥadīth*, where it evolved in the early centuries as the only method considered as leading to trustworthy knowledge if the transmitters of the chain were deemed reliable, sound and morally worthy, but also in the mathematical sciences. As a result, fewer and fewer texts by scholars of previous generations, centuries, and cultures were directly studied. Editions, paraphrases, commentaries, and super-commentaries of Euclid's *Elements*, for instance, replaced in many classes the study of the *Elements* themselves.

The goal of education in endowed teaching institutes was not to create critical or substantially new knowledge. New ways of looking on taught knowledge and asking questions about it, although lauded as marks of excellence, were considered exceptional. A major way of establishing a scholar's academic credentials, in particular when he was new to town, was to hold a public disputation (*munāẓara*). In such disputations a series of questions was asked to bring out the depth of knowledge of what was taught and discussed in the various disciplines. Defeat, evaluated by an arbiter who was either a well-established scholar or occasionally a ruler, vizier, or another high-ranking court official, was declared when a participant was not able to answer these questions satisfactorily. Thus, scholarly excellence did not consist in raising questions beyond the already established facts but in being broadly and substantively familiar with what was taught and discussed in scholarly circles. Such broad, encompassing knowledge was worthy of patronage and promotion.

Outcomes of patronage for the mathematical sciences

Outcomes of patronage for the mathematical sciences extant today include instruments and manuscripts, art objects, and architectural and technological monuments. Their character as products of patronage can be established through dedications, frontispieces, colophons, and marks of ownership. Other results of patronage were expeditions, measurements, observations, and oral performances as teachers, boon companions, and participants in sessions of serious debate or conviviality. These survived the centuries through reports and descriptions. Since it is impossible to describe here all known outcomes of patronage, or to summarize all the results of patronage relationships, a few examples will have to suffice.

According to King (2004, II 993–1020) and Charette (2006, 134, n1), over a thousand astrolabes and several hundred globes, quadrants, sundials, and other scientific devices made in Islamic societies survive. At least one hundred and fifty extant astrolabes, some twenty sundials, and about a dozen quadrants go back to pre-sixteenth-century workshops. Astronomical instruments were part and parcel of the Translation Movement. The prominent role of instruments resulted from the political and ideological functions that transformed the act of translating into a cultural movement sustained for almost two centuries by a fine net of patronage acts from Abbasid caliphs, courtiers, and practitioners of the translated knowledge. Gutas (1998, 41–50) convincingly argued that these functions rested on a concept of translation that was part of a pre-Islamic ideology of kingship from Sasanian Iran.[5]

Closely linked to a new, modified form of this ideology, which now was directed against the Byzantine Empire, was the organization of geodetic measurements and astronomical observations between 828 and 833 in Baghdad, Damascus, Mecca, and the desert of Sinjar near Mosul under the caliph al-Ma'mūn (Gutas 1998, 83–95). The geodetic observations were presented as a means of checking, comparing, and verifying data found in ancient texts and instruments in use in different regions of the Abbasid caliphate (King 2000, 215, 217–218, 223–234). The observation of a lunar eclipse in Mecca is described as a caliphal order to determine the qibla of Baghdad, the direction of prayer towards Mecca (King 2000, 214, 218–219). Other observations most likely served for astrological counselling (Charette 2006, 125). Astrology was probably not merely the practical but also the theoretical context of this astronomical programme, due to the rising impact of Aristotelian natural philosophy (Charette 2006, 135 n18).

Astrologers, instrument-makers, and a judge participated in these measurements and observations – constructing a series of instruments; determining appropriate sites; choosing the team members; supervising their work; witnessing, recording, and communicating the results to the caliphal patron. He in turn formulated successive research questions, informed himself about the reliability of the instruments and evaluated the results (King 2000, 215, 218–220, 223–224). The cultural and scientific results of this first major programme of empirical scientific activities had immediate and long-lasting consequences. They supported the radical shift from Indian and Sasanian models and parameters to Ptolemaic astronomy that took place in the following decades. The series of reports on this programme and its scientific results established astronomical observations and measurements as an important icon of courtly patronage. Elements such as specifically constructing instruments, forming a team of scholars, and inviting witnesses were repeated in later programmes under the Buyids, Saljuqs, and Ilkhanids.

5 I no longer subscribe to this view, see Brentjes 2023, 20.

Several princes of the Buyid dynasty either received an excellent education in the mathematical sciences or were presented with high quality writings on theoretical and practical mathematical problems. Berggren (unpublished) has evaluated some aspects of Buyid patronage of the mathematical sciences. He describes a deeply structured network, with patrons in the dynasty and the administration, and clients among the administrators, scholars, and instrument makers. The princely patron of this dynasty was first and foremost ʿAḍud al-Dawla. But several other princes also contributed, among them Rukn al-Dawla – one of the three founders of the dynasty and father to ʿAḍud – and ʿAḍud's sons Sharaf al-Dawla and Ṣamṣām al-Dawla. Rukn al-Dawla, an illiterate mercenary from Daylam on the Caspian Sea, ordered the measurement of the latitude of the city of Rayy and its longitudinal difference from the Abbasid capital. He provided an excellent education for his son by giving him his own vizier, Ibn al-ʿAmīd, as tutor. Ibn al-ʿAmīd collected manuscripts, commissioned a commentary on Book X of Euclid's *Elements* from Abū Yūsuf Yaʿqūb Rāzī, and himself excelled in mechanics. He used the latter for inventing new siege machines and is credited with constructing a mural quadrant used in the astronomical observations requested by Rukn al-Dawla (Sijistānī, *Ṣiwān al-ḥikma*, 321–324; Sayılı 1960, 104). Other Buyid viziers also patronized scholars of the mathematical sciences. Numerous scholars sponsored by Buyid patrons wrote works on geometry, number theory, algebra, and arithmetic in addition to texts on astrology and astronomy.

Works on higher geometry were of two types. One type continued and completed the research of ancient Greek geometers. The two main figures for identification and imitation were Archimedes and Apollonios. The scholars of the Buyid courts wrote on conic sections, loci, polygons, projections of the sphere on the plane, the construction of two mean proportionals, and the trisection of angles and studied the transformation of curves into equations and vice versa (Hogendijk 1981; Berggren 1986, 77–85). The other type began to modify and then to replace the methods and concepts of Greek mathematics with new theorems, procedures, and outcomes. Works on Menelaus' theorem and new theorems of plane and spherical trigonometry constitute one field of such innovation. Other fields of geometrical research that went beyond the classical heritage, or drew rather on Indian precedent, related to astronomical problems such as interpolation procedures for calculating tables. Approximate solutions were found to practical problems in architecture, surveying, accounting, and determining inheritance shares and legacies.

In number theory, arithmetic, and algebra, Buyid scholars also wrote important books which they dedicated to their patrons. In Karajī's *Fakhrī*, for instance, the author departs from the former understanding of algebra as a set of rules for solving quadratic equations and verifying their solutions, as defined by ninth-century scholars, some of whom had been patronized by Abbasid caliphs. Karajī now presented algebra as a discipline which applied the rules and procedures of arithmetic systematically to unknowns of the type x^n and $1/x^n$ and to polynomials (Rashed 1984/1994). Taken up a century later by the physician and mathematician al-Samaw'al, who wandered through Syria, Kurdistan, Azerbaijan, and other Islamic lands in search of patrons, Karajī's new algebra has had a deep impact on how the field was taught and studied by later generations.

In the period of endowed teaching institutions, patronage of the mathematical sciences took place, as discussed above, at courts, among the civil educated elite, and within the framework of the endowed institutions. The manuscripts, tables, and instruments that were produced in these diverse environments were primarily created within the context of teaching. Luxury specimens were fabricated mainly for princely education, as gifts for members of the ruling family, and as items held and displayed in princely libraries. At times the production of new knowledge, in the form of new solutions to standard problems, variations to extant solutions, and modifications of unsolved problems occupied scholars at court as well as in *madrasas*, *khānqāhs*, or mosques. Planetary theory and timekeeping are the two prominent fields of such innovative efforts. Debates over modifications to Ptolemaic models, and the relationships between geometrical models, physical properties, and philosophical principles, perhaps began as early as the ninth century. Major scholars patronized by rulers contributed to this debate, among them Ibn al-Haytham, Naṣīr al-Dīn Ṭūsī, al-ʿUrḍī, Quṭb al-Dīn Shīrāzī and ʿAlī Qushjī. In the second period of patronage, scholars linked to the endowed teaching institutions participated in it (Roberts and Kennedy 1959; Kennedy and Ghanem 1976; Saliba 1979, 1990, 1993; Langermann 1990; Ragep 1993, 2005). In addition to al-ʿUrḍī, Shīrāzī, and Qushjī who also lived in and taught at *madrasas*, Ibn al-Shāṭir, *muwaqqit* and muezzin in Damascus, Ṣadr al-Sharīʿa in Bukhara, and Shams al-Dīn Khafrī in Shiraz were important contributors to planetary theory (Kennedy 1957; Dallal 1995; Saliba 1994).

Muwaqqits, and people engaged in similar works and with similar professional links to mosques and madrasas, but named differently, start to appear in the late thirteenth and early fourteenth centuries, first in Mamluk Egypt and Syria. Later they also are known from al-Andalus, parts of North Africa, the Yemen, and the Ottoman Empire. As mentioned earlier, their task consisted in solving the astronomical and mathematical problems connected with determining prayer times, the direction of prayer towards Mecca, and the beginning of the new month (King 1993, 2004). These tasks had been tackled since the eighth century, often by some of the most brilliant scholars involved with the mathematical sciences. Independent of their religious beliefs they contributed to finding exact as well as approximate numerical and geometrical solutions to these problems. From this perspective it could well be claimed that a major outcome of courtly patronage for the mathematical sciences in both periods was the development of a rich field of methods for finding astronomical and mathematical solutions important to religious practices. When the *muwaqqit* emerged in the late thirteenth and early fourteenth century, such methods became acknowledged as an independent branch of the mathematical sciences called *ʿilm al-mīqāt*. In Charette's view (2006, 129) this included a broadening of content, as it united spherical astronomy, timekeeping, astronomical instrumentation, gnomonics, determination of the direction of prayer, chronology, and the prediction of the beginning of the new month with the first visibility of the lunar crescent. The *muwaqqit*s and the teachers of the new discipline developed new instruments, calculated multi-entry tables (often up to 40,000 entries and occasionally even to 415,000 entries), developed sophisticated tools for simplifying the necessary calculations, and created means for finding solutions valid for all latitudes.[6] The emergence of *'ilm al-mīqāt* also involved issues of legitimacy within the discursive framework set by al-Ghazālī in the eleventh century. Ibn al-Akfānī, a physician and madrasa professor in Cairo, declared *'ilm al-mīqāt* to be a discipline obligatory (*wājib*) for a Muslim (Witkam 1989, 59). Instrument making was mostly seen as socially beneficial (Charette 2006, 129). This discursive support for the new field stabilized its establishment within the teaching system and *'ilm al-mīqāt* indeed became a respectable mathematical discipline studied as part of a general education by numerous, but by no means all, students, including some of the later leading scholars in Mamluk society.

The outcomes of patronage of the mathematical sciences in Islamic societies described here are only a fraction of what was produced. Numerous instruments, more treatises, and wonderfully illuminated copies were produced over the centuries for courtly display in libraries and private settings, for men and women, rulers and their families, viziers, and emirs. Many professors, *muwaqqit*s, physicians, and other professionals who contributed to the mathematical sciences as teachers, researchers, observers, and instrument makers owed their positions directly or indirectly to military as well as civil patrons. Without patronage, the flourishing mathematical cultures of Islamic societies would have been impossible or at least much poorer.

Conclusion

Sources about patronage of the mathematical sciences in Islamic societies between the eighth and the nineteenth century are rich, but narrow in scope and uncertain in both the meaning and the reliability of their claims. It is possible and useful to collect all dedications of texts, instruments, paintings, and other relevant objects and analyse their rhetoric, focus, and function. Such research will broaden our knowledge about the mathematical fields supported by different kinds of patrons and can elucidate the meaning given to this support.

However, we should not consider dedicated works as the sole outcomes of patronage. They are, without doubt, the central elements in the exchange of benefits, honour, and gratitude that constituted and kept alive the patronage relationship. Patronage, however, is not reducible to the exchange of gifts between patron and client. It also was a relationship of work or, as the medieval authors preferred, of service, which bound the participants to each other in the many ways discussed in the previous sections, work that included studying, researching, observing,

6 King 2004; Charette 2003.

interpreting, and writing. Without continuous intellectual work the scholars may have lost their patrons, although in some cases sources indicate that wit and sociability were more valuable commodities than mathematical proficiency. The criticism uttered against the fifteenth-century scholar Kāshī, for instance – that he was a bore who had not mastered the refined protocol of Ulugh Beg's court and thus was frowned upon – reminds us that scholars of the mathematical sciences were participants in larger networks of social relationships, expectations, and behaviour. Excellence in the mathematical sciences while important was not sufficient for creating stable and comfortable connections between a client and a patron.

A second substantial field for research about patronage is offered by the rhetoric of patronage and its variations and changes over time and space. Islamic societies differed from each other while sharing a number of elements. The rhetoric of patronage embodies some of these differences and similarities. It reflects different degrees of violence, instabilities in client–patron relationships, the new opportunities provided by the endowed teaching institutes, the spread of patronage forms among the civil elites, and other changes. In this chapter I have tried to trace major differences, similarities, and changes. Focusing on shorter periods of time and smaller territories will help to uncover local and temporal particularities. Such particularities will offer possibilities for understanding why certain, but by no means all or even most, Islamic societies supported the mathematical sciences through patronage.

A third domain for further research is the study of the relative place ascribed to the mathematical sciences in the complex web of courtly patronage. The support given to individual mathematical fields by specific dynasties in relationship to other domains of culture such as medicine, history, law, or the arts and the social loci of these sciences defines their reputation, forms of practices, and various elements of their content. A clearer picture of the specifics of these relations and places will improve our understanding of the substantial changes in productivity, creativity, and focal points in the mathematical sciences in Islamic societies between the eighth and nineteenth centuries.

Bibliography

Asch, Ronald G. and Birke, Adolf M. (eds.), *Princes, patronage and the nobility: The court at the beginning of the modern age, c. 1450–1650*, Oxford: Oxford University Press, 1991.

Bayhaqī, Ẓāhir al-Dīn, *Tārīkh ḥukamā' al-Islām* (ed. Mamdūh Ḥasan Muḥammad), al-Qāhira: Maktabat al-thaqāfa al-dīniyya, 1996/1417.

Berggren, J. Lennart, *Episodes in the mathematics of medieval Islam*, New York: Springer, 1986.

Berggren, J. Lennart, 'Patronage', unpublished.

Berkey, Jonathan, *The transmission of knowledge in Medieval Cairo: A social history of Islamic education*, Princeton: Princeton University Press, 1992.

Bernards, Monique and Nawas, John (eds.), *Patronate and patronage in early and classical Islam*, Leiden: Brill Academic Publishers, 2005.

Blochet, E., *Catalogue de la collection de Manuscrits Orientaux de M. Ch. Schefer*, Paris: Ernest Leroux, 1900.

Brentjes, Sonja, 'Shams al-Din al-Sakhawi on *muwaqqits*, *mu'adhdhins*, and the teachers of various astronomical disciplines in Mamluk cities in the fifteenth century', in Emilia Calvo, Mercè Comes, Roser Puig, and Mònica Rius (eds.) *A sgared legacy Islamic science East and West, homage to professor J. M. Millàs Vallicrosa*, Barcelona: Publicacions i Edicions, 2008, pp. 129–150.

Brentjes, Sonja, 'Translation as an enduring and widespread cultural practice,' in Sonja Brentjes, Peter Barker, and Rana Brentjes (eds.) *Routledge Handbook on the Sciences in Islamicate Societies. Practices from the 2nd/8th to the 13th/19th Centuries,* London and New York: Routledge, 2023, pp. 15–24.

Burnett, Charles, 'The certitude of astrology: The scientific methodology of Al-Qabī ṣī and Abū Maʿshar', *Early Science and Medicine* 7/3 (2002), *Special Issue: Certainty, Doubt, Error: Aspects of the Practice of pre- and Early Modern Science In Honour of David A. King* (eds. Sonja Brentjes, Benno van Dalen, and François Charette), 198–213.

Charette, François, 'The locales of Islamic astronomical instrumentation', *History of Science*, 44 (2006), 123–138.

Charette, François, *Mathematical instrumentation in fourteenth-century Egypt and Syria: The illustrated treatise of Najm al-Dīn al-Miṣrī*, Leiden: Brill, 2003.

Dallal, Ahmad S., *An Islamic response to Ptolemaic astronomy*, Kitāb Taʿdīl Hay'at al-Aflāk *of Ṣadr al-Shari'a*, Leiden: Brill, 1995.

Danneskiold-Samsøe, J. F. C., *Muses and patrons: Cultures of natural philosophy in seventeenth- century Scandinavia*, Lunds: Lunds Universitet, 2004.

al-Dhahabī, Shams al-Dīn, *Al-'Ibar fī khabar man ghabar* (ed. Ṣalāḥ al-Dīn al-Munajjid and Fu'ad Sayyid), al-Juz' al-rābiʿ, Kuwayt: Da'irat al-Matbuʿat wa-al-Nashr, 1963 [748/1347].

Eddé, Anne-Marie, *La principauté ayyoubide d'Alep* (579–658/1260), Stuttgart: Franz Steiner Verlag, 1999.

Elger, Ralph, 'Review of Bernards and Nawas 2005', *Sehepunkte*, 7 (2007), www.sehepunkte.de/2007/07/12053.html.

Gutas, Dimitri, *Greek thought, Arabic culture: The Graeco-Arabic translation movement in Baghdad and early ʿAbbasid society (2nd–4th/8th–10th centuries)*, Routledge, 1998.

Halm, Heinz, *The Fatimids and their traditions of learning*, London and New York: Tauris in association with The Institute of Ismaili Studies, 1997.

Hogendijk, Jan P., 'How trisections of the angle were transmitted from Greek to Islamic geometry', *Historia Mathematica*, 8 (1981), 417–438.

Ibn Abī Uṣaybīʿa, *ʿUyūn al-anbā' fī ṭabaqāt al-aṭibbā'* (ed. Nizār Riḍā), Bayrūt: Dār Maktabat al-Ḥayat, 1965.

Ibn al-Fuwaṭī (642–723 AH), *Talkhīṣ majma' al-ādāb fī mu'jam al-alqāb* (ed. Muṣṭafā Jawād), Damascus: al-Maṭbaʿa al-Hāshimiyya, 1962–1965.

Ibn al-ʿIbrī: al-Malaṭī al-maʿrūf bi-Ibn al-ʿIbrī, Ghrīghūrīyūs, *Tārīkh Mukhtaṣar al-Duwal*, Bayrūt: Matbaʿa al-Kāthūlīkiyya, 1958.

Ibn al-Nadīm, *Kitāb al-Fihrist li-l-Nadīm* (ed. Reza Tajaddod), 2nd ed., Tehran: Marvi Offset Printing, 1971.

Ibn Wāṣil, Jamāl al-Dīn Muḥammad b. Sālim (d 697 AH/1298 ad), *Mufarrij al-kurūb fī akhbār Banī Ayyūb* (eds. Ḥasnīn Muḥammad Rabī and Saʿīd ʿAbd al-Fatāḥ ʿĀshūr), al-Qāhira: Maktabat Dār al-Kutub, 1977.

Kennedy, Edward S., 'Late medieval planetary theory', *Isis*, 57 (1957), 365–378.

Kennedy, Edward S. and Ghanem, Imad, *The life and work of Ibn al-Shatir, an Arab astronomer of the fourteenth century*, History of Arabic Science Institute, University of Aleppo, Aleppo: 1976.

Kettering, Sharon, *Patronage in sixteenth and seventeenth-century France*, Cambridge: Cambridge University Press, 2002.

Kettering, Sharon, *Patrons, brokers and clients in seventeenth-century France*, Oxford: Oxford University Press, 1986.

King, David A., 'al-Khalīlī's *qibla* table', *Journal of Near Eastern Studies*, 34 (1975), 81–122.

King, David A., 'On the role of the muezzin and *muwaqqit* in medieval Islamic societies', in F. Jamil Ragep, Sally P. Ragep, and Steven J. Livesey (eds.), *Tradition, transmission, transformation: Proceedings of two conferences on premodern science held at the University of Oklahoma*. Leiden: Brill, 1996, pp. 285–345.

King, David A., *Astronomy in the service of Islam*, Ashgate: Variorum, 1993.

King, David A., *In synchrony with the heavens: Studies in astronomical timekeeping and instrumentation in medieval Islamic civilization*, 2 vols, Leiden: Brill, 2004.

King, David A., 'Too many cooks . . . A new account of the earliest Muslim geodetic measurements', *Suhayl*, 1 (2000), 207–241.

Kraemer, Joel, *Humanism in the renaissance of Islam: The cultural revival during the Buyid age*, 2nd ed., Leiden: Brill, 1993.

Langermann, Y. Tzvi (ed. and trans.), *Ibn al-Haytham's On the Configuration of the World*, New York: Garland, 1990.

Loth, Otto, *A catalogue of the Arabic manuscripts in the library of the India Office*, London: Austin, 1877.

Matvievskaya, G. P. and Rozenfel'd, B. A., *Matematiki i astronomy musul'manskogo srednevekov'ya i ich trudy (VIII-XVII vv)*, 3 vols, Moscow: Nauka, 1985.

Mottahedeh, Roy, *Loyalty and leadership in an early Islamic society*, Princeton: Princeton University Press, 1980.

Petry, Carl, *The civilian elite of Cairo in the later Middle Ages*, Princeton: Princeton University Press, 1981.

Ragep, F. Jamil, ''Alī Qushjī and Regiomontanus: Eccentric transformations and Copernican revolutions', *Journal for the History of Astronomy*, 36 (2005), 359–371.

Ragep, F. Jamil, *Nasir al-Din al-Tusi's memoir on astronomy (al-Tadhkira fī 'ilm al-hay'a)*, 2 vols, New York: Springer, 1993.

Rashed, Roshdi, *Entre arithmétique et algèbre: Recherches sur l'histoire des mathématiques arabes*, Belles Lettres, 1984. English translation: *The development of Arabic mathematics: Between arithmetic and algebra*, Dordrecht: Kluwer Academic Publishers Group, 1994.

Rebstock, Ulrich, *Rechnen im islamischen Orient. Die literarischen Spuren der praktischen Rechenkunst*, Darmstadt: Wissenschaftliche Buchgesellschaft, 1992.

Roberts, Victor and Kennedy, Edward S., 'The planetary theory of Ibn al-Shatir', *Isis*, 50 (1959), 227–235.

al-Sakhāwī, Shams al-Dīn, *al-Ḍaw' al-lāmi' li-ahl al-qarn al-tāsi'*, Bayrūt, no date.

Saliba, George, 'Al-Qushjī's reform of the Ptolemaic model for Mercury', *Arabic Science and Philosophy*, 3 (1993), 161–203.

Saliba, George, *The astronomical work of Mu'ayyad al-Dīn al-'Urḍī (d 1266): A thirteenth century reform of Ptolemaic astronomy*, Beirut: Markaz Dirāsat al-Waḥda al-'Arabiyya, 1990.

Saliba, George, 'The original source of Quṭb al-Dīn al-Shīrāzī's planetary model', *Journal for the History of Arabic Science*, 3 (1979), 3–18.

Saliba, George, 'A sixteeenth-century Arabic critique of Ptolemaic astronomy: The work of Shams al-Dīn al-Khafrī', *Journal for the History of Astronomy*, 25 (1994), 15–38.

Sayılı, Aydın, *The observatory in Islam and its place in the general history of the observatory*, Ankara: Türk Tarih Kurumu, 1960.

Schöler, Georg, unter Mitarbeit von H.-C. Graf von Bothmer, T. Duncker Gökçen, H. Jenni, *Arabische Handschriften*, Teil II, Stuttgart: Franz Steiner Verlag, 1990.

al-Shahrazūrī, Shamsuddin Muḥammad b. Maḥmood [d. after 687 AH/1288 ad], *Nuzhat al-arwāḥ wa-rawḍat al-afrāḥ fī ta'rīkh al-ḥukamā' wa-al-falāsifah*, 2 vols (ed. Syed Khurshīd Aḥmed), Ḥaydarābād: The Dā'iratu l-Ma'ārifu l-Osmānia, Osmania University, 1976/1396.

al-Sijistānī, Abū Sulaymān al-Manṭiqī, *Ṣiwān al-ḥikma wa-thalāth rasā'il* (ed. 'Abd al-Raḥmān Badawī), Tehrān: Bunyād-i farhang-i Īrān, 1974.

al-Ṭabarī, *The history of al-Ṭabarī, volume XXX: The 'Abbāsid caliphate in equilibrium: The caliphates of Mūsā al-Hādī and Hārūn al-Rashīd* (transl. and annot. C. E. Bosworth), Albany: State University of New York Press, 1989.

Tasköprüzade, *Es-Saqa'iq en-No'manijje* (transl. Oskar Rescher) *Gesammelte Werke*, Abteilung IV, Band 1, Osnabrück: Biblio Verlag, 1978.

Witkam, J. J., *De egyptische arts Ibn al-Akfani (gest 749/1348) en zijn indeling van den wetenschappen*, Leiden: Ter Lugt Press, 1989.

5

COURTLY PATRONAGE OF THE ANCIENT SCIENCES IN POST-CLASSICAL ISLAMIC SOCIETIES

Historians of science, medicine and philosophy in Islamic societies will agree without hesitation that courtly patronage was of extraordinary importance for the introduction, spread and maintenance of the ancient sciences, as well as for the many new results that scholars achieved in these fields in different Islamic societies. Despite this generally held conviction, there are no studies of the phenomenon and its various forms in specific Islamic societies. A second conviction, albeit less firmly held, is the belief that one of the major factors that led to what is usually called the decline of the ancient sciences was the disappearance of courtly patronage at some unclear point in time.[1] This vagueness results from disagreement about when the decline commenced, and from a lack of clear statements about when courtly patronage ended. Opinions on the matter vary greatly, some seeing the eleventh century as the starting point, others the fifteenth or sixteenth century.[2] Given the importance of these historiographical problems it is surprising that very little research has been done assessing the evidence for the

1 See for instance Makdisi, G., *The Rise of Colleges: Institutions of Learning in Islam and the West*, Edinburgh, 1981, 77–78.

2 For the various positions on the issue of decline, its date, its conceptualization, and its scope see for instance Sayili, A., *The Observatory in Islam and its Place in the General History of the Observatory*, Ankara, 1960, 402–429; al-Hassan, A.Y., *Factors Behind the Decline of Islamic Science after the Sixteenth Century*, accessible online in the personal website of the author: www.history-science-technology.com/articles/articles%208.html (15–08–08); Saliba, G., *Islamic Science and the Making of the European Renaissance*, Cambridge, 2007; Sabra, A.I., "The Appropriation and Subsequent Naturalization of Greek Science in Medieval Islam," *History of Science*, 25 (1987), 223–243; Abdalla, M., *The Fate of Islamic Science between the Eleventh and Sixteenth Centuries: A Critical Study of Scholarship from Ibn Khaldun to the Present*, accessible online in the "Australian Digital Theses Program": http://www4.gu.edu.au:8080/adt-root/public/adt-QGU20040618.091027/index.html (15–08–08), 95–141.

DOI: 10.4324/9781003372455-6

disappearance of courtly support for all or some of the ancient sciences, and the link between this phenomenon (if it indeed can be shown to have happened) and changes in the content and innovative power of research that occurred in later Islamic societies.

In this paper I will focus on courtly patronage of the ancient sciences after 1200 in the territories between Egypt and India. I will show that courtly patronage of the ancient sciences did not disappear in this post-classical period. Several dynasties extended their support to scholars interested in a variety of ancient sciences. Such a claim finds clear substantiation in dedications and ownership marks attested in manuscripts; notes appearing in biographical dictionaries and historical chronicles about scholars at courts and their contacts with rulers, princes, emirs, viziers, other court officials and powerful women at courts; and courtly protocol and official honorific titles specified in administrative sources. The more challenging problems arise from the limitations of these materials and the need for non-trivial interpretations of the information they offer. Since courtly patronage continued after 1200 under several major and minor dynasties, the changes in scientific activities and the decrease in new results cannot be ascribed to lacking "state" support as such. The changes themselves, their character, scope, disciplinary, spatial and temporal occurrences as well as the modes in which they appear and the values they reflect, will not be discussed here. Neither will I offer suggestions about the factors contributing to such changes. All this goes far beyond the purpose and possibilities of this paper. But even more important, these issues are deeply steeped in prejudices and assumptions characteristic of scientific activities in our own days. There are no in-depth studies of specific cases that contextualize such changes in the value system of their times and places. In addition, the rigorous questioning of the assumptions underlying our judgments and of the suitability of our methods has only begun recently.

I use the term "ancient sciences" in this paper as a shorthand for the mathematical sciences (geometry, number theory, arithmetic, algebra, magic squares, mathematical cosmography and geography, astronomical handbooks, ephemerides, time-keeping and related issues), medicine, philosophy (metaphysics, logic, natural philosophy, economics, ethics, politics) and the occult sciences (astrology, alchemy, geomancy, numerology and related fields). The reason for doing this is that the use of the term "science" suggests a modern understanding and excludes medicine, philosophy and the occult sciences, while the use of the term "rational sciences" would exclude the mathematical sciences and include important religious disciplines such as *uṣūl al-dīn*, which I will not explore here. I will focus primarily, however, on the mathematical sciences and medicine, touching upon philosophy and the occult sciences, excluding astrology, only in a cursory manner. The reason for this limitation is found on the one hand in the lack of systematic surveys on the history of philosophy under most of the dynasties I will discuss in this paper (exceptions are the Safavids and the Mughals), and on the other in the fact that my research interests focus primarily on the mathematical sciences with some additional interest in a social history of medicine.

1 Courtly patronage of the ancient sciences in the classical period[3]

Courtly patronage of the ancient sciences in the classical period took place both in the form of personal relationships between a patron and a client and as an outgrowth of relations of loyalty between families. According to Mottahedeh, loyalties were the result of three cultural activities: providing benefits, concluding various types of formal commitment (oaths, vows, guarantees of safe conduct) and showing gratitude. Benefits (*niʿma*) given by a ruler to his subjects served to create stable, reciprocal ties. Generosity (*samāḥa*) was expected from the ruler when handing out benefits, but when combined with indulgence (*musāmaḥa*) could diminish the ruler's reputation.[4] Formal commitments regulated "duties and obligations that could be enforced without coercion."[5] Practitioners of the ancient sciences benefitted from the second type of formal commitment, the vow. The vow was made to God, and contained a promise to do some particular good work on the basis of a good intention. Repairs of irrigation channels and bridges, or the building of mosques and, later, *madrasas*, could be announced publicly in the form of such a vow. In combination with the actual work of repair or construction, the vow served to elevate the symbolic and real power of the ruler who made it. Practitioners of the mathematical sciences, in particular surveyors and *muhandisūn* (geometers, engineers, architects), were often involved in the repair and replacement work. The recipient of benefits was obliged to show gratitude (*shukr al-niʿma*), since masculine honor was based on acknowledging received benefits, reinforced by the Qurʾān's repeated reminder that the (true) believer is grateful for God's continuous benefits.[6]

Mottahedeh argued that patronage during Buyid rule resulted from this triangular net of benefits, formal commitments and gratitude.[7] In exchange for promotion, a patron expected his protégé to serve him in various ways as long as either of the two lived.[8] Many of the scholars who worked as astrologers or physicians at Buyid and other courts of the period stayed until they died in courtly service, often through the reign of more than one ruler. It is no exaggeration to claim that many of the most productive and innovative astronomers/astrologers, physicians and philosophers spent long years in courtly service. This means that the impressive results of nearly five centuries of mathematical, medical and philosophical research and teaching were produced through the institution of patronage at its major centers, namely, central and local courts.

3 This part is an abbreviated extract from a larger paper on patronage of the mathematical sciences in Islamic societies, in Robson, E. and Stedall, J. (eds.), *The Oxford Handbook of the History of Mathematics*, Oxford, 2008, 301–327.

4 Mottahedeh, R., *Loyalty and Leadership in an Early Islamic Society*, Princeton, 1980, 90–91.

5 *Ibidem*, 43.

6 *Ibid.*, 72–76.

7 *Ibid.*, 42.

8 *Ibid.*, 82–84.

The patronage relationships that rulers, viziers and other courtiers maintained with physicians, astronomers/astrologers and philosophers were discussed in a variety of terms. A patron extended honor (*ikrām*) and benefits, often in a generous manner, to the men who healed him, cast his horoscope or observed the stars. The physicians and occasionally also the astronomers/astrologers are said to have served (*khadama*) their patrons, often until their patrons' death. Honor paid to a client could take many forms, including precious robes, monetary gifts, privileged ranking in seating at court during official occasions, and inclusion in caliphs', sultans' or viziers' groups of boon companions (*nadīm*). Patrons also shared other more personal favors with their clients.

The scholars engaged in patronage relationships were expected to offer expertise in areas such as healing, observing the planets and stars, casting horoscopes, constructing instruments, writing books, making automata or repairing clocks, water wheels, channels, and other infrastructural components. An important professional service of physicians and astrologers was to accompany their patrons on military campaigns, pilgrimages, and other travels. Teaching sons of rulers, their relatives, and other courtiers was another service scholars of the mathematical sciences provided for their patrons. Scholars of all disciplines dedicated books and instruments to patrons from the early Abbasid period onwards. It is, of course, not always clear whether the dedication was an expression of gratitude for benefits received or a proposal to enter into a patronage relationship. The vocabulary used in histories of scholars and bibliographies includes translating, writing, composing or doing a work for a ruler, a vizier or a scholar. Instruments as well as texts were named after patrons and designed specifically for them.[9]

Other ways of talking about patronage relationships in the ancient sciences included terms like *ittaṣala bi* (to be bound to, to be connected with), *ghulām* (boy, youth, military slave, servant), *ḥamala* (to carry), *aʿẓama* (to invite), *ʿazzaza* (to elevate, to exalt, to make dear), *irtabaṭa* (to bind, to commit, to engage), *istakhdama* (to hire, to engage the service), *istaḥḍara* (to summon, to fetch), *ishtāqa ilā* (to covet, to desire ardently), *ḥaẓiya ʿinda* (to enjoy the favors, to be in the good graces) and *daʿā* (to summon, to call, to urge, to invite).[10] Some of these expressions, such as *irtabaṭa*, *istakhdama* and *daʿā*, were applied to both rulers and viziers, while others such as *ḥamala* were used only of rulers. *Irtabaṭa*, *istakhdama* and *daʿā* describe relationships in which power, although asymmetric, is shared because not only did scholars look for patrons, but rulers and viziers likewise tried to attract clients to their courts, occasionally going to great lengths to do so. But scholars did not always accept the prospective patron's offers, invitations or gifts. The Saljuq sultan Sanjar (r. 511–582/1117–1157) sent the enormous sum of 1,000 dinars to ʿAbd al-Raḥmān Khāzinī (fifth/twelfth century), a freed slave

9 Charette, F., "The Locales of Islamic Astronomical Instrumentation," *History of Science*, 44 (2006), 123–138, 133, tables 2a, 2b.

10 This list is by no means exhaustive.

of Byzantine origin. Khāzinī refused to take the money, saying that he possessed ten dinars, three of which sufficed to finance him for a year, plus a cat. He also rejected the same sizable monetary gift when it was offered by the wife of one of the Saljuq emirs.[11]

The term *ḥamala* reflects the often violent nature of medieval society and the role of scholars as pawns in conflicts between rulers, invaders, rebels, and other claimants to power. When Sebügtegin (d. 387/997) and his son Maḥmūd (r. 388–421/998–1030) conquered parts of Central Asia and Iran, they forced numerous scholars to flee westwards into the protection of the Buyid dynasty, or else carried them away to Ghazna. Among those who had to relocate after the final campaign in 408/1017 were the scientist Abū Rayḥān Bīrūnī (362–d. after 444/973–d. after 1053) and the physician Abū l-Khayr al-Ḥasan.[12] The Ghaznavids were not the only dynasty that forced scholars into their patronage. Various Khwarazmshahs brought scholars to their court in this manner, as did some dynasties in the post-classical period such as the Mongols in the thirteenth century and Timur (r. 771–807/1369–1404) and his family in the late fourteenth to fifteenth centuries.[13]

The historical sources paint a complex and rich picture of different forms and styles of patronage. Successful relationships that lasted a lifetime are found, as are those of a more fluid nature. Cases of stability over two or even three successive rulers occurred at several courts during this period. However, cases of conflict and failed continuation of patronage were far from exceptional. Important factors in the fluidity of courtly patronage included enmity among members of the patron's family and towards his clients, as well as military, economic, and political instability. Competition among different factions within a ruling family and among families of administrators, and the growing influence of the Turkish military slaves upon the choice of the next caliph made patronage an unreliable field of social relations. As a result, the figure of the itinerant, and often impoverished, scholar emerged. The asymmetry of power and the inherent instability of this imbalance are reflected in the various forms of punishment that a patron could heap on a client, either on his own behalf or on the behalf of another client. Extortion, loss of office and property, exile, imprisonment, and execution were widespread means of regulating access to power and funds among the civil elites, in particular on the level of viziers and their immediate entourage. Scholars of the mathematical sciences, even those who were powerful patrons in their own rights, also suffered under the arbitrariness of the patronage system. Several were exiled, incarcerated or killed.

Boundaries that separated patronage from other kinds of relationships, for instance those that existed in the realm of craftsmanship, are often blurred due to a lack of precise information in the sources. Ibn al-Nadīm (d. 380/990) used

11 Al-Bayhaqī, Ẓahir al-Dīn, *Ta'rīkh ḥukamā' al-Islam*, M.Ḥ. Muḥammad (ed.), Cairo, 1996/1417, 181.

12 *Ibidem*, 36; Matvievskaya, G.P. and Rozenfel'd, B.A., *Matematiki i astronomy musul'manskogo srednevekov'ya i ich trudy (VIII-XVII vv.)*, Moscow, 1985, 2, 264.

13 Al-Bayhaqī, *Ta'rīkh*, 36, 173.

the term *ghulām*, which in the tenth century primarily meant a military slave, to describe relationships involving scholars of the mathematical sciences and instrument makers. In most cases, the owner or master of a *ghulām* was either an astrologer or an instrument maker. Instrument makers like Aḥmad and Muḥammad, the sons of Khalaf (third/ninth century), often started their career as *ghilmān* (plural of *ghulām*) and later had *ghilmān* themselves. Both brothers had been *ghilmān* of the astrolabe maker ʿAlī b. ʿĪsā (third/ninth century) and then trained six other men as their *ghilmān*, all of whom became known as instrument makers.[14] Hence it is perhaps reasonable to interpret this relationship as one between a master and an apprentice rather than one of patronage.

2 Courtly patronage of ancient sciences in the post-classical period

Courtly patronage of ancient sciences in the post-classical period was characterized by two kinds of relationships to the classical period. On the one hand, it was institutionally similar to that of the classical period; on the other, it differed in two respects profoundly from the previous period. It was institutionally similar insofar as it took place first and foremost as a relationship between individuals. Rulers supported scholars, not disciplines, even if certain forms used for installing and legalizing patronage such as *waqf* (religious donation) apply an abstract, disciplinary rhetoric either with specifications for individuals or without. Rulers who decided to donate professorships for certain ancient sciences, in particular medicine, often did so because of a personal experience of need. In some cases, a vow made at an earlier moment of need preceded such a donation. Scholars praised rulers, not dynasties, in dedications and inscriptions. Historians, if they talked at all about activities relating to the ancient sciences, also mentioned them only in relationship to individual rulers, members of ruling families or courtiers. Hence when I speak in this paper of courtly or princely patronage, it is important to recall this individual character of patronage relationships. It is very rare that a dynasty as a whole or even only its heads followed one single, coherent approach towards the ancient sciences and their practitioners. The most that can be said in such rare cases is that several members of a dynasty supported a similar set of authors and texts and used them for similar purposes. The fact that the authors or texts that were patronized by various dynasties belonged to a limited number of disciplines, in particular medicine, astrology and to some extent magic, tends often to obscure the differences between individual rulers and other courtly patrons.

In the post-classical period, the rhetoric used for negotiating and representing patronage relationships lost much of its earlier richness. It shrank mainly to the three concepts of honor or grace, benefit and service. Most of the patronage relationships between an Ayyubid ruler or prince and a physician, for instance, were

14 Tajaddod, R. (ed.), *Kitāb al-fihrist li-l-Nadīm*, Tehran, 2nd ed., s.d., 343.

described using this terminology.[15] The rhetoric of patronage under the Mamluks continued to revolve around the same terms as in the Ayyubid period. Clients linked themselves with and served specific patrons, and these patrons honored or benefited them.[16] New terms such as *qarrara hu* or *fī* (to appoint s.o. to, place s.o. in), *istaqarra bi* (to settle down, to take a permanent place of living, i.e., a permanent position) and *ikhtaṣṣa bi* (to devote or dedicate an honor or service, to privilege s.o., to make s.o. special) reflect the demands and opportunities created by the system of endowed teaching institutes. Outside the Ayyubid-Mamluk domains, the impact of the endowed teaching institutes took longer to permeate the language of patronage. The continued importance of the courtly sphere for patronage of the ancient sciences in these other regions is reflected in the vocabulary used to describe relationships between patron and client. In addition to the prevalent rhetoric of service, honor and benefit, other terms like *ittaṣala bi* (to be bound to, to be connected with), *nasaba* (to link, to bring into relation) or *taʿalaqqa* (to be attached, to be dependent, to be devoted, to be connected, to belong) were used. They expressed the fact that a patron and a client were connected or bound to each other.[17]

These shifts in rhetoric point to the institutional differences that separate patronage activities during the post-classical period from those of the classical period. The reason for these differences can be found primarily in two changes that occurred in regard to the social loci available for the practice of patronage and the scientific disciplines. After Niẓām al-Mulk (1018–1092) made the creation of endowed teaching institutes a component of Saljuq cultural politics in the eleventh century, rulers, princes and powerful women of the Zangids, Ayyubids, Artuqids and other Sunni dynasties followed the Saljuq example in the twelfth century in northern Iraq, Anatolia, Syria and Egypt. This opened new domains for patronage outside the courtly sphere. However, in Saljuq times the courts remained major, if not the main social spaces for patronage of the ancient sciences, as suggested by the case of ʿAbd al-Razzāq al-Jazarī's (d. after 603/1206) work at the Artuqid court in Diyar Bakr or the presence of physicians in the entourage of the Atabeg Nūr al-Dīn Maḥmūd b. Zangī (r. 541–569/1146–1173). From the last decades of the thirteenth century onwards, princely patronage for *madrasas*, mosques and Sufi *khānqāh*s with their prescriptions for teaching posts, stipends and other positions, among them the post of a *muwaqqit*, were features that repeatedly occurred in the cultural politics of Mamluk sultans in Egypt and Syria. As is well known, Sultan al-Manṣūr Qalāwūn (r. 678–689/1279–1290), Sultan Lājīn (r. 696–98/1296–99) and Sultan al-Muʾayyad Shaykh (815–824/1412–1421) donated hospitals and

15 Ibn Abī Uṣaybīʿa, *ʿUyūn al-anbāʾ fī ṭabaqāt al-aṭibbāʾ*, N. Rīḍā (ed.), Beirut, 1965, 584–586, 589–591, 598–601, 635–637 *et passim*.

16 Al-Sakhāwī, Shams al-Dīn, *al-Ḍawʾ al-lāmiʿ li-ahl al-qarn al-tāsiʿ*, Beirut, s.d., 3, 296; 6, 112, 235; 8, 191; 9, 15; 10, 48.

17 Ibn al-Fuwaṭī, *Talkhīṣ majmaʿ al-ādāb fī muʿjam al-alqāb*, M. Jawad (ed.), Damascus, 1962–1965, 4, pt.1, 40, 132, 149; pt. 2, 513, 754; pt. 3, 513.

chairs for teaching medicine at mosques and *madrasa*s in Cairo and Damascus.[18] Sultan Ḥasan (r. 748–752 and 755–762/1347–1351 and 1354–1361) provided for a professor and six stipends in *ʿilm al-mīqāt* (the astronomical discipline for determining prayer times, the directions of prayer towards the Kaʿba in Mecca and the rise of the new moon as well as teaching the relevant geometrical, arithmetical, astronomical and instrumental knowledge) at his newly erected *madrasa* in Cairo. Other Mamluk sultans granted scholars of the ancient sciences professorial posts and posts as *muwaqqit*s or physicians at mosques, *madrasa*s and *khānqāh*s. Under the Mamluks, the social space of princely patronage shifted significantly. It moved from the courtly sphere to the sphere of the endowed teaching institutes and the hospitals. Only medicine with its offices of *raʾīs al-aṭibbāʾ* (head of the physicians) and the courtly *dīwān* remained embedded directly in the courtly sphere. The extent to which the Mamluks shifted their patronage from the court to the *madrasa* and hospital seems to be unique. No other dynasty in the Arab world, Anatolia, Iran, Central Asia or India established its patronage of the ancient sciences primarily outside the courtly sphere.

The Ottomans followed a mixed heritage by taking up at first examples set by the Saljuqs, the Rum-Saljuqs and other Turkic rulers in Anatolia. After the fifteenth century they also added components of Mamluk policy. As a result, they sponsored the ancient sciences within the courtly sphere, in particular medicine and astrology. Court physicians and court astrologers lived in the palace and were presided over by a head physician or head astrologer, two official courtly positions. The Ottomans also exercised a substantial amount of patronage outside the court within the realm of the endowed teaching institutes and hospitals. They sponsored hospitals, chairs for medicine at *madrasa*s, the teaching of several mathematical disciplines and the post of the *muwaqqit*. Most of the major mosques in Istanbul, Edirne and Bursa provided for such a post, and the large *madrasa* complexes often included a hospital and a position for teaching medicine.

The proliferation of endowed teaching institutes altered the landscape of princely, but also urban patronage in several ways. Teaching became a profession which guaranteed a stable income from the properties attached to the teaching institute. Once a client had been appointed to a position at a certain institute, regular visits with the patron were not necessary to assure funding. In this sense, the relationship became more distant and formal. However, the substantial regular income and the social prestige linked to major endowed institutes, in particular those endowed by rulers, their relatives, high ranking military leaders and major administrative officials, induced many professors to try to grab as many chairs as possible and to determine their own successors. This desire for chairs, money and influence resulted in all sorts of fights and trickery as these clients sought to defeat competitors and push colleagues from their posts. Patrons were now especially valuable to them for their capacity to offer protection against rivals from their own

18 Behrens-Abouseif, D., *Islamic Architecture in Cairo*, Leiden, 1989, 138–140.

ranks and to support claims in case an official court complaint became necessary. In this sense, the relationship between patron and client became more intense. It could assume criminal aspects, a feature that occurred in the classical period primarily among the major secretarial families. Such criminal aspects consisted mainly in activities undertaken to force an office holder out of his office, mostly a professorship at a *madrasa* and cognate teaching institute. They comprised spreading rumors, raising false accusations, ambushing, enforcing legal persecutions or cancelling an appointment by a governor or another official who formally was not responsible for the particular *waqf* (donation) and its stipulations.

The bonds created between patron and client depended on opportunities and circumstances, and thus varied not only among the dynasties, but among individual members of a given dynasty and the cities where they ruled. A rare case of an astrologer affiliated in customary Mamluk style to Sultan al-Manṣūr Qalāwūn was Ibrāhīm al-Ḥāsib al-Malikī al-Manṣūrī in the thirteenth century. His name indicates that he was a member of the sultan's household, perhaps even a man of the sultan's personal troops. He wrote one of the very rare astrological treatises composed in Mamluk times and territories. The sources Ibrāhīm claims to have used show his familiarity with the leading ninth- and tenth-century writers on the subject, among them Abū Maʿshar (d. 273/886), Yaʿqūb b. Isḥāq al-Kindī (d. ca. 260/873) and Abū Saʿīd Sijzī (ca. 334–ca. 411/ca. 945–ca. 1020).[19]

Timurids exercised a variety of patronage forms. For instance, in some cases they forced their patronage on unwilling clients who had been bound to weaker patrons through foster parent relationships characteristic of the slave armies of other dynasties, and exercised their princely prerogatives in regard to scholarly matters. Enforced patronage occurred mainly in the context of military conquest. Princely interference into scholarly matters took place via dismissals of office holders and occasionally through visits of classes. In other situations, Timurid princes took a less aggressive approach, stimulating competition between clients and raising intellectual challenges for them.

In 789/1387, when Timur stormed the Muzaffarid stronghold in Shiraz, he forced Sayyid Sharīf Jurjānī (741–816/1340–1413) to leave the city for Samarkand and to accept him as his new patron. Jurjānī apparently did not like this new arrangement, although he could not avoid it. But once Timur died, Jurjānī left Samarkand and returned to Shiraz. Ulugh Beg (r. 850–853/1447–1449) established different relationships with the scholars with whom he collaborated. He interfered in the scholarly prerogatives of Qāżīzāde Rūmī (d. 815/1412) as head of the *madrasa* Ulugh Beg had opened. But when Qāżīzāde Rūmī began a strike of disobedience in which he remained at home and refused to continue classes at school, Ulugh Beg inquired about what had elicited his strong response. The scholar told him that he would not work at a *madrasa* where someone other than

19 *Fihrist-i kitābhā-yi khaṭṭī-yi Kitābkhāna-yi Millī-yi Malik dābiste be-Āstān-i Quds Rīżāvī*, vol. 1, *Kitābhā-yi ʿarabī wa-turkī*, I. Afshar and M. Taqī Dānish Pazhūh (eds.), Tehran, 1352h., 1, 644.

the professors themselves decided what would be taught. According to Tashköprüzade (d. 969/1560), who transmitted this story, Ulugh Beg apologized and reinstated a teacher whom he had dismissed for his lack of mathematical knowledge.[20] While the Timurid prince clearly appreciated Qāżīzāde Rūmī as a scholar and a man, he disliked Ghiyāth al-Dīn Kāshī's (d. 833/1429) manners which apparently failed to comply with the rules of courtly etiquette and courtesy. He teased other scholars by sending them pages with mathematical problems which he asked them to solve.[21] ʿAlī Qushjī (d. 879/1474), whom he took under his wing as a young boy and provided with excellent education at his *madrasa*, was treated like a foster son, according to reports by later historians.[22] The academic knowledge that ʿAlī Qushjī acquired in Samarkand and the expertise he obtained by running the *madrasa* and the observation program after Kāshī's death endowed him with a reputation and status that made him an honored guest in the eyes of the Aq Koyunlu ruler Ūzūn Ḥasan (r. 871–883/1466–1478) and the Ottoman sultan Mehmet Fatih (r. 855–886/1451–1481), who invited him to courtly sessions of scholarly debate and conviviality.[23]

3 Differences and similarities between princely patronage patterns in the post-classical period

The possibility that anyone with sufficient funds could donate them for a pious purpose and open a teaching institute diminished the central role of princely patronage and created a broader spectrum of patronage opportunities outside the courts. The attitudes of rulers and their family members, as well as other members of the courtly sphere, towards continued support of scholars beyond endowing teaching institutes varied across different regions and periods. The personal entourages of many Ayyubid princes included scholars, among them physicians, geomancers, engineers and astrologers, with whom the princes spent time in meetings of debate and conviviality. Mamluk sultans and heads of Mamluk households invited scholars of the religious disciplines to sessions for reading the Qur'ān or *ḥadīth* collections. Scholars of the mathematical and occult sciences were called when a Mamluk needed explanations of natural events, counsel for future activities, particularly battles, or help with prayer times and the direction towards Mecca. Physicians and geomancers were called in cases of disease. But the entourages of high ranking Mamluks only rarely included scholars of the ancient sciences serving for a lifetime as had been common under the Ayyubids. Mamluk patronage for the ancient sciences, as noted, focused first and foremost on the endowed

20 Tasköprüzade, A.E., *Es-Saqa'iq en-No'manijje*, O. Rescher (transl.), Osnabrück, 1978, 8.

21 *Rijāl-i Kitāb-i Ḥabīb al-Siyār az jumle-yi mughūl tā marg-i Shāh-i Ismāʿīl-i avval*, ʿA. Navāʾī (ed.), Tehran, 1324h., 111–112.

22 *Ibidem*, 122–123.

23 *Osmanlz Matematik Literatürü Tarihi (History of Mathematical Literature During the Ottoman Period)*, Istanbul, 2001, 1, 26–28.

teaching institutes. This focus on appointing preferred scholars to chairs at *madrasas*, mosques or *khānqāhs* included *ʿilm al-mīqāt*, medicine and occasionally the philosophical sciences.

Turkoman and Turco-Mongol tribal dynasties in Iran, Central Asia and India pursued patronage policies in regard to the ancient sciences that resembled those of the Ayyubids. Courtly support among the Injus, Aq Koyunlus and Timurids for the ancient sciences seems to have focused primarily on scholars attached to the court. Support for scholars pursuing the ancient sciences at *madrasas* is presently known only as the exception to this rule. The Mughals and Safavids also confined most of their patronage activity to their courts. But in contrast to the other three dynasties, they also supported scholars at *madrasas*, including those with an interest in the ancient sciences. Dynasties of other ethnic and social origins, such as the Muzaffarids in southern Iran, sponsored scholars within the sphere of the courts and with positions at endowed teaching institutes. Thus, although all of these Turkoman and Turco-Mongol dynasties adopted what is generally termed Persianate culture, this culture did not produce uniform patronage patterns. Hence other factors need to be taken into consideration, among them specific beliefs and policies intended to justify and ensure the dynasty's rule and power. For the time being, the data I could collect are insufficient to draw explanatory links between these patterns and other factors for the Injus and the Aq Koyunlus. For the Timurids, art historians have shown that Timur's politics of legitimizing himself as heir to Chingizid rule shaped most of his military and cultural decisions.[24] In my view, Timurid interest in astrology, astronomical handbooks and planetary theory are similarly situated within this framework of justification. Timurid rulers and princes showed a particular interest in astronomical and astrological texts written by scholars who had worked at Ilkhanid courts. Several Timurids collected such texts and copied them for their libraries. At least three Timurid princes, if not more, also studied these texts. In many respects, Mughal patronage patterns continued the policies of their relatives, but incorporated additional elements, in particular with regard to the mathematical sciences, under the direct influence of scholars from Shiraz.

Other differences in the patronage patterns between the mentioned dynasties concern the relationship between the ancient sciences and the arts. When produced for an Inju, Timurid, Mughal, Safavid or Ottoman ruler, prince, governor, eunuch or other member of the courtly sphere, mathematical, astronomical, astrological, medical and philosophical texts were often carefully executed and finely illustrated with *shamses*, *sarlawhs*, *ʿunwāns*, gilded frames, calligraphic writings and in some cases even illuminated with miniatures and other paintings.[25] Such

24 Lentz, T.W. and Lowry, G.D., *Timur and the Princely Vision: Persian Art and Culture in the Fifteenth Century*, Los Angeles, 1989; Golombek, L. and Subtelny, M. (eds.), *Timurid Art and Culture: Iran and Central Asia in the Fifteenth Century*, Leiden, 1992.

25 Art historians who have worked on manuscripts with scientific content have focused primarily on the artistic aspects of their illumination and illustration, but rarely discuss aspects of patronage for the sciences these texts teach. Those who tried to include some reflections on the relevance of the

luxury books became objects of collection and display, and found themselves in the libraries of various rulers, their officials, wives or other relatives. Preferred texts for illustration or illumination were ʿAbd al-Raḥmān Ṣūfī's (290–375/903–986) *Kitāb ṣuwar al-kawākib al-thābita* (from now on called *Star Catalogue*), Zakariyyāʾ b. Muḥammad Qazvīnī's (599–682/1203–1282) *ʿAjāʾib al-makhlūqāt* and Dioscurides' (first century) *Materia Medica*. Ṣūfī's *Star Catalogue* was repeatedly copied in the Safavid period in particular, but copies were also produced at Timurid and Mughal courts as well as at the courts of other Indian Islamic dynasties.[26] Qazvīnī's cosmography was copied by artists either in its Arabic versions or in its Persian or Ottoman Turkish translations at the courts of the Injus, Jalayirids, Aq Koyunlus, Timurids, in the environment of the Safavid, Mughal and Ottoman courts and under dynasties in the Deccan.[27] Such activities took place in specific locations, not always necessarily in the capital of a dynasty. The Jalayirids and Aq Koyunlus, for instance, asked for copies of Qazvīnī's work primarily in Baghdad. The Jalayirids also ordered a copy of the earlier work by Salmān Ṭūsī (sixth/twelfth century) with a similar title and similar content, although more literary than scientific, but written in Persian.[28] This suggests that centers of painting experience and the availability of qualified painters were decisive factors for the illustration of scientific manuscripts. The fact that Salmān Ṭūsī's work, Dioscurides' *Materia Medica* and other scientific texts were illustrated and illuminated in Safavid Shiraz in the sixteenth and seventeenth centuries reinforces this claim.[29] However, as more manuscripts come to light and are carefully studied, our knowledge of where, when and for whom scientific manuscripts were illustrated and illuminated will continue to expand.

---

scientific content have usually relied on secondary sources rather than on an analysis of the content and context of the texts themselves. Because of these differences in interest and focus I have abstained from citing the studies of Qazvīnī's, Ṣūfī's and other authors' works by contemporary art historians.

26 Examples are MSS Paris, BnF, Arabe 4670, Iran, eleventh/seventeenth century; Arabe 6528, Gujarat, 1053/1643–4; Supplément persan 1551, Iran, late tenth/sixteenth century; Welch, A., *Collection of Islamic Art*, Geneva, 1972, 2, 69–70, mid-eleventh/mid-seventeenth century.

27 Examples are MSS Cambridge, University Library, Nn. 3.74, 974/1566; Dublin, Chester Beatty 2/2, 952/1545; Istanbul, Topkapı Sarayı, R 1660, Rabīʿ I 824/March 1421; London, British Library, Add. 23564, Shiraz, 845/1441, Or. 12220, Herat 910/1503–04; Manchester, John Rylands Library, Persian 2, 1029/1619, Persian 3, 1041/1632, Persian 37, ca. 1440; New York, Metropolitan Museum of the Arts, 34.109 (frontispiece of Istanbul, Topkapı Sarayı, R 1660; Paris, BnF, Smith-Lesouëf (oriental) 221; Supplément persan 334, second quarter ninth/second quarter fifteenth century, in the style of Shiraz; Supplément persan 2051, ca. 885/1480, Shiraz; Rampur, The Raza Library, A 4600, 923–4/1571–2 and A 4601, 1001/1591–2; Welch, *Collection of Islamic Art*, 1, 122, five pages of a late ninth/fifteenth-century manuscript; Robinson, B.W., *Persian Painting in the John Rylands Library: A Descriptive Catalogue*, London, 1980, 35, 273–290, 295–312, 351; Cagman, F. and Tanındı, Z., *Islamic Miniature Painting*, Istanbul, 1979, 20–21.

28 MS Paris, BnF, Supplément persan 332, 790/1388.

29 MS Paris, BnF, Supplément persan 332, beginning of the tenth/beginning of the sixteenth century.

Another important factor that determined the kind of manuscripts that were copied and illustrated was the image a ruler or prince wished to portray. The stories told about Ulugh Beg or Iskandar Sulṭān (r. 811–818/1409–1414), for instance, suggest that they wanted to project the image of a scientifically educated prince. Hence, they invested money, manpower and time in the production of high-quality illustrated copies of scientific texts for their libraries and for their personal use.

Professionally decorated manuscripts made in Iran or India include astronomical handbooks such as the *Ilkhanid Tables* and *Zīj-i jadīd-i gurgānī*, works on planetary theory such as Shams al-Dīn Khafrī's (d. 957/1550) *al-Takmila fī l-Tadhkira*, Ibn Sīnā's (d. 430/1037) philosophical and medical encyclopedias *Kitāb al-Shifāʾ* and *al-Qānūn fī l-ṭibb*, medical writings of other authors, books on theoretical geometry such as Euclid's *Elements*, geographical works such as Abū Isḥāq Ibrāhīm b. Muḥammad Iṣṭakhrī's (d. ca. 340/951) *al-Masālik wa-l-mamālik*, treatises on astrology and works of mixed content. Mostly, the painters and calligraphers remained anonymous as did the recipients of the manuscripts. In a few cases, either some or all of this information is given. Examples are a manuscript of Ibn Sīnā's *Kitāb al-Shifāʾ* that was written and illustrated for the Ottoman sultan Mehmet III (r. 1003–1012/1595–1603) and two manuscripts of the Persian translation of Ṣūfī's *Star Catalogue* made by the court astrologer Ḥasan b. Saʿd Qāʾinī for the governor of Mashhad Abū l-Fatḥ Manūchihr (d. 1045/1636).[30]

In contrast, Zangid, Ayyubid and Mamluk illustrated or illuminated scientific manuscripts are less numerous and cover a much smaller spectrum of themes and titles. Mechanics, natural history, astrology, medicine, the art of war and a broadly construed knowledge of horses and other useful animals were their main domains. At least one Zangid copy of Dioscurides' *Materia Medica* is preserved in Istanbul.[31] Copies of al-Jazarī's book on automata are extant from the Mamluk period, as are copies of Qazvīnī's *ʿAjāʾib al-makhlūqāt* and Ṣūfī's *Star Catalogue*.[32] Several illuminated manuscripts on horses reflect Mamluk experience with Mongol war tactics that posed a grave challenge to Mamluk cavalry.[33] Other texts sponsored by Mamluk sultans dealt with curing sick horses and the diseases of other animals.[34] Such texts show that Mamluk patronage of the ancient

30 *Fihrist-i kitābhā-yi khaṭṭī-yi Kitābkhāna-yi Millī-yi Malik dābiste be-Āstān-i Quds Riżāvī*, 1, 463–464.

31 MS Istanbul, Topkapı Sarayı, Ahmet III 2127, 626/1228.

32 Welch, *Collection of Islamic Art*, 1, 28–32; MS Paris, BnF, Arabe 2490, 922/1516; Atıl, E., *Art of the Arab World*, Washington, 1975, 102–114.

33 See for instance Muḥammad b. ʿĪsā al-Āqsarāyʾī's (d. 749/1348) *Nihāyat al-suʾl*, in a copy produced in Mamluk Syria or Egypt in 1371, MS London, British Library, Add. 18866. Another work of this genre of *furūsiyya* is Muḥammad b. Yaʿqūb b. Khazzām al-Khuttalī's (eighth/fourteenth century) *Kitāb al-makhzūn fī jāmiʿ al-funūn*. See *De Bagdad à Ispahan: Manuscrits islamiques de la filiale de Saint-Pétersbourg de l'Institut d'Études Orientales, Académie des Sciences de Russie*, Musée du Petit Palais, 14 octobre 1994–8 janvier 1995, Lugano-Paris-Milan, 1995, 170–176.

34 An example is Abū Bakr b. Badr (or: Mundhir) al-Bayṭar's (c. 741/1340–1) *Kāmil al-ṣinaʿatayn*. MS Bursa, Kharajī Oghlu, no. 1122.

sciences and related knowledge did not include scholars only, but extended to members of their army and their stables. In addition, there exist single copies from the Mamluk period of each of the following: a work on war machines for the Ayyubid Ṣalāḥ al-Dīn (r. 564–589/1169–1193), Abū Maʿshar's *Kitāb al-mawālīd*, Iṣṭakhrī's *al-Masālik wa-l-mamālik* and Ibn al-Durayhim's (712–761/1312–1360) *Kitāb manāfiʿ al-ḥayawān*.[35]

Another difference between the post-classical dynasties that sponsored the ancient sciences at their courts or through their endowment of teaching institutes concerns the disciplines that were privileged. In most cases, the only way to determine which disciplines were favored at court is to collect information about dedicated texts, instruments and other material objects such as medical bowls, amulets or miniatures. Although such dedications do not necessarily reflect the interests and tastes of the dynasties and their courts, they indicate at least perceptions among scholars, instrument makers and other craftsmen about what kind of discipline could expect an open ear and a friendly gesture. Zangid and Ayyubid princes received sundials, globes and medical bowls with magic squares and incantations, some of which are obvious fakes.[36] Ṣalāḥ al-Dīn also received medical treatises and the aforementioned work on war machines. One of the Ayyubid viziers, Ibn al-Qifṭī (d. 646/1248), was the recipient of a work on toponyms and other geographical information.[37]

Thanks to Ibn Abī Uṣaybiʿa's (592–670/1194–1270) dictionary on physicians and other scholars of the ancient sciences, our knowledge about Ayyubid patronage of these sciences is comparatively better than that of other Islamic dynasties in the same period.[38] As Ibn Abī Uṣaybiʿa's book proves, Ayyubid rulers enjoyed surrounding themselves with several physicians to whom they were often linked for a lifetime. In many cases a court physician only moved to the next Ayyubid court after a patron had died. In contrast, Ayyubid rulers did not consider astrologers a necessary scholarly group at their courts. Only a minority of Ayyubid princes patronized astrologers. Those who did apparently did not employ more than one astrologer at a time.

More interest seems to have existed for philosophical debates at the courts in Hama and Damascus. In the first decades of the thirteenth century, scholars with strong interests in logic, metaphysics and *kalām* fought at Muʿaẓẓam ʿĪsā's (r.

35 MSS Paris, BnF, Arabe 2583; London, British Library, Or. 5305; Escorial, Ar. 898; Ruiz Bravo-Villasante, C., *El Libro de las utilidades de los animales de Ibn al-Durayhim al-Mawsili*, Madrid, 1981. There are several other illustrated or illuminated scientific manuscripts that are either not dated or do not mention the place of their production, which are ascribed by art historians to Ayyubid or Mamluk Syria or Egypt. For a debate on Mamluk illustrated and illuminated manuscripts see Atıl, E., "Mamluk Painting in the Late Fifteenth Century," *Muqarnas*, 2 (1984), 159–171.

36 Savage-Smith, E. and Maddison, F., *Science, Tools & Magic*, London, 1997.

37 *Fihrist-i kitābhā-yi khaṭṭī-yi Kitābkhāna-yi Millī-yi Malik dābiste be-Āstān-i Quds Rīżāvī*, 1, 698.

38 Somewhat surprisingly, Ibn al-Qifṭī has little to say about Ayyubid patronage of the ancient sciences in his dictionary.

615–624/1218–1227) court in Damascus about Ibn Sīnā's philosophical heritage and Fakhr al-Dīn Rāzī's (544–606/1149–1209) evaluation of this oeuvre. Sayf al-Dīn al-Āmidī (d. 631/1233), who had been a client of the prince in Hama until the latter's death, received an invitation to come to the Damascene court. There he excelled through his intelligence and eloquence.[39] His strong self-esteem brought him many enemies, among them Shams al-Dīn Khusrawshāhī (d. 652/1254), his rival in the rational sciences and the teacher of Muʿaẓẓam ʿĪsā's son and successor Naṣīr al-Dīn Dāʾūd (r. 624–626/1227–1228), but more importantly Ibn al-Ṣalāḥ al-Shahrazūrī (577–643/1181–1245), one of the leading scholars of *ḥadīth* and a sworn opponent of logic, which he had unsuccessfully studied with Kamāl al-Dīn b. Yūnus (d. 640/1242) in Mosul. In addition to alienating leading *ʿulamāʾ* in Damascus, Sayf al-Dīn apparently also violated some of the rules of patronage by allowing the ruler of Amid, his native town, to negotiate with him about leaving Ayyubid patronage without actually taking up the offer. When the beleaguered lord of Amid answered a taunt by the head of the Ayyubid family al-Kāmil (r. 615–635/1218–1238), which asserted that he had no scholar worth his name in his town, with the proud rebuke that Sayf al-Dīn had accepted his offer to become the town's new chief judge, the scholar's fate was sealed. Al-Kamil's brother al-Ashraf (r. 626–635/1228–1237), who had conquered Damascus one year after his brother Muʿaẓẓam ʿĪsā's death and forced his nephew al-Naṣīr Dāʾūd to leave for the fortress town Karak together with his teacher Khusrawshāhī and his physician Ibn Abī Uṣaybīʿa, had developed a deep dislike for Sayf al-Dīn. After the confrontation involving Amid, he dismissed the scholar from his chair at the ʿAzīziyya *madrasa*, according to a claim akin to that of Sibṭ b. al-Jawzī (d. 654/1256): "Him who teaches something other than *tafsīr* and *fiqh* and turns to the doctrine of the philosophers I will expel."[40] Although it is not known whether al-Ashraf asked Ibn al-Ṣalāḥ for legal support, the scholar backed in principle the ruler's breach of patronage conduct when he issued a *fatwā* against teaching any other discipline but *ḥadīth*, *tafsīr* and *fiqh*.[41] However, al-Ashraf surrounded himself with other scholars of the ancient sciences, some of whom he explicitly invited to join his entourage.[42] This is one of the reasons why I read the story of Sayf al-Dīn's dismissal by al-Ashraf primarily as a story of conflicts in patronage relationships between a brilliant and self-confident scholar, his peers and his patrons.[43]

39 Brentjes, S., *"Orthodoxy", Ancient Sciences, Power, and the Madrasa ("College") in Ayyubid and Early Mamluk Damascus*, Berlin, 1997, Preprint 77, 23, footnote 67; 24, footnote 70; 27, footnote 76.

40 *Ibidem*, 23.

41 *Ibid.*, 18. For a brief summary of the traditional reading of Sayf al-Dīn's house arrest see for instance Madelung, W., "The Search for God's Law: Islamic Jurisprudence in the Writings of Sayf al-Dīn al-Āmidī by Bernard G. Weiss," *Islamic Law and Society*, 4 (1997), 122–125, 122.

42 Brentjes, *"Orthodoxy"*, 24, footnote 69.

43 There are other points that support seeing Sayf al-Dīn's repeated troubles as resulting directly from his breach of courtly etiquette. Sibṭ b. al-Jawzī, for instance, told a story of Sayf al-Dīn angering Muʿaẓẓam ʿĪsā by coming too late to a session and not showing his respect to the ruler. This

In contrast, philosophy was not a discipline that attracted much interest among the Timurids. First and foremost, they patronized history and geography, which traditionally were not considered scientific disciplines, except for mathematical geography.[44] Above and beyond this focus on history and geography that occupied an important position in Timurid self-representation, astrology and astronomy drew most of their attention. The casting of horoscopes was a regular component in Timurid intellectual culture. Iskandar Sulṭān and Ulugh Beg patronized astronomy, geometry and arithmetic. The observatory and *madrasa* for the mathematical sciences built by order of Ulugh Beg in Samarkand are well known, as are the anthologies composed at Iskandar Sulṭān's courts in Shiraz and Isfahan. Several of the surviving manuscripts from Iskandar Sulṭān's courts contain scientific treatises on astrology, astronomy, geometry, mathematical geography, alchemy and medicine.

Several Timurid rulers and princes, beginning with Timur himself, were dedicatees of texts which included scientific topics or were exclusively devoted to one of the ancient sciences. Some manuscripts of a Persian translation of the *Rasāʾil ikhwān al-Ṣafāʾ* contain a dedication to Timur.[45] After the death of Iskandar Sulṭān, Ghiyāth al-Dīn Kāshī, who had worked at his court in Shiraz and written several astronomical texts for the prince, dedicated at least two of his mathematical works to his new patron Ulugh Beg. Other important collaborators in the program of astronomical observations patronized by the Timurid prince also dedicated works to him, but in contrast to Kāshī's treatises, the texts dedicated by Qāżīzāde Rūmī and ʿAlī Qushjī to Ulugh Beg were of an elementary nature. The overall pattern of such texts in the mathematical sciences seems to be that *Zijes*, surveys and introductions were considered appropriate either as offers to Timurid rulers and princes in search of patronage, or as expressions of appreciation and gratitude for received support and benefits. Also belonging to this group are the texts dedicated to Abū l-Qāsim Bābur Bahādūr Khān (r. 851–862/1447–1457) and Abū Saʿīd Gürāgān (r. in Samarqand 855–863/1451–1459 and Herat 864–874/1459–1469). In addition to writings on the mathematical sciences, astronomical instruments also were dedicated to Ulugh Beg.

Medicine and mineralogy were among the other disciplines considered by scholars to be worthwhile subjects for the production of treatises dedicated to Timurid rulers and princes. The well-known and often-copied anatomical work *Tashrīḥ al-badan* by Manṣūr b. Muḥammad b. Aḥmad (fl. ca. 794/1390) was dedicated by its author to Pīr Muḥammad Bahādur Khān (807–809/1404–1406);

---

indiscretion allegedly induced Sibṭ b. al-Jawzī to interfere in favor of his colleague. See Brentjes, *"Orthodoxy"*, 23, footnote 67.

44 Ettinghausen, R., "An Illuminated Manuscript of Hafiz-i Abru in Istanbul," *Kunst des Orients*, 1 (1959), 30–44; Soudavar, A., *Art of the Persian Courts*, New York, 1992, 62–65; Ferrier, R.W. (ed.), *The Arts of Persia*, New Haven & London, 1989, 205–206.

45 *Fihrist-i nuskhahā-yi khaṭṭī-yi kitābkhāna-yi Imām-i Ṣādiq-i Qazvīn*, M. Tāyār Marāghī (ed.), Qazvin, 1378, 1, 225–226, n.° 190.

the commentary on Najīb al-Dīn Muḥammad b. ʿAlī Samarqandī's (d. 618/1221) *al-Asbāb wa-l-ʿalāmāt* by Burhān al-Dīn Nafīs Kirmānī (fl. 841/1437–8) was dedicated to Ulugh Beg; and the *Jawāhir-nāme* by Zayn b. Ḥājjī was dedicated to Shāhrukh (r. 808–851/1405–1447).[46] As indicated earlier, Mamluk sultans and emirs primarily supported *ʿilm al-mīqāt*, medicine and *furūsiyya* (the knowledge of horses and the art of cavalry). Some also patronized astrology and geomancy, but there was no official position for an astrologer or geomancer at court. The wealthy and powerful Mamluk Amīr Yashbak min Mahdī al-Dawādār (d. 885/1480) commissioned a treatise on geomancy.[47] As with veterinary medicine and cavalry, Mamluk sultans paid special attention to administrative practice. For instance, in 850/1446–7 Sultan Qāytbāy (873–901/1468–1495) received a treatise, written in beautiful *thulūth*, on the fiscal year, its relation to the lunar calendar and its use in official documents.[48] An exceptional case in the Mamluk period was that of the Ayyubid scholar-prince Abū l-Fidāʾ, the governor of Hama. As is well known, he wrote his own substantial works on mathematical geography and history. In 710/1310–11 he also commissioned a commentary by ʿUmar b. Dāʾūd b. Sulaymān al-Fārisī (fl. ca. 710/1310–11) on Naṣīr al-Dīn Ṭūsī's *Tadhkira*, which discusses planetary theory.[49]

The Jalayirids and the Aq Koyunlu supported music, considered part of the mathematical sciences, as well as mineralogy, natural history and philosophy. Manuscripts dedicated to rulers of these two dynasties include the *Jawāhir-nāme* by one Muḥammad b. Manṣūr for Ḥasan Bahādur Khān (Ūzūn Ḥasan) and his son Khalīl Bahādur Sulṭān, and two works dedicated to Yaʿqūb Bahādur Khān (r. 884–896/1479–1490): Khwāja Fażl Allāh b. Rūzbehān Khūnjī Iṣfahānī's (ninth/fifteenth century) Persian translation of the philosophical story of Ḥayy b. Yaqẓān, and Idrīs b. Ḥusām al-Dīn Munshī's *Rabīʿ al-abrār*, in which he debated the natural properties of each season.[50] A single text on geometry, Ilḥāq b. Abī Isḥāq's commentary on Naṣīr al-Dīn Ṭūsī's edition of Euclid's *Elements*, was dedicated to this same Aq Koyunlu ruler in 886–7/1481–2.[51] The connection of Jalāl al-Dīn Davvānī (830–907 or 8/1426–1501 or 2) first with Timurid, then with Aq Koyunlu rulers is well known. He dedicated several of his writings on philosophy and

46 *Fihrist-i nuskhahā-yi khaṭṭī-yi Kitābkhāna-yi ʿUmūmī-yi Āyatullāh al-ʿUẓmā Marʿashī Najafī*, Qom, 12, 1315sh, 144, n.° 4569; 14, 1366sh, 148, n.° 5364; *Fihrist-i nuskhahā-yi khaṭṭī-yi kitābkhāna-yi Imām-i Ṣādiq-i Qazvīn*, M. Ṭāyār Marāghī (ed.), Qazvin, 1378, 1, 151 n.° 117.

47 MS Dublin, Chester Beatty, n.° 3120.

48 King, D.A., *A Survey of the Scientific Manuscripts in the Egyptian National Library*, Cairo, 77, C 83.

49 *Ibidem*, 60, C21.

50 *Fihrist-i nuskhahā-yi khaṭṭī-yi Kitābkhāna-yi ʿUmūmī-yi Āyatullāh al-ʿUẓmā Marʿashī Najafī*, 16, 1368sh, 222, n.° 6234; 19, 1369sh, 379–380, n.° 7574.

51 MS London, British Library, Or. 1514. Although I have not seen them, there are two other copies of this text extant in the Archaeological Museum and the Süleymaniye Library in Istanbul. *Osmanlı Matematik Literatürü Tarihi*, 1, 31–32.

related topics to princes of these two dynasties.[52] Asked by Ūzūn Ḥasan for commentaries on philosophical texts, he edited Naṣīr al-Dīn Ṭūsī's ethics and wrote treatises on Shihāb al-Dīn Suhravardī's teachings.[53] In 887/1482 he wrote an introduction to Ilḥāq's unfinished commentary on the *Elements*, recommending it and expressing his hope for its speedy completion.[54] Commentaries on astronomical texts, such as Ṭūsī's *Sī faṣl* on the compilation of ephemerides, also are attributed to Davvānī.[55]

The rulers of the Inju dynasty were interested in encyclopedias, astrology, medicine and magic. The Inju ruler Abū Isḥāq (r. 744–754/1344–1353) sponsored some of the scholars who left the Ilkhanid capital Sultaniyeh due to the upheaval and uncertainties caused by the death of Abu Saʿīd Bahādur Khān (r. 716–736/1316–1335). Muḥammad b. Maḥmūd Āmulī (d. ca. 753/1352) was one of these scholars. Around 741/1340, he wrote his encyclopedia of the sciences, which included substantial material on a broad range of the ancient sciences, both their so-called fundaments and their branches. A manuscript of an unidentified text on Iranian history which included numerous miniatures was produced in the environment of the Inju court in Shiraz in about 741/1340. Welch has argued that it is a Persian version of the same type of astrological history as that encountered in Māshā'allāh's (d. 200/815) and Abū Sa'id Sijzī's Arabic works.[56]

The Muzaffarids exhibited interest in the same fields as the rival Injus, and apparently also in geography. They began as a local dynasty in Maybud, starting with Mubāriz al-Dīn Muḥammad (r. 713–759/1313–1357). In 748/1347, a geographical work with the classical title *Ṣuwar al-aqālīm* was dedicated to him in Yazd. In 754/1353, Mubāriz al-Dīn conquered Shiraz. His son Shāh Shujāʿ (r. 759–786/1357–1384) patronized ʿAlī b. Muḥammad al-Sayyid al-Sharīf Jurjānī (741–816/1340–1413), whom he appointed in 778/1377 to a teaching position at the *Dār al-Shifā'* in Shiraz. Jurjānī wrote a commentary on an epitome of the *Elements*, a copy of which is extant in Mashhad.[57] He may also have been the author of an encyclopedia called *Maqālīd al-'ulūm fī l-ḥudūd wa-l-rusūm*, a copy of which is extant at the British Library.[58] It is dedicated to Shah Shujaʿ. Shāh Shujāʿ's need for magical support is indicated by a mirror with magic squares and other formulas that was dedicated to him in Muḥarram 777/June 1375.[59]

52 *Fihrist-i nuskhahā-yi khaṭṭī-yi Kitābkhāna-yi ʿUmūmī-yi Āyatullāh al-ʿUẓmā Marʿashī Najafī*, 28, 1379sh/1421, 280, n.º 483.

53 Cooper, J., "al-Dawani, Jalal al-Din (1426–1502)," in C. Edward (ed.), *Islamic Philosophy. From the Routledge Encyclopedia of Philosophy*, accessible on-line in www.muslimphilosophy.com/ip/rep/H038.htm (15–08–08).

54 MS London, British Library, Or. 1514, ff 1a–b.

55 *Fihrist-i nuskhahā-yi khaṭṭī-yi Kitābkhāna-yi ʿUmūmī-yi Āyatullāh al-ʿUẓmā Marʿashī Najafī*, 21, 231.

56 Welch, *Collection of Islamic Art*, 3, 42.

57 Matvievskaya and Rozenfel'd, *Matematiki*, 2, 476.

58 MS London, British Library, Or. 3143.

59 Soudavar, *Art of the Persian Courts*, 46–47.

Safavids, Ottomans and Mughals sponsored astrology and medicine by establishing regular courtly offices for their practitioners. The majority of texts on the ancient sciences dedicated to Safavid rulers, princes, governors and viziers fall into these two disciplines. They include surveys on diseases and remedies, discussions of the properties of individual remedies, ephemerides, horoscopes and surveys of astrological themes and methods.[60] Characteristic of Safavid interest in medicine is the apparently widespread incorporation of magic in the form of engraved bowls.[61] In addition to these two disciplines, works on astronomical instruments, in particular the astrolabe, on star constellations, arithmetic, geometry, geomancy, *fāl* (an art of prognostication) and philosophy were either dedicated to or commissioned by Safavid courtly patrons, in particular the shahs ʿAbbās I (r. 985–1038/1587–1629), ʿAbbās II (r. 1052–1077/1642–1666), Sulaymān (r. 1077–1105/1667–1694) and Ḥusayn (r. 1105–1135/1694–1722).[62]

An important element in Safavid patronage of the ancient sciences was their interest in and support for translations of Arabic texts into Persian. In the course of the seventeenth century, several Arabic medical, geographical and astronomical texts were translated at Safavid courts in Isfahan and Mashhad, among them Abū l-Fidāʾ's (672–732/1273–1331) *Taqwīm al-buldān*, ʿAbd al-Raḥmān Ṣūfī's *Star Catalogue* and Ibn Jazla's (d. 495/1100) *Taqwīm al-abdān* The Safavid reputation for having scientific works translated from Arabic into Persian is emphasized by the ascription of the translation of Abū l-Fidāʾ's work to Khalīfa Sulṭān (d. 1064/1653), a son-in-law of ʿAbbās I and great vizier to three Safavid rulers, who is said in some manuscripts to have translated the geography on Ṣafī's (r. 1038–1052/1629–1642) order in 1050/1640.[63] In addition to scientific works, Ibn al-Qifṭī's history of scholars of the ancient sciences was also translated into Persian for a Safavid shah, namely Sulaymān.[64]

Mughal rulers and princes, in particular Akbar (r. 963–1014/1556–1605), Jahāngīr (r. 1014–1037/1605–1627), Shāh Jahān (r. 1037–1068/1628–1658) and Dārā Shikūh (10251069/1615–1659), patronized the translation of Sanskrit mathematical, astronomical, philosophical and religious works into Persian. In 995–6/1587 the poet Fayżī lent his name to the translation of Bhāskara's (d. after 1183)

60 See, for instance, Tourkin, S., "The Horoscope of Shah Tahmāsp," in J. Thompson and Sh.R. Canby (eds.), *Hunt for Paradise: Court Arts of Safavid Iran 1501–1576*, Milan, 2003, 327–331.

61 See Savage-Smith, E., "Safavid Magic Bowls," in Thompson and Canby (eds.), *Hunt for Paradise*, 241–247.

62 For a survey on mathematical, astronomical and astrological works dedicated to Safavid patrons see Brentjes, S., "The Mathematical Sciences in Safavid Iran: Questions and Perspectives," in F. Speziale and D. Hermann (eds.), *Muslim Cultures in the Indo-Iranian World during the Early-Modern and Modern Periods*, Berlin: Klaus Schwarz Verlag, Tehran: Institut Français de Recherche en Iran, 2010, 325–402. For some of the lavishly illustrated manuscripts on *fāl* made for Ṭahmāsb see Arthur, M. *Sackler Gallery, Smithsonian Institution*, Washington, DC, S1986.253a-b, S1986.254.

63 *Fihrist-i kitābhā-yi khaṭṭī-yi Kitābkhāna-yi Millī-yi Malik dāt iste be-Āstān-i Quds Rīżāvī*, 2, 129.

64 *Ibidem*, 2, 129; *Fihrist-i nuskhahā-yi khaṭṭī-yi farsī*, A. Munzāvī (ed.), Tehran, s.d., 6, 3943.

Līlāvatī for Akbar.[65] In about 1074–5/1663, a Hindu scholar wrote a Persian commentary on the translated text. In 1044/1634–5, ʿAṭāʾ Allāh Rashīd b. Aḥmad Nadīr, the eldest son of the architect of the Taj Mahal, translated Bhāskara's *Bījagaṇita* for Shāh Jahān.[66] In addition to these two mathematical texts, various *Upanishads* and the *Mahābhārata* were translated into Persian.[67]

Ottoman sultans and viziers, among them Mehmet Fatih and Ahmet Fazil Köprülü (1045–1087/1635–1676), sponsored translations of scientific works from Greek into Arabic, and from Persian, Arabic, Latin, Italian and French into Ottoman Turkish. Examples are Ptolemy's, (Pseudo-)Ibn Ḥawqal's and Blaeu's works on geography, Qazvīnī's book on cosmography and natural history, Naṣīr al-Dīn Ṭūsī's little survey on the astrolabe, Čaghmīnī's introduction into mathematical cosmography, works written by *muwaqqit*s in Mamluk Egypt and Syria, as well as astronomical and astrological works mostly by French writers of the seventeenth and eighteenth centuries. Scholars responded to the changing cultural conditions at court by dedicating translations into Persian or Ottoman Turkish to sultans, viziers or religious dignitaries. This could include the translation of treatises earlier dedicated in another language to a different patron, mostly outside the Ottoman realm. Examples are ʿAlī Qushjī's rededications of his introductory works on arithmetic and mathematical cosmography, first dedicated to Ulugh Beg, to Mehmet Fatih after translating them from Persian into Arabic. This practice, whereby works formerly offered to one patron were rededicated to another, also can be observed with other dynasties, and could occur with works other than translations. It reflects the difficulties arising from loss of patronage due to the death of a patron. The quantity of works translated under the Ottomans was considerable and has not yet been systematically investigated.

4 Honors, benefits and services

The main services scholars of the ancient sciences were expected to provide their patrons consisted of medical treatment in case of illness, preparation of remedies and drugs, casting horoscopes, determining auspicious days,

65 Winter, H.J.J. and Mirza, A., "Concerning the Persian Version of Lilavati," *Journal of the Asiatic Society*, 18 (1952), 1–10, 2–3.

66 *Catalogue of the Arabic and Persian Manuscripts in the Oriental Public Library at Bankipore*, vol. IX *(Persian MSS.), Philology and Sciences*, M. Abdul (ed.), Patna, 1925, cc. 1112–1113; Rahman, A., Alvi, M.A., Khan Ghori, S.A. and Samba Murthy, K.V. (eds.), *Science and Technology in Medieval India: A Bibliography of Source Materials in Sanskrit, Arabic and Persian*, New Delhi, 1982, 391.

67 Göbel-Groß, E., *Sirr-i akbar: Die persische Upaniṣadenübersetzung des Mogulprinzen Dara Šukoh; Eine Untersuchung der Übersetzungsmethode und Textauswahl nebst Text der Praîna-Upaniṣad Sanskrit-Persisch-Deutsch*, Marburg, 1962; Athar Ali, M., "Translations of Sanskrit Works at Akbar's Court," *Social Scientist*, 20 (1992), 38–45; D'Onofrio, S., "A Persian Commentary to the Upaniṣads: Dārā Shikōh's Sirr-i akbar," in Speziale and Hermann (eds.), *Muslim Cultures*, 533–564.

compiling ephemerides and surveys on good and bad days, teaching children, and dedicating surveys, introductions and occasionally more substantive treatises on individual scientific topics to their patrons. Dedications were expected to include praise for the recipient, the specification of his titles and the expression of good wishes for his rule, health and afterlife. Loyalty, as discussed for Sayf al-Dīn al-Āmidī above, was another item a patron expected of his client. To what extent clients were expected to offer new ideas for modelling the trajectories of the planets or arguing about movement of physical bodies, or to present unique approaches to solving old problems such as the determination of the direction for prayer, is difficult to ascertain. Certain rulers, princes and viziers in the post-classical era, among them Mehmet Fatih, Iskandar Sulṭān, Ulugh Beg, ʿAlī Shīr Navāʾī, Humāyūn, Akbar and Dārā Shikūh, however, took note of theoretical questions, received at least an introductory education in the theoretical fundaments of astronomy and asked their clients for surveys of such matters. Stories told in biographical dictionaries and historical chronicles suggest that in addition to being able to address everyday necessities, such as determining the auspicious day for a particular activity or healing a sick person, scholarly clients were expected to be conversant in all themes discussed at the endowed teaching institutes and beyond. Knowing more than was regularly taught was particularly important when visitors from other regions and countries came, displayed superior knowledge, and defeated local scholars in a *munāẓara*. Patronized debates with strangers were akin to battles. Poor performance on the part of the local scholars could entail a loss of face, and hence of reputation, for the participating patron. Defeating a stranger on the other hand, was honored by material and pecuniary rewards and could entail the start of a successful patronage relationship, with a continued rise through courtly functions and offices and an ever-increasing salary. On one occasion, in the second half of the ninth/fifteenth century, a scholar of the occult sciences from North Africa came to Istanbul. Sultan Mehmet Fatih arranged for a public disputation in his presence between the scholars of Anatolia and this visitor. The local scholars could not answer the visitor's questions about the occult sciences, which were unfamiliar to them. The sultan was very upset, since he felt he had lost face. He ordered that a scholar be found with excellent knowledge of the occult sciences. A certain Mullā Khiżr Bey, who, despite his youth, had studied all possible disciplines and taught in a provincial town, was recommended and called to the capital. In a further disputation this Anatolian scion of the sciences managed to defeat the visitor from the West due to the latter's lack of knowledge outside the occult disciplines. The sultan was delighted, honored the young scholar greatly, appointed him to the teaching post at the *madrasa* founded by his grandfather in the former Ottoman capital Bursa, increased his daily salary several times and promoted him to other prestigious and profitable offices.[58]

68 Tasköprüzade, *Es-Saqaʾiq*, 54–55.

The story about Mehmet Fatih and Mullā Khiżr Bey illustrates some of the actions a patron undertook for a cherished client. He fostered the client's career through offices or incorporation into the household, protected him from rivals and other obstacles, and expressed his gratitude, appreciation and respect by means of gifts and payments. A patron needed to be generous in his rewards for services rendered, but had to avoid overtaxing his personal treasury or that of the dynasty. As Tashköprüzade's story about Ulugh Beg and Qāżīzāde Rūmī, summarized above, suggests, a patron ideally did not interfere in scholarly activities or his client's academic responsibilities. In practice, however, as Qāżīzāde Rūmī's reply to Ulugh Beg implies, a patron did interfere more often than not. The lines of obligation became blurred when patrons were also scholars, and when scholars obtained influential offices and became powerful patrons in their own rights.

5 Court libraries

Libraries were mostly bound to individuals, although endowed teaching institutes and particularly important shrines and mosques acquired a steadily growing number of donated manuscripts that became attached to the institute rather than to its professor(s). While the originals of dedicated works usually were destined for the library of a ruler, vizier or other member of a court, most manuscripts that filled such a court library were copied by scribes of the court's workshop, bought at markets or from owners of libraries, received as gifts or taken as booty. Various scientific manuscripts exist today that were copied explicitly for a courtly library. The following examples illustrate this point: two manuscripts of ʿAbd al-Raḥmān Ṣūfī's *Star Catalogue*, in the Bibliothèque nationale de France and the Library of Congress respectively, were made for Ulugh Beg's library; Iṣṭakhrī's *al-Masālik wa-l-mamālik* in the British Library was made for a Mamluk patron; and ʿAyn al-Zamān Abū ʿAlī b. ʿAlī Marwazī's (d. 548/1153) *Keyhān Shenākht* in the Public Library of Āyatollāh Marʿashī in Qom and Ulugh Beg's *Zij-i jadid-i gurgānī* were made for Ulugh Beg's library. The second is kept today in the Freer and Sackler Gallery in Washington, DC. Al-Jazarī's book on the automata was made in 755/1354 for the library of the Mamluk emir Naṣir al-Dīn Muḥammad b. Tulak al-Ḥasanī al-Malikī al-Ṣaliḥī, and came later to the library of the Ottoman sultan Ahmet III (r. 1115–1143/1703–1730).[69] Some of the more precious manuscripts such as this last work ended up on the international art markets, cut to pieces and sold to museums and private collections in the US and Europe.

Mughal, ʿĀdil Shāhī and Ottoman rulers and other members of the court seem to have searched for and bought scientific manuscripts not only because of the fame of their authors and the importance of their content, but also because of the antiquity of a given copy and its artistic execution. A partial copy of Bīrūnī's astronomical *magnum opus* dedicated to the Ghaznavid Sultan Maṣʿūd b. Maḥmūd

69 Welch, *Collection of Islamic Art*, 1, 31–32; Soudavar, *Art of the Persian Court*, 67–69.

(r. 421–432/1030–1040) was acquired by a courtier of Shāh Jahān's court in 1060/1649. A copy of Ghiyāth al-Dīn Kāshi's *Miftāḥ al-ḥisāb* dedicated to Ulugh Beg was bound for Tippu Sulṭān's (r. 1163–1214/1750–1799) library.[70] Other manuscripts changed hands as a result of wars, as exemplified by a copy of Ṭūsī's edition of Euclid's *Elements*, which came from the library of Muḥammad ʿĀdil Shāh (r. 1035–1070/1626–1660) in Bijapur in 1069/1659 into the library of the Mughal ruler Awrangzīb (r. 1069–118/1659–1707) after he had defeated the last ʿĀdil Shāh in 1097/1686.

To some extent, it is possible to reconstruct collections of scientific works that were once in court libraries in the post-classical period. A systematic survey of ownership marks and seals in scientific manuscripts can contribute to a better understanding of courtly patronage of the ancient sciences by showing which types of texts and which authors were collected, in which times the copies were produced or, occasionally, where and from whom copies were bought or looted. A comparison of texts produced and reproduced in the lifetimes of these libraries can indicate to what extent their composition reflects the contemporary scientific life. While such a reconstruction is beyond the scope of this paper, the information provided by catalogues of the Āstān-i Quds Rīżāvī Library in Mashhad about the library of the Afshar ruler Nādir Shāh (1148–1160/1736–1747) will serve as an example. The biography of Nādir Shah does not indicate whether he received in his childhood, youth or adulthood a substantive education in the intellectual traditions of his Safavid predecessors. Nevertheless, his library was rich in manuscripts, with texts on the ancient sciences as the following table shows.

Table 5.1 Scientific texts from Nādir's Library[71]

Author	*Title*	*Discipline*
ʿAbd al-Sattār b. Qāsim	Tadhkirat al-ḥukamāʾ	history of scholars
Abū l-Fatḥ Shahrastānī	Kitāb al-milal wa-l-niḥal	history of philosophy and religious sects
Ḥunayn b. Isḥāq (translator)	al-Muʿālajāt al-buqrāṭiyya	medicine
ʿAlī b. ʿĪsā Kamāl	Tadhkirat al-kaḥḥālīn	ophthalmology
Muḥammad b. Zakariyyāʾ Rāzī	Mufīd al-khāṣṣ	medicine

(*Continued*)

70 Blochet, E., *Catalogue de la collection de manuscrits orientaux de M.Ch. Schefer*, Paris, 1900, 87; Schöler, G., unter Mitarbeit von H.-C. Graf von Bothmer, T. Duncker Gökçen, H. Jenni, *Arabische Handschriften.* Teil II, Stuttgart, 1990, 172; Loth, O., *A Catalogue of the Arabic Manuscripts in the Library of the India Office*, London, 1877, 215–216, 220.

71 I ordered the titles as taken from the catalogues according to disciplinary affiliations starting with two books on history of scholars and beliefs followed by medical, mathematical, occult and philosophical sciences according to the understanding of the period

Table 5.1 (Continued)

Author	*Title*	*Discipline*
Ibn Sīnā	al-Qānūn[72]	medicine
Ibn Sīnā	al-Qānūn	medicine
Ibn Sīnā	al-Qānūn	medicine
Fakhr al-Dīn Rāzī	Sharḥ al-Qānūn	medicine
	Dustūr-i sākhtan-i tiryāq	
	Dhākhira-yi Iskandariyya	
Ismāʿīl b. Ḥusayn Jurjānī	Dhākhira-yi Khwārazmshāhī	medicine
Ibn al-Nafīs al-Qurashī	Sharḥ al-Qānūn	medicine
Abū Yazīd Ṣahār Bakht	Miftāḥ al-khazāʾin	medicine
Ḥusayn b. Bistām Wāsiṭī	Ṭibb al-aʾimma	medicine
Abū l-Fidāʾ	Taqwīm al-buldān	mathematical geography
Ḥusām al-Dīn Sālār	Risāla dar istikhrāj-i samt-i qibla	mathematical geography
Ḥusam al-Dīn Sālār	al-Risāla fī l-khuṭūṭ al-mutawāziya	geometry
ʿAbdallāh Ḥāsib	al-Fawāʾid al-Bahāʾiyya	arithmetic
Naṣīr al-Dīn Ṭūsī	Ḥisāb	arithmetic
	Manẓūma dar riyāżī	mathematics
Sayyid Sharīf Jurjānī	Sharḥ-i Tadhkira	astronomy
Niẓām al-Dīn Birjandī	Sharḥ-i Bīst bāb	astronomy
Bahāʾ al-Dīn ʿĀmilī	Risāla dar asṭurlāb	astronomy
Abū Maʿshar	al-Mudkhal	astrology
ʿAlī Shāh Muḥammad Khwārazmī	Risāla-yi tasyīrāt-i nujūmi	astrology
Ḥusayn Waʿiẓ Kāshifī	Lavaʾi al-qamar	astrology
	Fālnāma	occult science
Aristotle (ascribed)	al-Samāʾ wa-l-ʿālam	natural philosophy
Ibn Sīnā	al-Shifāʾ	philosophy
Qāżī Sirāj al-Dīn	Laṭāʾif al-ḥikma	philosophy
Naṣīr al-Dīn Ṭūsī	Risāla dar ṣudūr-i mavjūdāt az mabdāʾ	philosophy
Mubārak Shāh Bukhārī	Sharḥ-i Ḥikmat al-ʿayn	philosophy
Qāżī Ḥusayn Maybudī	Sharḥ Hidāyat al-ḥikma	philosophy
Shams al-Dīn Maḥmūd Iṣfahānī	Sharḥ-i qadīm-i Tajrīd	metaphysics
Ibn Sīnā	Manṭiq, al-Shifāʾ	logic
Quṭb al-Dīn Rāzī	Sharḥ al-Shamsiyya	logic
Sayyid Sharīf Jurjānī	Risala-yi kubrā	logic

72 The three manuscripts mentioned here contain different parts of the work.

The titles preserved today in Mashhad represent a puzzling mixture. Among them are books that were repeatedly copied in the tenth/sixteenth and eleventh/seventeenth centuries and most likely studied in Safavid *madrasas*, such as the work on arithmetic *al-Fawāʾid al-Bahāʾiyya*, Quṭb al-Dīn Rāzī's commentary on the *Shamsiyya* and Qāżī Ḥusayn Maybudī's commentary on *Hidāyat al-ḥikma*. Ibn Sīnā's *Kitāb al-Shifāʾ*, as well as his *Qānūn*, were also copied more than once and could be found in *madrasas* in Isfahan in the eleventh/seventeenth century. Other titles, such as *al-Risāla fī l-khuṭūṭ al-mutawāziya* by Ḥusām al-Dīn Sālār, are rarities and not known to have been part of the *madrasa* education in the Safavid period. Since titles such as the one on parallel lines were not only a rarity in Nādir's time, but also find no mathematical context in his library, at least insofar as it was given to Āstān-i Quds Rīżāvī Library, it is highly unlikely that the Afshar ruler acquired the manuscript on his own as a result of specific interest in a seventh/thirteenth-century discussion of the parallel postulate.

Because of these structural features, I suggest viewing the extant manuscripts on the ancient sciences from Nādir's library as a remainder of the library of the Safavid shahs. It must be emphasized, however, that this constitutes only a minor part of what this library may have included. With the exception of a single title, namely Bahāʾ al-Dīn ʿĀmilī's treatise on the astrolabe, no works dedicated to a Safavid shah, and no other of Bahāʾ al-Dīn's mathematical and astronomical writings – so famous in the eleventh/seventeenth century in Iran, India and the Ottoman Empire – can be found among Nādir's manuscripts. This suggests not only that a substantial quantity of manuscripts once present in the palatial library were dispersed or destroyed when the Afghans sacked Isfahan in 1136/1722. It also implies that while Nādir obviously was interested in collecting a range of texts on the ancient sciences, he also was eager to present himself as an admirer of the intellectual luminaries studied in the Safavid period.

Reflections

Patronage of the ancient sciences in post-classical Islamic societies is a field worthy of study, and rich in material evidence in both libraries and manuscripts. This paper only touches in a preliminary manner on some of the most obvious issues that patronage involved. Issues not addressed here due to a lack of data include, for instance, princely patronage of writers outside the scholarly sphere, the similarities and differences between patronage for scholars and instrument-makers, and patronage exercised by minor dynasties and courts of governors. A considerably more meticulous examination of dedications, ownership marks and other types of information would help greatly to expand and clarify the available evidence. This material could then serve as the basis for analyzing the types and forms of relationships that a patron and a client could engage in, and for understanding the specific details of patronage activities as exercised by individual patrons and clients. Applying theories from other human and social sciences would be helpful for interpreting connections between social structures, cultural rituals and the presence or absence of patronage activities relating to individual ancient sciences. As for the link between courtly patronage of the ancient sciences and their increasing loss of innovative potential, further study is needed to establish links between courts, princely education and courtly libraries, on the one hand, and *madrasa*s, their educational profiles, public scholarly debates and the content of texts produced for a patron, in comparison with those written for students or colleagues, on the other. Once we understand more fully the similarities and differences between the intellectual atmospheres at courts and *madrasa*s, we may be better equipped to answer which of the two social loci of the ancient sciences has contributed causally to the increasingly elementary and repetitive character of the content of newly written works in these disciplines.

6

THE LANGUAGE OF 'PATRONAGE' IN ISLAMIC SOCIETIES BEFORE 1700

Introduction

In the first centuries of Islamic societies, 'clients' were called *mawla*s. This described an adoption of someone by an Arabic tribe. It was a pre-Islamic custom which was now applied to the newly converted. Such a *mawla*ship was legally regulated. It could be inherited or transferred. This custom became so prestigious that in the 8th and 9th centuries, some men invented their pedigrees as *mawla*s, while others rewrote their ancestry immediately as that of an Arabic tribe.

A new concept rose to prominence under the Abbasid caliphs in the 9th century and even more so under the Buyid *umara'* in the 10th century: *istina'*. It signified the rearing and training of (slave) soldiers. According to Mottahedeh, the relationship of *istina'* was of a more informal kind than the *mawla*ship. He also saw it as "a surprisingly formal and serious relationship" where a man expected from his protégé (*sani'*) a lifelong commitment.[1] Hence, Mottahedeh identified *istina'* with 'patronage'.[2] I could not find, however, a single case where *istina'* or its verbal forms were used for scholars. Nor did the triangle that Mottahedeh described as the defining feature of Buyid patronage, namely benefit (*ni'ma*), gratitude (*shukr al-ni'ma*) and obligation (*'uhda*, such as oath, vow or the guarantee of safe conduct) appears in texts about scholars and their relationship to rulers or courtiers.[3]

Does this mean that there was no patronage for the sciences or for scholarship in general? This is barely plausible. Too many mathematical, astronomical, astrological and other scholarly texts as well as instruments with dedications exist. Lately I even happened to come across a copy from 705/1306 of Qutb al-Din Shirazi's

1 R. Mottaheded, *Loyalty and Leadership in an Early Islamic Society*, Princeton, Princeton University Press, 1980, pp. 82–84.
2 Ibidem.
3 Ibidem, pp. 70–79.

DOI: 10.4324/9781003372455-7

(634–710/1236–1311) encyclopedic work *Durrat al-Taj* dedicated to an emir in northern Iran that appears to have been illustrated by a painter in an art workshop. On its first folio is a circular motif (*Shamse*) that claims that Qutb al-Din himself had transcribed this copy for the library of the *Great Amir*.[4] Thus, since relationships between scholars, rulers and courtiers existed that we usually name 'patronage', in which aspects did these relationships differ from those characteristics for the military and where did they agree? How did people in different Islamic societies before 1700 speak about such relationships? Which features changed over time? These and similar questions are at the heart of my paper. In section 1 I will discuss the linguistic similarities and differences between military and civilian 'patronage relationships', mainly in regard to physicians and astrologers, but occasionally also to scholars of geometry and other mathematical sciences, and the activities described by such terms. In section 2 I will discuss problems with the expression most often used in the case of civilian relationships in two Islamic societies between the late 12th and the early 16th centuries, namely *khidma*, meaning service and later also employment. In section 3 I will talk about some changes in civilian "patronage relationships" as the result of the spread of the madrasa as the major symbolic institution of education. My last section is concerned with the arts and their usage as markers of 'patronage' for the sciences.

1 Linguistic similarities and differences between military and civilian patronage relationships in the case of physicians and astrologers

Basic terms that were used to talk about relationships between rulers and courtiers on the one hand and physicians or astrologers on the other come from the vocabulary used for relationships between the caliphs or the *umara'* and their troops. Examples are terms like benefit (*ni'ma*), honor, grace (*karama, ikram*), stipend (*jamakiyya, jiraya*) or *iqta'* (tax farm). Paying a stipend to or bestowing a tax farm on a scholar is mentioned occasionally already in the classical period, in particular for the Abbasid caliphs. It became more regular for the Ayyubids, a dynasty of Kurdish mercenaries that rose to power by overthrowing the Fatimid caliphate in Egypt and Syria in 1171. According to Eddé the increased transfer of social and economic privileges of the military to leading religious scholars reflected the need of this upstart family to win civilian support in the main cities of their new realm.[5] In addition to religious scholars, the Ayyubids also granted some of their physicians a tax farm, while paying them a stipend was the rule.[6]

4 MS Istanbul, Süleymaniye Library, Fazil Ahmad Pasha 867, f 1a.

5 A.-M. Eddé, *La principauté ayyoubide d'Alep (579–658/1260)*, Stuttgart, Franz Steiner Verlag, 1999, p. 280.

6 S. Brentjes, 'Ayyubid princes and their scholarly clients from the ancient sciences,' in A. Fuess, J.-P. Hartung (eds.), *Court Cultures in the Muslim World: Seventh to Nineteenth Centuries*, SOAS/Routledge Studies on the Middle East, London, Routledge, 2010, pp. 326–356.

Civilian terminology expressing different layers or perspectives of a relationship between a scholar and male or female members of a dynasty, a vizier or another courtier was fairly rich and variegated. In the classical period it included, as I have explained elsewhere, words that express either the perspective of the 'client' or that of the 'patron'. A 'client' was someone who was bound to or connected with someone else (*ittasala bi*) and who enjoyed the favor or was in the good graces of another person (*haziya inda*). A 'patron', in contrast, bound (*irtabata*), engaged (*irtabata*), elevated (*'azzaza*), summoned (*da'a*), fetched (*istahdara*), carried away (*hamala*) or hired (*istakhdama*) someone else. This terminology contains at least one element of Mottahedeh's triangle, namely obligation (*ittasala bi, irtabata, 'azzaza*). It indicates asymmetric, but shared power, i.e., 'clients' were not powerless: they were coveted or desired ardently (*ishtaqa ila*), could be invited (*da'a*) instead of ordered and, if successful, were favored or graced (*haziya ila*).[7] Indeed, Ibn Abi Usaybi'a's description of the relationships between Ayyubid princes and their physicians shows beyond doubt that the physicians were active participants in forming such a relationship – they chose, they refused, they had alternatives, they had skills, they were rich, they had preferences. At times, they were willful, spoiled or overbearing. In addition to honor (*ikram*) and benefit (*ni'ma*), Ibn Abi Usaybi'a also used more or less frequently terms like favor (*huzwa*), respect (*ihtiram*) and companionship (*suhba*).[8] The first two are well known from Mottahedeh's study. Favor given by the patron complements the abovementioned verb describing the 'client's' status. It underlines that despite the more or less regular monthly payment of stipends and the regular daily visits of the royal patrons by their physicians the relationship was not an employment regulated by a written contract and remunerated by a salary. Respect, however, indicates something new, if not in regard to such relationships in previous Islamic societies, so at least in the discussions about such relationships. It implies that the relationships between Ayyubid princes and their physicians also contained an element of merit. Merit, mostly described as rank (*martaba*), was appealed to in negotiations about the character of the relationship and the amount of the remuneration. In 1207/8, the Ayyubid Sultan al-'Adil informed his vizier that he wished to add a second physician to the medical service of the troops who would cooperate with the current office holder Muwaffaq al-Din 'Abd al-'Aziz (d. 604/1208). The vizier recommended his own physician Muhadhdhab al-Din whom he informed about the promotion: "I praised you in front of the sultan and these thirty Nasiri dinar are for you every month in the service." Al-Muhadhdhab refused: "O my Lord, the physician Muwaffaq al-Din 'Abd al-'Aziz has every month one-hundred

7 S. Brentjes, 'Patronage of the mathematical sciences in Islamic societies. Structure and rhetoric, identities and outcomes,' in E. Robson, J. Stedall (eds.), *The Oxford Handbook of the History of Mathematics*, Oxford, Oxford University Press, 2008, pp. 301–328.

8 Ibn Abi Usaybi'a, *'Uyun al-anba' fi tabaqat al-atibba'*, N. Rida (ed.), Beirut, Dar Maktabat al-hayat, 1965, pp. 581f, 589, 591, 600, 630, 635, 637, 639, 646, 652, 661, 683f, 698, 718 et passim.

dinar and a [further] payment equivalent to them. I know my rank in the science and I do not serve below this settlement."[9] When al-Muhadhdhab had begun his career as a young doctor, the vizier Safi al-Din b. Shukr had taken him in his service paying him a comfortable income (*jamakiyya*) since "he knew . . . his rank in the medical arts."[10]

Companionship finally describes a particularly close relationship which demanded for instance a high degree of mobility. This meant that the physician was expected to travel with the prince wherever the latter went. If Ibn Abi Usaybi'a can be trusted, this special relationship demanded the consent of the client and led to no negative consequences if rejected, since several Ayyubid physicians refused to enter into such a companionship.[11] It is unclear though which other features separated companionship from a standard 'patron'–'client' relationship.

2 The problem of *khidma*

The differences between military and civilian terminology became increasingly blurred when forms of *khadama* (to serve, to be at service) pushed alternative parlance to the fringes. As *istakhdama* (to hire, to engage the service) shows it was already used in the classical period. Service (*khidma*) and its cognates became central terms apparently first with the smaller military dynasties in northern Iraq, Syria and perhaps Anatolia. The regional spread of this terminology as the dominant language of 'patronage' needs further study. It is by no means clear when and how this preponderance came into being. It is, however, evident that writers in Baghdad and cities of Ilkhanid Iran continued to use a much more diversified language.

What kind of relationships can *khidma* and its cognates describe? In a text on arithmetic needed by scribes and other practitioners Abu l-Wafa' Buzjani (994–998), a courtier of the Buyid emir 'Adud al-Dawla (r. 978–983) in Baghdad and a well-known scholar of the mathematical sciences, wrote that everybody who had knowledge (of something) approached the Buyid emir for "*khidmatihi*" trusting in his (giving) him charitable gifts and being honest towards him.[12] Here, "*khidmatihi*" cannot possibly mean the emir's service to the applicants in a literal sense nor does the grammatical construction seem to allow a translation like 'they approached him for being taken into his service'. Thus, it might be possible that "*khidmatihi*" referring to the emir signifies here his exercise of 'patronage' which then is specified by charitable gifts and honesty or integrity. Immediately afterwards, however, Abu l-Wafa' applies the verb *khadama* to himself in the sense of a service of a 'client' to his 'patron': "I had served him with composing a book . . ."[13]

9 Ibidem, p. 729.

10 Ibidem.

11 Ibidem, pp. 673, 696.

12 A. Salim Sa'idan, *Tarikh 'ilm al-hisab al-'arabi*, 'Amman, al-Jami'a al-urduniyya, 1971, p. 64.

13 Ibidem, p. 64.

In later sources, *khidma* seems to cover all sorts of 'services', military as well as civilian ones, religious as well as secular types. Its increasing application in Ayyubid and Mamluk sources to relationships between rulers and courtiers on the one hand and scholars, including physicians, astrologers and *muwaqqit*s (scholars who determined prayer times, the directions of prayer and the beginning of the new moon with mathematical and astronomical means and tools) on the other poses serious problems for the understanding of what constituted 'patronage' and what belonged to other categories of social relationships, if we do not wish to inflate the usage of the term "patronage" to any and sundry kind of *khidma*.

Ibn Abi Usaybiʿa's *suhba* (companionship) seems to indicate that *khidma* was a generic term for a relationship between two people in a hierarchy, while *suhba* describes a relationship of closer personal ties with more obligations. The Ayyubid historian and scholar of the mathematical and philosophical sciences Ibn Wasil (604–697/1208–1289), for instance, described both his relationship to his father as well as that to his Ayyubid royal 'patron' as *khidma*.[14] As I have suggested elsewhere, in Ibn Abi Usaybiʿa's portrayal of Ayyubid conditions *khidma* described a relationship that consisted of single professional acts in a case of sickness, but excluded long-term, personal obligations of shared time beyond the immediate treatment of the disease.[15] The example of Muhadhdhab al-Din quoted above shows that *khidma* positions were offered, not only sought. In the case of al-ʿAdil, his vizier acted as the broker, the HR director, so to speak. In cases of *suhba* Ibn Abi Usaybiʿa's language seems to suggest that the invitation came directly from the "patron" to the "client". When *khidma* positions were on offer in the Ayyubid period, they could be negotiated. Since invitations to *suhba* were also negotiated, negotiation seems to have been a characteristic element in the relationships between Ayyubid princes and their elite doctors. The doctors' potential for negotiation depended on their professional skills, rank and information. Muhadhdhab al-Din could not have refused the offered position, if his professional self-confidence had not been well developed nor could he have challenged the offered salary if he had not known the salary of his colleague. No retaliation from the vizier against his unwilling and ungrateful 'client' apparently occurred. Muhadhdhab al-Din rather became a very successful physician of several Ayyubid princes.

The problem to understand what kinds of relationship were subsumed under the term *khidma* continues for the Mamluk period. Four examples from Shams al-Din al-Sakhawi's biographical dictionary illustrate the range of difficulties.

a. Was the Christian secretary, employed or appointed by a Mamluk officer, his 'client' or his (well-paid) 'servant'?

   Ibrahim b. ʿAbd al-Razzaq al-Iskandari, known as Ibn Ghurrab; his origin was in Cairo in a family of Coptic secretaries; he bound himself to the service

14 S. Brentjes, *op. cit.*, 2010, p. 344.
15 Ibidem, p. 330.

for Mahmud al-Ustadar (majordomo) and distinguished himself in it and was promoted until he was the head of administration of the private (properties/lands) before he was 20 years old replacing Sa'd al-Din Abu l-Faraj b. Taj al-Din Musa in Dhu l-Hijja 798 h. . . . he continued to rise becoming the head of the secret correspondence and related matters; finally Sultan al-Zahir Barquq (1382–89, 1390–99) held him in high prestige and his son al-Nasir Faraj gave him the long-term (?) post (*istaqarra bi*) of supervising the army in addition to the private (properties/lands) and other things . . . and (appointed him) together with his brother to the vizierate . . .[16]

b. A scholar employed (?) and promoted a scholar (?)

Ibrahim b. 'Umar al-Baqa'i, an inhabitant of Cairo, then of Damascus . . ., he studied a bit of astrology, . . . he took *fiqh* (law) in Cairo with a number of scholars, among them al-Sakhawi's teacher Ibn Hajar al-'Asqalani; . . . he earned his living as a witness for one of his sheikhs, namely al-Fakhr al-Asyuti, and others and by copying (manuscripts), teaching small children and other (things). He travelled in the service of my sheikh (i.e., Ibn Hajar) to Aleppo where he took from the sheikhs of transmission (i.e., the transmission of *hadith*) as well as elsewhere . . . My sheikh promoted him by nominating him in the lifetime of al-Zahir Jaqmaq (1438–53) to the reading of *hadith* in the fortress . . .[17]

c. Scholars and the wife of a Mamluk officer employ (?) a scholar and, occasionally, also his wife

Ahmad b. 'Ali al-Qahiri al-Hanafi, *khadim* (domestic servant, attendant, 'client'?) of al-Amin al-Aqsaray'i . . . "he rose in the service of the sheikh and in his attendance (*mulazama*) in the *hajj*, the closeness at the two *haramayn* (Mecca and Medina) and other things. He was present in his lectures and he did not betray (?, *anfaka*) him until he died after he had given him the permission to teach and to give *fatwās* according to what was said. He became very wealthy through his affiliation to him and his appointments to many posts. He attended the service at the *Ashrafiyya* (madrasa) as a deputy, and he desired his independence from it after the death of its holder. After his sheikh (had died), he joined the sister of al-Muhibb, who was the wife of the *majordomo* in the year 87. He was entrusted with the external affairs and his wife with the internal affairs so that the two did not interfere in each other's business until she died. Likewise, he attended the service of al-Burhani al-Karaki, the Imam, until he became entrusted, in the days of his disappearance, with the revenues of his domains, and he wrested them from him as tenure (? *wa-ntaza'aha minhu al-mulk*).[18]

16 Shams al-Din al-Sakhawi, *Al-Daw' al-lami' li-ahl al-qarn al-tasi'*, Cairo, s.d., vol. 1, p. 65.

17 Ibidem, pp. 101–102.

18 Ibidem, vol. 2, p. 7.

d. A Muslim scholar worked (?) for a (Christian?) secretary in the Mamluk administration: "Ahmad b. Muhammad al-Qahiri al-Shafiʿi, was deputy judge, served Ibn al-Kuways, the *katib al-sirr* (secretary of the secret correspondence), then Ibn Mazhar, and he acquired reputation and landed property."[19]

These four examples indicate that the boundaries between 'patronage' and 'servitude' had become fluent. I know, however, too little about the social and economic differences between labor, employment, service and other kinds of dependencies in Islamic societies for offering more than the following observations. The duration of either type of relationship obviously was not a primary criterion of differentiation between them. The rank of the superior person per se with whom the relationship was formed also does not clarify which type of social format the relationship possessed. High ranking Mamluk officers, their wives, the sultans themselves, but also scholars were served and promoted those who served them. The "servants" could come from well-regarded families or could be poor. Becoming rich and the owner of property/land, however, was certainly an important ingredient of scholarly 'service' in the higher echelons of Mamluk society. The rejection of service for the powerful that can be found so often among religious scholars either stressed as a praiseworthy behavior or indeed exercised had apparently lost its appeal. 'Clients' were praised when proving trustworthy, but as example d shows also looked for enriching themselves as did 'patrons'.

3 'Patronage' and the madrasa

Madrasas, mosques, houses of *hadith*, Sufi *khanqah*s and cognate institutes plus hospitals were the only institutional recipients of 'patronage' in the way they were set up, namely by a religious donation (*waqf*). Whether observatories should or could be regarded as another such institutional recipient depends on the formal status, i.e., whether a *waqf* certificate was signed when they were founded or extended. In Ilkhanid Maragha this seems to have been the case. Institutional 'patrons', on the other hand, did not exist. The 'patronage' for institutions contained in the way how the institutions were run always an element of a person-to-person patronage. The act of founding such an institution by *waqf* included donating professorships, posts for physicians, *muwaqqit*s and other personnel, depending on the purposes and intentions of the donor. These positions needed to be filled. The appointment procedures for new office holders were regulated, but in practice not always followed. The concept of the *waqf* prescribed that the donor or an administrator chosen by him or her alone had the right of appointing office holders. After his or her demise, a successor, also chosen by the donor, would take over this function. In particular in this period after the death of the original donor,

19 Ibidem, p. 102.

other powerful people either from the ruling dynasty or from among the scholars often interfered in this process not merely by recommending or choosing office holders against the wishes of the succeeding administrator/s, but, as is well known in particular for the Mamluk period, by more violent means of interference such as bribery, denunciation, theft or even murder.[20] All these kinds of meddling with the legal stipulations of the *waqf* gave many opportunities to people engaged in powerful and well-functioning 'patron'-'client' relationships. Outsiders, however, were not the only people who tried to influence the choice of a new professor or other office holder. Those who actually held a post as a professor undertook various activities to ensure that the post remained either in their families or in the hands of a preferred student. They also engaged in numerous efforts, legal and illegal, to amass themselves as many posts as possible. As example c above shows they hired deputies who took over their teaching obligations. Insofar we can understand the relationship of *khidma* between two scholars as one of deputizing of a younger or less successful scholar for an older one who had been more successful in the struggle for posts. The reasons for this massive social networking among the professorial elite and those who wanted to enter it were the monetary and cultural gains that these posts offered. The donation certificate stipulated a regular payment and its amount in addition to certain types of work the office holder was supposed to carry out such as reading the Qur'an, teaching a certain subject to a certain number of students, determining prayer times etc. Thus, a position at a madrasa or another one of the previously mentioned institutions seems to have been much more stable and formally closer to a contractual employment than a position at court or in the administration. The more positions a professor, a *muwaqqit*, an Imam or a physician could acquire the higher his income and the greater his fame, bargaining power and personal network. *Muwaqqit*s as well as physicians did not hold only positions in their specialization, but could combine them with professorships in the various law-schools.[21] Hence, they too were beneficiaries and exploiters of the madrasa system and its opportunities of "patronage" relationships. This is made visible too by the fact that the biographical entries which were dominated by a vocabulary reflecting the studying and teaching purposes of the scholarly system in many cases also include elements of the "patronage" language and that in addition to rulers, courtiers, the military and their wives or daughters also scholars acted now in the capacity of 'patrons' and took other scholars not only as adjuncts or junior companions, but as 'servants', i.e. most likely 'clients', as examples b and c above imply.

20 M. Chamberlain, *Knowledge and Social Practice in Medieval Damascus, 1190–1350*, Princeton, Princeton University Press, 1992, pp. 91–107; S. Brentjes, 'Reflections on the role of the exact sciences in Islamic culture and education between the twelfth and the fifteenth centuries,' in M. Abattouy (ed.), *Etudes d'histoire des sciences arabes*, Casablanca, Fondation du Roi Abdul-Aziz Al Saoud pour les Etudes Islamiques et les Sciences Humaines [*sic*], 2007, pp. 15–33, in particular pp. 29–30.

21 Al-Nu'aymī, *Al-Dāris fī tārīkh al-madāris*, J. al-Ḥasanī (ed.), Beirut, Maktabat al-Thaqāfa al-Dīniyya, 1988, 2 vols., vol. 1, p. 519, vol. 2, p. 131.

In addition to the new pecuniary and cultural possibilities for the educated to devote themselves to their studies and writings and to form allegiances and power centers, the language of 'patronage' also grew richer. New terms seem to have been for instance to appoint s.o. to, to place s.o. in (*qarrara hu or fi*), to settle down, to take or give a permanent place of living, i.e., a permanent position (*istaqarra bi*) and to devote or dedicate an honor or service, to privilege s.o., to make s.o. special (*ikhtassasa bi*). But as example a above indicates they were also applied to the secretarial 'class'. This means more research about the role of the teaching institutions and their relationships to the members of the secretarial 'class' is needed for a more precise and reliable understanding of the 'language of patronage'.

4 The arts as markers of patronage for scholars

In addition to dedications, biographical dictionaries and relevant remarks in historical chronicles, the quality of the handwriting, in particular if executed in one or more styles of calligraphy, and various kinds of illustrations serve as markers that not only the text, but the particular copy or manuscript may have been an element of a 'patronage'-relationship. As art historians have argued since the early twentieth century, the translation of Greek, Syriac and Middle Persian scientific manuscripts included also the transfer of frontispieces, miniatures and other art forms.[22] Although no illustrated scientific manuscripts are extant from the translation period (8th–10th centuries) except for the rare folios from Dioscurides' *Materia Medica*, some of the few extant exemplars from the twelfth and thirteenth centuries were undeniably produced in a courtly context. One of these splendidly illustrated works is a manuscript of Ibn al-Razzaq al-Jazari's book on automata and other mechanical devices (1205/6).[23] Another one is the Pseudo-Galenic *Book of the Antidotes* (*Kitab al-Diryaq*) of which two manuscripts exist in Paris (date 1198/9) and Vienna (13th c.), the calligraphy and painting of which are superb (in both cases in Seljuq 'style').[24] From the Ilkhanid dynasty onwards, more material has survived to establish a safer basis for the importance of art as a marker

22 K. Weitzmann, 'The Greek sources of Islamic scientific illustration,' in G. Miles (ed.), *Archaeologica Orientalia in Memoriam Ernst Herzfeld*, Locust Valley, New York, J.J. Augustin, 1952, pp. 244–266; R. Ettinghausen, *Arab Painting*, Geneva, Skira, 1962; R. Ettinghausen et al., *The Art and Architecture of Islam*, New Haven, Yale University Press, 1987; E. Grube, 'Materialien zum Dioskurides Arabicus,' in R. Ettinghausen (ed.), *Aus der Welt der islamischen Kunst, Festschrift für Ernst Kühnel*, Berlin, Gebr. Mann, 1959, pp. 163–93; E.R. Hoffman, 'The author portrait in thirteenth-century Arabic manuscripts: A new Islamic context for a late-antique tradition,' *Muqarnas*, vol. 10 (1993), pp. 6–20, pp. 6, 8–9; E.R. Hoffman, 'The beginnings of the illustrated Arabic book: An intersection between art and scholarship,' *Muqarnas*, vol. 17 (2000), pp. 37–52.

23 D.R. Hill (ed. and transl.), *The Book of Ingenious Mechanical Devices*, Dordrecht, D. Reidel, 1974.

24 MS Paris, BnF, Arabe 2964; MS Vienna, ÖNB, Codex A.F. 10 See also J. Kerner, 'Art in the name of science: The Kitab al-Diryaq in text and image,' Abstract, SOAS Conference: *Arab Painting: Text and Image in Illustrated Arabic Manuscripts*, 2004, pp. 13–14; www.qantara-med.org/qantara4/public/show_document.php?do_id=1496&lang=en, accessed 2 June 2011.

for "patronage" for the mathematical and other sciences. In particular the great and powerful dynasties like the Timurids, the Safavids, the Mughals and the Ottomans sponsored art workshops at their courts where many excellently illustrated scientific manuscripts were produced. Examples are Arabic, Persian and Ottoman Turkish versions of ʿAbd al-Rahman al-Sufi's (903–986) *Book of the Constellations of the Fixed Stars* (*Kitab suwar al-kawakib al-thabita*) written for the already mentioned Buyid emir ʿAdud al-Dawla in the tenth century, Arabic and Persian versions of Dioscurides' *Materia Medica*, astronomical handbooks like the *Ilkhanid Tables* produced by a team of scholars at the Ilkhanid court of Maragha in the thirteenth century or the so-called *New Royal Tables* or *Tables of Ulugh Beg* produced by another team of scholars with some participation of the Timurid prince Ulugh Beg (1394–1449) in the early fifteenth century in Samarkand, works on planetary theory like Qutb al-Din Shirazi's *The Royal Gift* (*al-Tuhfa al-shahiyya*) written for Taj al-Din Muʿtazz b. Tahir, the vizier of Amir-Shah Muhammad b. Sadr al-Saʿid (13th c.) or Shams al-Din Khafri's (d. c. 1550) commentary on Nasir al-Din Tusi's (1201–1274) astronomical textbook *The Memoirs* (*al-Tadhkira*), both produced in Safavid Iran in the sixteenth and seventeenth centuries.[25] Many of these illustrated scientific texts were originally products of 'patronage' for the sciences. Now they were favored outlets for such activities as elements of education, research, cultural policy or political ambitions. Nasir al-Din Tusi, Ulugh Beg, the Safavid governor of Mashhad, Manuchihr Khan (d. 1636), and the Safavid governor of Azerbaijan in this period, Mirza Muhammad Ibrahim (d. 1699), contributed as patrons and the latter also as a scholar to the production of copies of Sufi's book on the star constellations in its entirety or partially (only the images).[26] The copies made for Ulugh Beg and Manuchihr Khan, today extant in the National Library of France in Paris, the New York Public Library and the Dar al-Kutub in Cairo, belong to the most splendidly illustrated ones of this work.[27]

25 MS Paris, BnF, Arabe 5036; P. Kunitzsch, 'The astronomer al-Ṣūfī as a source of Uluġ Beg's Star Catalogue (1437),' in Ž. Vesel, H. Beikbaghan et B. Thierry de Crussol des Epesse, *La science dans le monde iranien à l'époque islamique*, Tehran, Institut Français de Recherche en Iran, 1998, pp. 41–47, in particular pp. 42–3; reprinted in P. Kunitzsch, *Stars and Numbers, Astronomy and mathematics in the Medieval Arab and Western Worlds*, Variorum CS 791, Aldershot, Burlington, Ashgate, 2004, xii; Ž. Vesel, 'Science and scientific instruments,' in N. Pourjavady (gen. ed.), *The Splendour of Iran*, 3 vols.; vol. iii: C. Parham (ed.), *Islamic Period: Applied and Decorative Arts: The Cultural Continuum*, London, Booth-Clibborn Editions, 2001, pp. 268–307, 38 colored images; A. Soudavar, *Art of the Persian Courts*, New York, Rizzioli, 1992, p. 67.

26 B. Schmitz, *Islamic Manuscripts in the New York Public Library*, with contributions by Latif Khayyat, Svat Soucek, Massoud Pourfarrokh, New York, Oxford, Oxford University Press and The New York Public Library, 1992, pp. 55, 123–124, figure 127; D.A. King, *A Survey of the Scientific Manuscripts in the Egyptian National Library*, Cairo, The American Research Center in Cairo, Inc., 1986, pl. iii; Kunitzsch, *op. cit.*, 1998, p. 43; F. Richard, *Splendeurs persanes. Manuscrits du XII e au XVIIe siècle*, París, Bibliothèque Nationale de France, 1997, p. 78.

27 B. Schmitz, *op. cit.*, figure 127; MS Cairo, Dār al-Kutub, MMF9, D.A. King, *op. cit.*; MS Paris, BnF, Arabe 5036, http://mandragore.bnf.fr/jsp/rechercheExperte.jsp, Richard, *Splendeurs persanes*, p. 78.

A number of these manuscripts and a few others on astrology and technology were illustrated with paintings that show a close relation to the styles developed in Isfahan. These manuscripts include the undated *Royal Gift* by Qutb al-Din Shirazi mentioned above; three copies of ʿAbd al-Rahman Sufi's *Book of the Constellations of the Fixed Stars*, two of which were made on order of Manuchihr Khan between 1630 and 1634 in Mashhad, while one is undated, but placed by Anthony Welch in the middle of the seventeenth century; at least two copies of the Arabic text of this work, one produced possibly around 1630, the other being undated; a new Persian translation of Dioscurides's *Materia Medica* ordered by Shah ʿAbbas I (r. 1588–1629); at least three Persian copies of the *Materia Medica* made in 1645 and 1657; and a work about the astrological meanings of each of the 360° degrees of the sky made in Isfahan in 1663.[28] In addition to these copies, there are other illustrated scientific manuscripts produced in seventeenth-century Safavid Iran that may or may not be related to the art of the court.[29] A further feature of this triangular relationship between 'patrons', the arts and scholars was the emergence of pictures of scholars linked in 'patronage' relationships with the Ilkhanid, Timurid or Safavid courts. Examples are miniatures of Nasir al-Din Tusi, the vizier of *waqf* at the court of the first Ilkhanid ruler of Iran, Hülägü Khan (r. 1256–1265), and director of the observatory in Maragha, and the Safavid head physician Muzaffar b. Muhammad al-Hasani al-Shifaʾi.[30]

Examples of this role that courts and their art workshops played from the thirteenth to the seventeenth centuries in the production of luxury versions of scientific texts and that involved in several cases the participation of a court astronomer/astrologer or other scholars as a translator, commentator or copyist can also be found in connection with the courts of other dynasties than the Ilkhanids, Timurids and Safavids. Hence, a closer attention of historians of science in Islamic societies to illustrated manuscripts as suggested by Vesel[31] and well argued for by her in her chapter 'Science and Scientific Instruments' in the *Splendor of Iran* and the *Images of Islamic Science* is urgently needed.[32]

28 B. Schmitz, *op. cit.*, p. 61; A. Welch, *Collection of Islamic Art: Prince Sadruddin Aga Khan*, Ginebra, Chateau de Bellerive, 1972, 4 vols., vol. 2, pp. 60–61, 69–70, 122–123; Ž. Vesel, *op. cit.*, 2001, pp. 293–297.

29 See, for instance, B. Schmitz, *op. cit.*, p. 61.

30 For Tusi see F. Richard, 'Les "portraits" de Naṣīr al-Dīn Ṭūsī,' in N. Pourjavady, Ž. Vesel (eds.), *Naṣīr al-Dīn Ṭūsī, Philosophe et Savant du XIII Siècle*, Bibliothèque Iranienne 54, Tehran, Institut Français de Recherche en Iran, Presses Universitaires d'Iran, 2000, pp. 199–206, 4 images on pp. 202–206; for Shifaʾi see for instance MS Paris, BnF, Supplément persan 1572, f 3.

31 Ž. Vesel, *op. cit.*, 2001, pp. 268–307; Ž. Vesel, S. Tourkin & Y. Porter (eds.), avec la collaboration de F. R. et F. Ghasemloo, *Images of Islamic Science: Illustrated Manuscripts from the Iranian World*, Tehran, UNESCO-IFRI-Fondation M. Van Berchem-Azad University, 2009.

32 Este trabajo se ecuadra en el proyecto de Excelencia de la Junta de Andalucía código P07-HUM02594, "Estudios sobre Historia y Filosofía de las Ciencias Físicas y Matemáticas".

7

THE STUDY OF GEOMETRY ACCORDING TO AL-SAKHĀWĪ (CAIRO, 15TH C) AND AL-MUḤIBBĪ (DAMASCUS, 17TH C)

1 The geometrical disciplines, texts, and authors listed by Shams al-Dīn al-Sakhāwī

Shams al-Dīn al-Sakhāwī (830–902/1427–1497) was born and raised in a prominent family of scholars who originally came from Sakhā, a town in the central Nile Delta, and had settled in the Fatimid district of Cairo. He studied *ḥadīth*, i.e., the Prophetic traditions, and related Islamic disciplines with several prominent teachers, the most famous among them being Ibn Ḥajar al-ʿAsqalānī (773–852/1373–1449). He also studied other disciplines of what is considered today as part of the formal education of Sunni scholars such as *fiqh*, *uṣūl*, grammar, rhetoric, or metrics, although not every student indeed engaged in their study.[1] In the entry dedicated to his own biography in his major work, the biographical dictionary *al-Ḍawʾ al-lāmiʿ li-ahl al-qarn al-tāsiʿ* (*The Shining Light of the People of the Ninth Century*), al-Sakhāwī reports that he studied further fields that are not considered canonical in modern scholarship but attracted a stable clientele in the Mamluk period such as history, logic, arithmetic, or *ʿilm al-mīqāt* (astronomical timekeeping and related issues). He took the last two disciplines together with *ʿilm al-farāʾiḍ* (calculation of inheritance shares) in Cairo with several teachers, among them one of the leading experts in these fields, Aḥmad b. Rajab b. Ṭaybughā, Shihāb al-Dīn, known as Ibn al-Majdī (767–850/1365–1447).[2]

1 Petry, C.F., ""al-Sakhāwī". *Encyclopaedia of Islam*. Edited by: P. Bearman, Th. Blanquis, C. E. Bosworth, E. van Donzel and W. P. Heinrichs. Brill, 2007. Brill Online. 23 January 2007 <www.brillonline.nl/subscriber/entry?entry=islam_SIM-6503>

2 Shams al-Dīn al-Sakhāwī, *al-Ḍawʾ al-lāmiʿ li-ahl al-qarn al-tāsiʿ*. Bayrūt, s.d., vol. 8, pp. 3–4, vol. 9, p. 10.

DOI: 10.4324/9781003372455-8

In addition to the dictionary about famous male and female scholars of the fifteenth century, al-Sakhāwī wrote about history, biography, historiography, *ḥadīth*, and various other religious topics to which he added some treatises on geography, grammar, arithmetic, and prophetic medicine.[3] He held several posts at prestigious educational institutions in Cairo. Al-Sakhāwī travelled widely in the Mamluk domain and undertook several pilgrimages to Mecca and Medina. He spent the last years of his life in Medina, where he died.

Al-Sakhāwī's biographical dictionary contains some ten thousand biographical entries. Only a minority of them include any of the mathematical disciplines known at the period. The disciplines most often studied are *ʿilm al farāʾiḍ*, *ʿilm al-ḥisāb* (arithmetic), *ḥisāb al-jabr wa-l-muqābala* (algebra), and *ʿilm al-mīqāt*. Geometrical disciplines, whether theoretical, i.e., *ʿilm al-handasa*, or practical, i.e., *ʿilm al-misāḥa* (surveying), appear less often. This general picture fits well with the manuscripts written by the scholars and their students mentioned by al-Sakhāwī, among them Jamāl al-Dīn al-Māridānī (d. 809/1406), Shihāb al-Dīn Ibn al-Hāʾim (c. 756–815/1355–1412), and Badr al-Dīn al-Māridānī,[4] known as Sibṭ al-Māridānī, and extant today in libraries in Cairo, Damascus, Istanbul, Berlin, Paris, Oxford, and elsewhere. Only one of these leading representatives of the four mathematical sciences ranking highest in al-Sakhāwī's dictionary is listed as the author of a text on (practical) geometry in Rosenfeld and Ihsanoğlu (2003), namely, Ibn al-Hāʾim.[5] Neither did al-Sakhāwī list geometrical writings by any of them.[6] This observation is confirmed by the manuscripts extant today in Cairo's major manuscript library and catalogued by King (1986).[7] But given the amount of geometrical knowledge and skill needed for the study of timekeeping and the construction of astronomical instruments it is highly unlikely that no classes were taught at least on an elementary level.

A closer look into the biographies compiled by al-Sakhāwī shows that the access that extant manuscripts provide us to the past is partial and incomplete. The glance into the past through manuscripts is not merely handicapped because manuscripts have been lost and destroyed. Al-Sakhāwī's dictionary suggests that not all subjects taught by scholars of his period were codified through teaching material written by them or their students. Certain themes were obviously taught on the basis of texts by other authors, which may have been annotated in class rather than copied or dictated. Such an interpretation of the lack of geometry among the extant

3 al-Sakhāwī (see footnote 2), vol. 8, pp. 16–19.

4 I chose this spelling in accordance with al-Sakhāwī (see footnote 2), vol. 5, p. 19 who explained it as a *nisba* related to the mosque al-Māridān.

5 Rosenfeld, Boris A.; Ihsanoğlu, Ekmeleddin: Mathematicians, Astronomers & Other Scholars of Islamic Civilisation and their Works (7th–19th c.). Istanbul, IRCICA, 2003, pp. 261, 263–264, 276–277, 293–298.

6 al-Sakhāwī (see footnote 2), vol. 1, pp. 300–302; vol. 2, pp. 157–158; vol. 5, p. 19; vol. 8, pp. 35–36.

7 King, David A.: A Survey of the Scientific Manuscripts in the Egyptian National Library. Winona Lake, Indiana, Eisenbrauns, 1986, C47, 58, 62, 97.

texts on mathematical sciences written by teachers and students of inheritance mathematics, arithmetic, algebra, and timekeeping is supported when al-Sakhāwī occasionally writes that someone studied with a teacher, while arithmetic was studied by listening to another teacher.[8] The fact that King's catalogue (1986) shows not even copies of standard texts of theoretical geometry from the ninth/fifteenth century, but only copies of such texts from earlier (seventh/thirteenth century) or later (eleventh–thirteenth/seventeenth–nineteenth centuries) times, poses an interesting puzzle. To dismiss it by simply declaring this lack as the result of physical losses over time is not very satisfying. A more systematic analysis of extant manuscripts and their local affiliations, if given, is needed to understand whether this lack of copies of writings about theoretical geometry reflects a social phenomenon linked to the kind of persona that teachers of mathematical sciences adhered to in fifteenth-century Cairo. It is possible that we find here a reversal of the usual situation where theoretical treatises, or, as Høyrup (2003) called them, the books of (great) mathematicians, were preserved thanks to their specific social environment which included reification through writing, while the mathematical knowledge and skills of those who earned a living through practicing such knowledge and skills outside the classroom are lost due to its primarily oral transmission.[9] Theoretical geometry seems to have functioned in fifteenth-century Cairo as propaedeutic knowledge that did not need to be written down.

The number of scholars reported by al-Sakhāwī to have studied geometrical texts with a teacher is very small.[10] Almost all of them favored theoretical geometry. Al-Sakhāwī's entries testify to this choice either by listing *handasa* as one of their fields of study or by mentioning explicitly a few authors and texts they read with their teachers, namely Euclid (c. 365–c. 300 BCE), Naṣīr al-Dīn Ṭūsī's (599–674/1201–1274) *Taḥrīr* of the *Elements*, Shams al-Dīn al-Samarqandī's (second half seventh/thirteenth century) *Ashkāl al-ta'sīs*, and Qāżīzāde al-Rūmī's (d. c. 845/1440) commentary on al-Samarqandī's treatise. While it has to be kept in mind that al-Sakhāwī was primarily interested in *'ilm al-ḥadīth* (transmission of the sayings of the prophet and his companions) and history, strove to delineate his own place in Cairene and Meccan intellectual history of the ninth/fifteenth century, and aimed to define the members of the intellectual elite in urban centers of Egypt, al-Shām (Syria), and al-Ḥijāz, the slight differences he made in the terminology for and presentation of the fields, authors, and texts for different Islamic societies and regions should not be considered as meaningless. Some of these little differences correspond well with the geographical accumulation of authors and disciplines as represented by manuscripts extant today. Others challenge historical

8 al-Sakhāwī (see footnote 2), vol. 1, p. 175.

9 Høyrup, Jens: Practitioners – –school teachers – –'mathematicians': the division of pre-Modern mathematics and its actors. Talk given at Les Treilles 2003, p. 8.

10 See appendix I for those entries that list geometrical texts.

images presented by Islamic historians of the past and often accepted too willingly by us today.

Examples of the first type are al-Sakhāwī's exclusive use of the term *riyāḍiyyāt* (mathematics) in studies or teachings of Iranian scholars and his tendency to be slightly more specific when talking about authors and texts studied in a number of disciplines, among them *handasa*, *hayʾa*, and *falsafa* or *ḥikma* by students from the larger Iranian world.[11] In biographical entries about teachers and students from the Arab world, al-Sakhāwī focuses his eye for details on arithmetic and timekeeping, their various subgenres, and the local texts studied in this context. These differences confirm the impression of a greater regularity and quantity of study in respect to the theoretical branches of mathematics and philosophy in the larger Iranian world in comparison to Egypt, Syria, or the Arabian Peninsula after 1200, an impression gleaned from manuscripts and chronicles. They also imply that in al-Sakhāwī's view the scholarly communities of Cairo, Alexandria, Damascus, or Mecca thought differently about how the domain of mathematical studies was constituted. They did not see this domain as one coherent discipline neatly structured as a set of boxes filling a closet one after the other in their appropriate places, each box containing its own proper set of well-arranged boxes. They did not teach a standardized set of rules, methods, skills, and concepts, but rather learned them in the same kind of empirical way as we teach today history, cultural studies, or other disciplines of the humanities – by reading authoritative texts. Since individual teachers may have had different understandings of these texts, it was fully appropriate to study the same texts with different teachers. As a result, the teachers in al-Sakhāwī's dictionary teach almost always the same set of texts on the basics of arithmetic, algebra, or timekeeping and the students study them with more than one teacher, even if the one teacher was the acclaimed authority of the period.

An example of the second type is al-Sakhāwī's biography of the Maghribī scholar Muḥammad b. Muḥammad [. . .] al-Zawāwī al-Bijāyī (821 or 22–c. 864/1418–1460).[12]

In contrast to Ibn Khaldūn's (734–809/1332–1406) claim of the sharp decline in the study of the theoretical sciences in the Maghrib, this scholar undertook a broad range of studies in logic, dialectics, philosophy, geometry, mathematical cosmography, mechanics, number theory, theoretical music, optics, burning mirrors, and the theoretical fields of the religious and philological disciplines in Tlemcen and also in Bijāya.[13]

The texts on theoretical geometry taught and studied according to al-Sakhāwī during the ninth/fifteenth century were primarily Euclid's *Elements*, al-Samarqandī's *Ashkāl al-taʾsīs*, Qāḍīzāde al-Rūmī's commentary on the latter text, and possibly commentaries by other authors on this text. They were taught

11 See for instance al-Sakhāwī (see footnote 2), vol. 10, pp. 7, 48.

12 al-Sakhāwī (see footnote 2), vol. 9, pp. 180–188.

13 al-Sakhāwī (see footnote2), vol.9, pp. 181–182.

in Herat, Isfahan, possibly Shirvan, Cairo, Alexandria, and possibly in Damascus. The professors who taught the *Elements*, *Ashkāl al-taʾsīs*, and commentaries on the latter in Cairo were Jamāl al-Dīn al-Māridānī, Ibn al-Majdī, and al-Jamāl al-Kūrānī (9/15th c.). Jamāl al-Dīn al-Māridānī and Ibn al-Majdī were leading teachers of *ʿilm al-farāʾiḍ*, arithmetic, algebra, and *ʿilm al-mīqāt*, the latter having studied timekeeping and its affiliated branches with the former.[14] Ibn al-Majdī is ranked highly by al-Sakhāwī as someone "with whom the nobles of each *madhhab* (law school), one rank after the other, took (classes)", cherished as *al-ustādh* (the master), and commemorated as an epithet in the name of one of his successful students, al-Sayyid ʿAlī Tilmīdh Ibn al-Majdī (c. 808–870/1405–1465).[15] In addition to the fields for which Ibn al-Majdī left manuscripts behind, he did not only teach theoretical geometry and mathematical cosmography, but also philosophy (*ḥikma*), *Shāfiʿī* law, and Arabic. His teacher Jamāl al-Dīn al-Māridānī taught in addition to the major branches of mathematics mentioned above *ʿilm al-falak* and *ʿilm al-hayʾa*, being praised for his competence in the latter branch.[16] Al-Sakhāwī confirms that his teacher Ibn Ḥajar al-ʿAsqalānī studied with Jamāl al-Dīn al-Māridānī *ʿilm al-mīqāt* and *ʿilm al-hayʾa*.[17] Other entries in al-Sakhāwī's dictionary mention the study of *handasa* without specifying the texts or authors read. An example for this almost silent ritualistic presentation of a teacher's academic profile in the theoretical disciplines of mathematics is a scholar often mentioned by al-Sakhāwī as a teacher of religious disciplines and fields of *adab* – Muḥammad b. Sulaymān al-Kāfiyājī (before 790–c. 879/1388–1474).[18] He was one of al-Sakhāwī's teachers and ranked high among the scholars of his time, but is basically ignored by historians of mathematics.[19] Al-Sakhāwī informs us that al-Kāfyiājī was a leading scholar of his age in several religious and philological disciplines, logic, *ḥikma*, medicine, and several mathematical disciplines, namely *ʿilm al-hayʾa*, *handasa*, spherics, optics, and burning mirrors.[20] Al-Kāfiyājī taught some of these disciplines, including theoretical geometry, in Cairo to at least four students, among them Jalāl al-Dīn al-Suyūṭī (849–911/1445–1505), but we hear nothing by al-Sakhāwī of the texts or authors al-Kāfiyājī read in his classes.[21] Among al-Kāfiyājī's own works, which al-Sakhāwī estimates as more than one hundred, the biographer lists one astronomical text, a commentary on Maḥmūd Čaghmīnī's (fl. 620/1223) *al-Mulakhkhaṣ fī l-hayʾa*.[22] This text is registered by Rosenfeld and Ihsanoğlu (2003), who also list one text on geometry and surveying

14 al-Sakhāwī (see footnote 2), vol. 1, p. 300.
15 al-Sakhāwī (see footnote 2), vol. 1, p. 300; vol. 4, p. 165; vol. 5, p. 285.
16 al-Sakhāwī (see footnote 2), vol. 2, p. 152; vol. 4, p. 19.
17 al-Sakhāwī (see footnote 2), vol. 4, p. 19.
18 This is the spelling of al-Sakhāwī (see footnote 2), vol. 8, p. 26.
19 al-Sakhāwī (see footnote 2), vol. 8, p. 26.
20 al-Sakhāwī (see footnote 2), vol. 7, p. 261.
21 al-Sakhāwī (see footnote 2), vol. 4, p. 164; vol. 9, pp. 24, 179.
22 al-Sakhāwī (see footnote 2), vol. 7, p. 260.

extant in Cairo, *Ḥall al-ishkāl fī mabāḥith al-ashkāl*.[23] According to King (1986), it is not al-Kāfiyājī's treatise that is extant in Cairo, but the fragment of a commentary on this text.[24] Rosenfeld and Ihsanoğlu (2003) characterize al-Kāfiyājī as a Turkish theologian and astronomer. While terminological clarity is admittedly difficult to achieve, al-Sakhāwī's dictionary shows the much broader scope of madrasa-based education in Cairo during the fifteenth century. Al-Kāfiyājī taught Quranic exegesis, *uṣūl al-dīn* (the fundamentals of religion), *uṣūl al-fiqh* (fundamentals of law), logic, *ʿilm al-kalām* ('rational theology'), several philological disciplines, among them rhetoric, mathematical cosmology, and theoretical geometry. He embraced apparently almost all those domains of knowledge available in Islamic societies of the Middle East and thought of by many historians and historians of science even nowadays as separated, if not even isolated from each other.

The last point that can be made on the basis of al-Sakhāwī's dictionary concerns the combination of disciplines with which geometry was taught and studied. This aspect can be studied from a broader and a narrower perspective. The broader perspective would situate the study of geometry in the entire context of taught fields, texts, and authors, an approach that goes far beyond the boundaries of this paper. The narrower perspective delineates the immediate environment in which al-Sakhāwī has placed the geometrical fields. This immediate environment is marked by linguistic and classificatory means as lists of fields studied with one or more teachers. The individual entries are collected in Appendix 2. Occasionally I added fields of study which al-Sakhāwī separated formally from the immediate environment of geometry because they too were seen as belonging or related to the numerous fields of mathematics. Examples are geomancy, letter magic, or the construction of curves on astronomical instruments.

The entries in al-Sakhāwī's dictionary that list *handasa* or *misāḥa* as fields of education are more numerous than those that mention specific texts and authors. Although this is no surprise since this style of reporting on education dominates all the fields of study in biographical dictionaries of the period, it shows that the study of both kinds of geometry in classes set apart from timekeeping, arithmetic, algebra, and other mathematical and astronomical themes was much more widespread in ninth/fifteenth-century Arab cities than the reference to specific texts and authors indicates. Since scholars such as Jamāl al-Dīn al-Māridānī or Jamāl al-Dīn al-Kūrānī are not reported among the students of *handasa*, but later taught this topic, it is clear that more people got at least some training in geometry than is made explicit by the biographer.[25] This visible lack of comprehensiveness is an argument against using biographical dictionaries for gauging the quantity of involvement with the mathematical sciences. Their richness in qualitative information, on the other hand, can barely be overestimated. The disciplines

23 Rosenfeld, Ihsanoğlu (see footnote 5), p. 291.
24 King (see footnote 7), C 71, p. 416.
25 al-Sakhwī (see footnote 2), vol. 3, pp. 48–49.

that were combined with geometry as fields of study by students and teachers in al-Sakhāwī's dictionary are first and foremost *farā'iḍ*, arithmetic, and timekeeping. The affiliation of classes teaching the *Elements* or al-Samarqandī's *Ashkāl* with scholars who taught *farā'iḍ*, arithmetic, and timekeeping corresponds well with the fact that geometry as a generic field of study was also mostly offered by the same group of teachers and their students once they themselves became teachers. The entries also indicate a much greater focus on theory than on practical geometry. This probably results from the function geometrical studies had within the larger field of astronomical disciplines taught by most of the teachers. Such an interpretation is supported by the fact that *hay'a* is presented in al-Sakhāwī's dictionary in its relationship to *handasa* as an equal to *farā'iḍ*, arithmetic, and timekeeping. Minority positions are occupied by combinations with philosophy, some of the occult sciences, medicine, and *taṣawwuf* (Sufi teachings). They point to the wider spectrum open for educational interests in geometry and the fact that not everybody had the field's applicability in mind when going to classes.

2 The geometrical disciplines and activities listed by Muḥammad Amīn b. Faḍl Allāh al-Muḥibbī

Muḥammad Amīn b. Faḍl Allāh al-Muḥibbī (d. 1111/1699) was born and raised in Damascus in a prominent family of scholars. He studied for some time in Beirut, where his father had moved in 1076/1666, and in Bursa, financed by a friend of his father who served as military judge in various towns of Anatolia. His patron procured a position for al-Muḥibbī in Edirne. When he fell ill and moved to Istanbul, al-Muḥibbī accompanied him and cared for him until his death in 1091/1681. Thereafter, he returned to Damascus. He compiled a substantial collection of poems and the biographical dictionary that is the basis for this discussion called *Khulāṣat al-athar fī a'yān al-qarn al-ḥādī 'ashar* (*The Essence of the Tradition on the Notables of the Eleventh Century*).[26] It contains some 1,300 biographies of scholars, poets, and men of the military and administration. Relatively late in his life, al-Muḥibbī received official appointments, first as a deputy judge, then as a teacher at a *madrasa* in Damascus. He died in his hometown in Jumādā 1111/ November 1699.

Fifty-nine biographical entries of the *Khulāṣat al-athar* contain references to one or more mathematical sciences. A few of the scholars dealt with in these entries lived either before the eleventh/seventeenth century or outside the Ottoman Empire.[27] In contrast to al-Sakhāwī, al-Muḥibbī speaks regularly of either

26 Brockelmann, C., "al-Muḥibbī". *Encyclopaedia of Islam*. Edited by: P. Bearman, Th. Bianquis, C. E. Bosworth, E. van Donzel, and W. P. Heinrichs. Brill, 2007. Brill Online. 02 February 2007 <www.brillonline.nl/subscriber/entry?entry=islam_SIM-5431>

27 Muḥammad Amīn b. Faḍl Allāh al-Muḥibbī, *Khulāṣat al-athar fī a'yān al-qarn al-ḥādī 'ashar*, Bayrūt, s.d., 4 vols., vol. 1, p. 373; vol. 4, p. 31.

al-ʿulūm al-riyāḍiyya (the mathematical sciences) or simply *al-riyāḍī* (mathematical) as a generic term of the field.[28] This shift of terminology corresponds well with his usage of other terms of Greek origin such as *al-arithmāṭīqī* and the greater visibility of the philosophical disciplines.[29] Another difference to al-Sakhāwī's way of talking about mathematics is al-Muḥibbī's stricter adherence to a disciplinary language. While al-Sakhāwī often would refer to titles of mathematical texts or parts of them when he described what teachers taught or students studied al-Muḥibbī chose the language of disciplinary classifications. As a result, his report is more homogeneous and less colorful than that of his Egyptian predecessor. The dominant disciplines in al-Muḥibbī's dictionary are the same as those in al-Sakhāwī's work – *ʿilm al-farāʾiḍ*, *ʿilm al-ḥisāb*, *ḥisāb al-jabr wa-l-muqābala*, and *ʿilm al-mīqāt*. Several of the texts in *ḥisāb* and *mīqāt* of the ninth/fifteenth century continue to be studied, in particular texts by Ibn al-Hāʾim and Ibn al-Majdī.[30] Of the latter a collection of his texts was compiled for teaching.[31] Mathematical cosmography and astrology continue to be taught and studied and the relevance ascribed to them by the two authors does not seem to differ significantly.

Another result of al-Muḥibbī's preference of disciplinary classifications over individual titles is the light this language sheds on the inter-disciplinary linkages which scholars of his time and location pursued and accepted. Somewhat surprisingly and again in contrast to what his predecessor emphasized, eleventh/seventeenth-century teachers and students of the mathematical sciences in Damascus saw these disciplines either linked to the rational sciences or to the occult sciences. The trend to combine studies of arithmetic or algebra with that of magic squares, other forms of licit magic, or simply the occult sciences in general has become more prominent.[32] But the difference to al-Sakhāwī's representation is not only one of quantity. It appears also to be one of perspective. In at least one case, al-Muḥibbī not merely links the mathematical sciences to the occult, but calls them occult sciences.[33] In addition to this explicit identification, the person whose scholarly biography is narrated in this instance is a *muwaqqit* at the sanctuary of the Prophet and epistolary secretary of the Sharīf Saʿd b. al-Sharīf Zayd al-Aʿlam (c. 1041–1077/1631–1666), the ruler of Medina.[34] Al-Muḥibbī's integration of the mathematical sciences in another case within the spectrum of the rational sciences may not cause too much attention among modern readers, since there is the widespread assumption that the term 'rational science' is identical with the tern

28 al-Muḥibbī (see footnote 27), vol. 1, p. 44; vol. 2, pp. 142, 147, 149, 300, 437, 456; vol. 3, pp. 322; vol. 4, pp. 28, 330.

29 al-Muḥibbī (see footnote 27), vol. 1, p. 88; vol. 4, p. 207.

30 al-Muḥibbī (see footnote 27), vol. 1, p. 88; vol. 2, p. 456; vol. 3, pp. 130, 304.

31 al-Muḥibbī (see footnote 27), vol. 3, p. 130.

32 al-Muḥibbī (see footnote 27), vol. 1, pp. 44, 96, 175, 373; vol. 3, pp. 39, 314, 322.

33 al-Muḥibbī (see footnote 27), vol.1, p. 178. For the text of the example see appendix 3.

34 al-Muḥibbī (see footnote 27), vol.1, p. 178.

'ancient science'.[35] Biographical dictionaries and historical chronicles contradict, however, such identification as shown by the following list of disciplines making up the rational sciences cited by al-Muḥibbī in another entry: the two *aṣl*, logic, rhetoric, prosody, grammar, and inflection.[36] Neither was it standard fare among writers of treatises on disciplinary classifications in eleventh/seventeenth-century Ottoman lands to include either all of the mathematical sciences or some of them under the rubric of the rational sciences.

This double singularity in al-Muḥibbī's rhetorical treatment of the mathematical sciences does not reflect a lack of knowledge and familiarity with these disciplines. He studied some of them, in particular *handasa*, with several scholars.[37] Neither is it possible to attribute the singularity to processes of professional reorientations within clusters of previously connected disciplines of knowledge. The only trend that is recognizable over the two centuries that separate the two authors is the increasing presence of kinds of knowledge previously not studied and taught at madrasas, mosques, or Sufi convents in the educational practices exercised at these institutions. Apparently, it was processes of inclusion and extension and not processes of closure and boundary drawing that allowed for the integration of formerly contested domains of knowledge into the main loci of religious and secular education and for multiple forms of talking about the connections between well-established disciplines in the center of education, peripheral disciplines, and newcomers to the canon of institutionalized education. Such a perspective finds further support in a third form used by al-Muḥibbī for situating the mathematical sciences – a triad that evokes philosophical habits, namely *al-ʿulūm al-riyāḍiyya wa-l-ḥikmiyya wa-l-ṭabīʿiyya* (the mathematical, philosophical and natural sciences).[38]

In contrast to the surprising lack of treatises or copies of treatises on geometry in ninth/fifteenth-century Cairo discussed above, al-Muḥibbī's dictionary indicates that in eleventh/seventeenth-century Damascus, scholars wrote works on *handasa* and *misāḥa*.[39] The focus remained, however, on works on *mīqāt* and arithmetic. The Damascene teachers of geometry and other theoretical disciplines such as *hayʾa* are often linked to the standard religious disciplines; but occasionally a *muwaqqit* taught *handasa* too.[40] Strangers such as Bahāʾ al-Dīn al-ʿĀmilī (954–1030/1547–1621) also taught these theoretical disciplines when visiting Damascus, Cairo, Jerusalem, or Mecca.[41] An interesting and challenging aspect of al-Muḥibbī's report on al-ʿĀmilī's teaching of *hayʾa* and *handasa* in Jerusalem is the latter's condition of secrecy before he agreed to the teaching requested by a local scholar. Since no context is given for al-ʿĀmilī's demand, it is impossible

35 al-Muḥibbī (see footnote 27), vol. 4, p. 31. For the text of the example see appendix 3.
36 al-Muḥibbī (see footnote 27), vol.1, p. 252.
37 al-Muḥibbī (see footnote 27), vol. 2, p. I 02; vol. 4, p. 330.
38 al-Muḥibbī (see footnote 27), vol. 4, p. 28.
39 al-Muḥibbī (see footnote 27), vol. 1, p. 277; vol. 2, p. 300; vol. 4, p. 11.
40 al-Muḥibbī (see footnote 27), vol. 2, p. 300; vol. 3, p. 178.
41 al-Muḥibbī (see footnote 27), vol. 3, pp. 440–443.

to determine whether the secrecy applied to any of the *ʿulūm* he was asked to teach, i.e., to the act of teaching per se, or specifically to the two disciplines he then taught.

Al-Muḥibbī mentions one text on *handasa*, a super-commentary on the commentary on *Ashkāl al-taʾsis*, and a few texts on practical geometry that were either read or written by students or teachers.[42] Texts taught in *hayʾa* appear slightly more often in his book, mostly the introductory text by Čaghmīnī.[43] Al-Muḥibbī makes it clear, however, that he knew the basic school texts of geometry and astronomy by referring to the *Elements*, the *Middle Books*, and the *Almagest* in one of his biographical entries.[44] One of al-Muḥibbī's teachers in geometry did not teach the field with books alone. He caught the attention and admiration of his student by visualizing the drawings of the book through precisely constructed models which he had inherited from his own teacher.[45]

3 Conclusions for the historiography of mathematics in later Islamic societies

The comparative analysis of the two biographical dictionaries undertaken here with the goal to clarify one specific point, namely the teaching and studying of geometry in Arabic cities during the ninth/fifteenth and the eleventh/seventeenth centuries, has resulted in several new insights which cannot be achieved through the study of mathematical manuscripts and scientific instruments alone. This analysis does not merely add to our factual knowledge about which mathematical disciplines were pursued. It shows that these fields were connected with larger frameworks and trends within the educational spheres of these cities. On the basis of manuscripts, only connections with teachers of inheritance mathematics, arithmetic, algebra, and timekeeping will be expected and visible. Shifts in rhetoric and attitudes remain widely opaque. The biographical dictionaries indicate that a second group of scholars was involved in teaching some of the ancient sciences, in particular their theoretical fields – scholars who mainly taught religious sciences such as *ḥadīth* and *tafsīr*, but combined them continuously with major rational sciences, in particular the two *aṣl*, logic, and some parts of *kalām*. They also suggest a new attitude towards the ancient sciences, in particular theoretical mathematical and philosophical disciplines, in eleventh/seventeenth-century Ottoman Damascus. Although there is no indication in this kind of source that the new attitude also yielded new results, the cultural shift is interesting in its own rights and deserves closer investigation. Furthermore, the study of the two dictionaries elucidates that educational opportunities and approaches as well as movements between different

42 al-Muḥibbī (see footnote 27), vol. 1, p. 277; vol. 2, p. 300; vol. 4, pp. 11, 371.
43 al-Muḥibbī (see footnote 27), vol. 1, pp. 12; vol. 2, p. 458; vol. 3, p. 348.
44 al-Muḥibbī (see footnote 27), vol. 3, p. 306.
45 al-Muḥibbī (see footnote 27), vol. 4, pp. 330–331.

regions of the Islamic world described by them do not agree with information and evaluation given by other, often highly trusted authors such as Ibn Khaldūn. Ibn Khaldūn's comments on the decline of teaching mathematics and philosophy are not born out by the biographies presented by al-Sakhāwī. Maghribī scholars travelling to Egypt and the Holy Cities are often acknowledged as having received a solid education in North African towns. Occasionally their education in mathematics, astronomy, and philosophy sounds broader than the one offered in Cairo, Alexandria, Damascus, Jerusalem, or Mecca. Intellectual journeys towards Egypt, Syria, and the Hijaz continue to be undertaken from the West and the East. The transfer of mathematical, astronomical, and philosophical knowledge takes place in different directions. During the ninth/fifteenth century, there is a clear influx of such knowledge from the greater Iranian world through scholars going for pilgrimage to Mecca and later visiting or settling in Cairo. In the same time, parts of the predominantly taught mathematical sciences, in particular arithmetic and timekeeping, were exported from Cairo to Mecca and Medina. In the eleventh/seventeenth century, this double movement seems to be much weaker, although texts by Iranian authors continue to be studied and highly appreciated. The focus of scholarly travelling is now Anatolia and in particular Istanbul. The establishment *of muwaqqit*s in the Holy Cities may owe as much to the older transfer of this particular kind of knowledge from Cairo southwards as to the Ottoman acceptance of and support for this profession. A number of other observations are interesting in their singularity such as the usage of models for teaching geometry or in their potential for a different understanding of social relationships among Muslim scholars if proven to be more widespread. Reading biographical dictionaries thus yields a number of valuable insights for a much denser depiction of the cultural context of teaching and learning mathematics in specific localities than available so far. Reading biographical dictionaries also has the potential to replace the tedious and highly loaded debate about intellectual decline by some other categories and questions, among them some of those highlighted in this paper.

APPENDIX 1

Texts and authors on theoretical geometry mentioned by al-Sakhāwī:

1 Aḥmad b. Muḥammad [. . .] al-Qāhirī, the preacher, known as Ibn al-Qardāḥ, and perhaps he was called al-Qurdāḥ [. . .]. He was born after the eighties or perhaps at their end [. . .] and he served al-ʿIzz b. Jamāʿa in the arts such as music and others, and he took timekeeping and other (things) from al-Jamāl al-Māridānī and *ʿilm al-falak* from al-Shams b. Ayyūb, the head of the ʿAmrī Mosque in Miṣr [. . .]. And I heard that he studied Uqlīdis in (its) entirety with Ibn al-Majdī.[46]

2 ʿAlī b. Aḥmad [. . .] ai-Qurashī, al-Qalqashandī originally, al-Qāhirī [. . .] but he also took (classes) about them (i.e., Arabic and *farāʾiḍ*) and about arithmetic and algebra with Ibn al-Hāʾim and equally with Jamāl al-Dīn al-Māridānī together with some timekeeping. But he read with him Uqlīdis.[47]

3 ʿAlī b. ʿAbd al-Qāḍir [. . .] al-Shāmī originally, al-Qāhirī al-Azharī al-Faraḍī [. . .] and known as al-Sayyid al-Faraḍī. He was born in circa 808 in Cairo and grew up there. [. . .] he served Ibn al-Majdī in *farāʾiḍ*, arithmetic, algebra, and other (fields) over a long period until he became as mentioned. He took with him the reading or the listening *of Ashkāl altaʾsīs* on *handasa* and he asked him about everything the understanding of which he could press out of him in order to verify it for him. That is why he excelled.[48]

4 ʿAlī b. Muḥammad [. . .] al-Jurjānī, [. . .] ʿĀlim al-Sharq and known as al-Sayyid al-Sharīf. [. . .]. He studied in his country [. . .]. Then he came to Cairo [. . .]. Then he went to Bilād al-Rūm (Anatolia) and then he returned to Bilād al-'Ajam (Iran). [. . .]. He wrote works and it is said that they surpassed fifty. The son of his grandson told me that among them are [. . .] a commentary on Naṣīr al-Dīn al-Ṭūsī's *Tadhkira* and al-Čaghmīnī on mathematical

46 al-Sakhāwī (see footnote 2), vol. 2, p. 142.
47 al-Sakhāwī (see footnote 2), vol. 5, p. 161.
48 al-Sakhāwī (see footnote 2), vol. 5, p. 242.

cosmography, [. . .] a commentary on the text of *Ashkāl al-taʾsīs*, [. . .] on al-Ṭūsī's *Taḥrīr* of Euclid [. . .][49]

5 Muḥammad b. ʿAbd al-Wāḥid [. . .] al-Sīwāsī originally, then al-Qāhirī [. . .], known as Ibn al-Humām. Born in 790, probably, as I read in his own hand, but al-Maqrīzī said in his *ʿUqūd:* in the year 88 or 89, in Alexandria. His father, who was the judge of Alexandria, died when he was circa ten years old. [. . .] he took [. . .] Uqlīdis with Ibn al-Majdī [. . .].[50]

6 Muḥammad b. Qāsim [. . .] al-Ghazzī, then al-Qāhirī [. . .], known as Ibn al-Gharābīlī. He was born in (the month of) Rajab circa in the year 859 in Ghazza [. . .]. He came to Cairo in Rajab of the year 81. [. . .] He read [. . .] with al-Jamāl al-Kūrānī (something from) *Sharḥ Ashkāl al-taʾsīs*.[51]

7 Muḥammad b. ʿAwaḍ [. . .] al-Sikandarī [. . .], known as Junaybāt. He was born in Alexandria in 788. He read there the Qurʾān [. . .] he also studied with al-Muʿizz timekeeping and the beginnings of Uqlīdis [. . .].[52]

8 Muḥammad b. Muḥammad [. . .] al-Sikandarī, then al-Qāhirī [. . .] he was born close to the year 790 [. . .] and he took Uqlīdis from Jamāl al-Dīn al-Māridānī and was excellent.[53]

9 Muḥammad b. Marāhim [sic] al-Dīn al-Shams al-Shirwānī, then al-Qāhirī [. . .]. He was born circa in the 80s. [. . .] He did not occupy himself with *ʿilm* before he had reached his twentieth year. Then he took from al-Sayyid Muḥammad b. al-Sharīf al-Jurjānī and Qāżīzāde al-Rūmī, the author of the commentary on *Ashkāl al-taʾsīs*. He favored him over the former in *riyāḍiyyāt* (the mathematical disciplines).[54]

49 al-Sakhāwī (see footnote 2), vol. 5, pp. 328–329.
50 al-Sakhāwī (see footnote 2), vol. 8, p. 127.
51 al-Sakhāwī (see footnote 2), vol. 8, p. 286.
52 al-Sakhāwī (see footnote 2), vol. 9, pp. 272–273.
53 al-Sakhāwī (see footnote 2), vol. 10, p. 27.
54 al-Sakhāwī (see footnote 2), vol. 10, p. 48.

APPENDIX 2

Combinations of (theoretical or practical) geometry with other disciplines

1 Aḥmad b. Ibrāhīm [. . .] al-Dimashqī, then al-Dimyāṭī, [. . .] known as Ibn al-Naḥḥās. [. . .]. He had a complete knowledge of *farā'iḍ* and *ḥisāb* which he acquired through his capability to solve the rarities of *ḥisāb* through *handasa*. He wrote on it together with very good knowledge of law and shared (knowledge) of other fields.[55]

2 Aḥmad b. Rajab b. Taybughā al-Majdī [. . .], known as Ibn al-Majdī as a relation to his grandfather. [. . .] He became the leader of the people in the branches of *ḥisāb* and *handasa* and *hay'a* and *farā'iḍ* and *'ilm al-waqt* without doubt.[56]

3 Aḥmad b. Sulaymān [. . .] al-Bilqāsī, then al-Qāhirī al-Azharī [. . .]. Then he served [. . .] and Ibn al-Majdī in *farāiḍ* and *ḥisāb* and *mīqāt* and *hay'a* and *handasa* [. . .].[57]

4 Aḥmad b. Ṣadaqa [. . .] al-'Asqalānī al-Makkī originally, then al-Qāhirī [. . .], known as al-Ṣīrafī [. . .]. And he took [. . .] and *falak* and *muqanṭarāt* and algebra and *handasa* and *hay'a* and *ḥikma* and Arabic from al-Khawāṣṣ and al-Qalqashandī and Ṭāhir [. . .].[58]

5 Aḥmad b. 'Ubaydallāh [. . .] al-Sijīnī, then al-Qāhirī al-Azharī [. . .] al-Faraḍī [. . .] and his interest grew stronger in serving Ibn al-Majdī in law and its fundaments and Arabic and *farā'iḍ* and *ḥisāb* and *misāha* (surveying) and algebra and *handasa* and *mīqāt* and the best of his disciplines in which he excelled. He limited himself to him by repeatedly taking much from these disciplines from him more than once and he became great through his benefit from him.[59]

55 al-Sakhāwī (see footnote 2), vol. 1, p. 203.
56 al-Sakhāwī (see footnote 2), vol. 1, p. 300.
57 al-Sakhāwī (see footnote 2), vol. 1, p. 310.
58 al-Sakhāwī (see footnote 2), vol. 1, p. 317.
59 al-Sakhāwī (see footnote 2), vol. 1, p. 376.

6 Aḥmad b. Muḥammad [. . .] al-Bījūrī originally, al-Qāhirī, [. . .] he took [. . .] *farāʾiḍ* and *ḥisāb* from al-Ḥijāzī and Abū l-Jūd and al-Būtījī, [. . .] and medicine from al-Jazarī, and *al-mīqāt* from al-Shams al-Ṭuntadāʾī, the resident of the Baybarsiyya, and *al-jayb* (literally: the Sine; but standing for al-Māridānī's treatise on the Sine Quadrant) from al-ʿIzz al-Wanāʾī, and calligraphy from al-Zayn b. al-Ṣāʾigh. He acquired the art of ink and similar things, shooting with bow and arrow from the lines, [. . .], (?) rarities of the game with the needle and the lance from the second of the two, and *al-mīqāt* from al-Shams al-Shāhid, the brother of the Friday preacher Dirāya, and al-Shāṭir Shūmān, and the art of the point (*ṣināʿat al-nuqṭa*) and the rules of surveying from Aḥmad b. Shihāb al-Dīn. He occupied himself in manifold ways with what I mentioned with others until he excelled in copper casting, filing, bloodletting with a feather and brocade embroidery so that I don't know now who joined him (i.e., studied with him) and he is not a master in many crafts but, (mastered) some in theory and with this he is badly known in comparison to others of those who were inferior to him. He turned to reading at the Azhar [. . .] and read there books on the disciplines (*funūn*). He went for pilgrimage more than once and passed through Medina, the city of the Prophet, in the year 56[60] and read there too books on the disciplines. [. . .] he commented something on what he knew of the needle and the lance, abbreviated *Maṣābiḥ al-Zullām fī l-Minfāq* (?) and added to it things he had picked up from his shaykh. He also abbreviated *Kitāb al-manāzil* by Abū l-Wafāʾ al-Būzjānī as one *manzila*, which is on surveying. He added to it things from (the treatise on) surveying by al-Tabrīzī. [. . .].[61]

7 Aḥmad b. Muḥammad [. . .] al-Bijāyī al-Tūnisī [. . .], known as Abū l-ʿAbbās b. Kuḥayl. [. . .] he took [. . .] *handasa* by being present and by listening from Ibn Marzūq, but he heard in his sessions more than what he read with him on the particular sciences, and also from Abū l-Qāsirn al-ʿUqbānī, [. . .].[62]

8 Aḥmad b. Muḥammad [. . .] al-Qusantīnī originally,[63] born in Alexandria, grown up in Cairo, [. . .] he took [. . .] medicine from al-Sharns Muḥammad al-Balādurī who held the highest rank in it, and *al-Khazrajiyya* on metric and *Fuṣūl* by Ibn al-Hāʾim on *farāʾiḍ* and *al-Nuzha* on arithmetic with the pen and the two treatises by al-Māridānī (i.e., on the Sine Quadrant and the Almucantar Quadrant) from Nāṣir al-Dīn al-Bārinbārī, and *handasa* and *hayʾa* through his reading and arithmetic through listening to Ibn al-Majdī, and logic through his reading and *Ādāb al-baḥth* from Abū Bakr al-ʿAjamī, the physician [. . .].[64]

60 The date meant here is of course 856 h.

61 al-Sakhāwī (see footnote 2), vol. 2, pp. 66–67.

62 al-Sakhāwī (see footnote 2), vol. 2, p. 136.

63 This means his family came from Constantine.

64 al-Sakhāwī (see footnote 2), vol. 2, p. 175.

9 Ḥasan b. ʿAlī [. . .] al-Ṭalkhāwī, then al-Qāhirī, [. . .] he came to al-Azhar [. . .] and memorized [. . .] *Alfiyyat al-farāʾiḍ* by Ibn al-Hāʾim, *al-Lamba* by al-ʿAfīf on medicine, [. . .]. He took *farāʾiḍ* and arithmetic and *mīqāt* and *hayʾa* and *handasa* and algebra and *ḥall al-shams* by the method of *al-Durr al-yatīm* (a treatise by Ibn al-Majdī) from al-Shihāb al-Sijīnī. Perhaps he consulted al-Sharfī b. al-Jayʿānī about *farāʾiḍ* and arithmetic and *hayʾa* together with *al-waqʿiyyāt* (construction of various curves on astronomical instruments) from al-Muḥibb b. al-ʿAṭṭār; and *al-waqʿiyyāt* alone from Ibn Walī al-Dīn, a son-in-law of al-Ghumarī, and *al-mīqāt* alone from Nūr al-Dīn al-Naqqāsh and his son and al-Badr al-Māridānī, and *al-ḥarf* (letter magic) from Nāṣir al-Dīn b. Qurqmās, and *al-raml* (geomancy) from Muḥammad al-Naḥrīrī [. . .].[65]

10 Ḥusayn b. ʿAlī [. . .] al-Bayḍāwī al-Makkī [. . .] al-Faraḍī, al-Ḥāsib, [. . .], known as al-Zamzamī. [. . .] he paid attention to *farāʾiḍ* and arithmetic and he took that from al-Shihāb b. Ẓuhayra and al-Burhān al-Burullusī al-Faraḍī, inhabitant of Mecca, and thought about both (fields). Then he increased in excellence after he took that from al-Shihāb b. al-Hāʾim by reading with him in Mecca some of his works. He took *ʿilm al-falak* in Cairo from al-Jamāl al-Māridānī and did not stop to add and pay attention until he became the imām, the knowledgeable, the excellent, the perfect whom the people knew in *farāʾiḍ* and *hayʾa* and arithmetic and *ʿilm al-khaṭaʾayn* and algebra and *handasa* and *falak* and ephemerides. The leadership in this knowledge came to him finally in the region of the Hijaz, Mecca, Medina and Yemen. He wrote about it. His brother al-Burhān, mentioned previously, profited from him in it.[66]

11 Ḥusayn b. Muḥammad [. . .] al-Ḥillawī, a *nisba* to the town Ḥilla, then al-Makkī [. . .] and known as Ibn ʿUlayf [. . .] he read with al-Ḥussām b. Ḥasan al-Abīwardī *al-Mufaṣṣal* by al-Zamakhsharī and he took from him the two *aṣl* (fundaments of religion and fundaments of law) and arithmetic in its branches and *misāḥa* and *taṣawwuf* (Sufi teachings) [. . .].[67]

12 Zakariyyāʾ b. Muḥammad [. . .] al-Qāhirī al-Azharī [. . .] al-Qāḍī. [. . .] he took *ʿilm al hayʾa* and *handasa* and *mīqāt* and *farāʾiḍ* and arithmetic and algebra and other (fields) from Ibn al-Majdī and he read with him things from his works and also *farāʾiḍ* and arithmetic from al-Shams al-Ḥijāzī and al-Būtījī; and also from Abū l-Jūd al-Yanabī with whom he read *al-Majmūʿ* and *al-Fuṣūl*, and *al-ḥikma* from al-Shirwānī and Jaʿfar, mentioned previously, and medicine from al-Sharaf b. al-Khashshāb, and metric from al-Warūrī and *ʿilm al-ḥarf* from Ibn al-Qurqmās al-Ḥanafī, and *taṣawwuf* from Abū ʿAbdallāh al-Ghumarī and al-Shihāb Aḥmad al-Adkāwī and Muḥammad al-Fawī [. . .].[68]

65 al-Sakhāwī (see footnote 2), vol. 3, p. 115.
66 al-Sakhāwī (see footnote 2), vol. 3, pp. 151–152.
67 al-Sakhāwī (see footnote 2), vol. 3, p. 155.
68 al-Sakhāwī (see footnote 2), vol. 3, p. 235.

13 ʿAbd al-Raḥmān b. ʿAlī [. . .] al-Shaybānī al-Zabīdī [. . .], known as Ibn al-Daybaʿ [. . .] and its meaning is in Nubian language the white. [. . .] he worked on arithmetic, algebra, *handasa*, *farāʾiḍ*, *fiqh* and Arabic with his uncle whom we already mentioned (Abū l-Najā Muḥammad, the physician) [. . .].[69]

14 ʿAbd al-Raḥīm b. Ibrāhīm [. . .] al-Qāhirī [. . .], our neighbour [. . .] and he also took from al-Shirwānī *al-uṣūl* and logic and from al-Kāfiyājī *hayʾa*, *handasa* and other (things) and *farāʾiḍ* and arithmetic together with algebra from al-Sayyid ʿAlī Tilmīdh Ibn Majdī and metric from al-Abadī or someone else [. . .].[70]

15 ʿAlī b. ʿAbd al-Qādir [. . .] al-Qāhirī al-Mīqātī. He participated in the lectures of al-Walī al-ʿIrāqī and took *mīqāt* and *handasa* from Ibn al-Majdī and engraving from the husband of his mother. He excelled in all of them. He earned his living with engraving in a shop of the gold market. He held the leadership at the Muqassī Mosque and at the Jamāliyya Sāḥibiyya (*Madrasa*) and others than the two such as the Ashrafiyya Tomb of Īnāl, but he taught the field at several places [. . .].[71]

16 ʿAlī b. Muḥammad [. . .] al-Qurashī al-Andalusī al-Bastī, [. . .], known as al-Qalaṣāwī [sic]. [. . .] He was born in 815 in the town of Basta and read there the Qurʾān [. . .], then he studied with Muḥammad al-Qusṭurlī arithmetic and read with the *faqīh* Jaʿfar on it and on *farāʾiḍ* and *fiqh* and with the *faqīh* Abū Bakr al-Bayyāz Arabic and the *Manẓūma* of Ibn Barrī in a useful reading [. . .]. Then he travelled to the town of al-Mankab [. . .], then to Tlemcen in the year 40. [. . .] Then he served the shaykh Aḥmad b. Zāghū [. . .] and Qāsim al-ʿUqbānī [. . .] and Muḥammad b. Marzūq. He studied with him *tafsīr*, *ḥadīth*, *farāʾiḍ* and grammar and with al-ʿUqbānī *tafsīr*, *ḥadīth*, *fiqh* and the two *aṣl* and with Ibn Zāghū *tafsīr*, *ḥadīth*, *fiqh*, *farāʾiḍ*, arithmetic, *handasa*, grammar, rhetoric and with ʿĪsā b. Amizzabān [. . .], *farāʾiḍ*, arithmetic and logic [. . .].[72]

17 Qāsim b. Aḥmad [. . .] al-Ḥalabī al-ʿAyntābī al-Kutubī, [. . .]. Our shaykh said in his . . . *Inbaʾ* after his uncle: he was one of the excellent in arithmetic, *handasa*, grammar, talismans, *ʿilm al-ḥarf* with a surplus of intelligence.[73]

18 Muḥammad b. Aḥmad [. . .] al-Tūnisī [. . .], inhabitant of the two sanctuaries and known as al-Wānnawghī. [. . .] He was born, it is believed, in 759 in Tunis, grew up there, and listened to its transmitter and reciter Abū l-Ḥasan b. al-ʾAbbās al-Baranī [. . .] and to Ibn ʿUrfa. He profited from him in *fiqh*, *tafsīr*, the two *aṣl* and the science of arithmetic and *handasa* [. . .].[74]

69 al-Sakhāwī (see footnote 2), vol. 4, p. 104.
70 al-Sakhāwī (see footnote 2), vol. 4, pp. 164–165.
71 al-Sakhāwī (see footnote 2), vol. 5, p. 242.
72 al-Sakhāwī (see footnote 2), vol. 6, pp. 14–15.
73 al-Sakhāwī (see footnote 2), vol. 6, p. 178.
74 al-Sakhāwī (see footnote 2), vol. 7, p. 3.

19 Muḥammad b. Aḥmad [. . .] al-Bisāṭī, then al-Qāhirī, the scholar of the age, [. . .], known as al-Bisāṭī. [. . .] He never stopped acquiring education in the disciplines and demanding that what is expressible from them and what is understandable until he achieved preeminence in *fiqh*, the two *aṣl*, Arabic, grammar, rhetoric, logic, *ḥikma*, algebra, medicine, *hay'a*, *handasa* and arithmetic and became the imam of the age. [. . .].[75]

20 Muḥammad b. Aḥmad [. . .] al-Sarāy'ī [. . .] al-ʿAjamī originally, al-Qāhirī, [. . .], the grandson of al-Shams al-Āqsarāy'ī [. . .]. He also took from al-Basāṭī and the method of the people (i.e., the jurists) from al-Zayn al-Khawāfī and researched about *handasa* with Ibn al-Majdī and followed the Qur'ān with Abū ʿUmar and ʿAlī al-Zayn Ṭāhir al-Mālikī, although he was older than him.[76]

21 Muḥammad b. Sulaymān [. . .] al-Mibyawī [. . .] al-Rūmī [. . .], known as al-Kāfiyājī. He was born in Kukja kī (Kökçe koy) in Sarūkhān (Saruhan) of the realm al-Rūm of Ibn ʿUthmān (the Ottomans) before circa 790. [. . .] He took from al-Shams al-Fanarī and al-Burhān Amīr Ḥaydar al-Khāfī, one of the students of al-Taftāzānī, and Wājid and ʿAbd al-Wāḥid al-Kūtā'ī and others. He read very much *al-Kāfiya* by Ibn al-Ḥājib and read it until he was linked to it by adding him (to it) as is the custom of the Turks with respect to origin. [. . .] And in total, he became the arch-scholar of eternity and the unique of the century and the rarity of the era and the pride of this time. Moreover, he was the master in (the following disciplines) the two *aṣl*, *tafsīr*, grammar, inflection, rhetoric, logic, *hay'a*, *handasa*, *ḥikma*, dialectics, the spheres, (burning) mirrors and optics with a good share (of knowledge) in *fiqh* and medicine.[77]

22 Muḥammad b. ʿAbd al-ʿAzīz [. . .] al-Shīrāzī originally, al-Makkī, al-Zamzamī, [. . .], inhabitant of Cairo [. . .]. He was born in 846 or 847 in Mecca and grew up there. [. . .] he and his brother Abū Bakr who came earlier (in this dictionary) studied histories (or: calendars) night after night in al-Masjid al-Ḥarām (the main mosque of Mecca); he memorized *al-Minhāj* and other (works); he took there (in Mecca) *al-falak* from Nūr al-Dīn al-Zamzamī. He went to Cairo in 65 and stayed there for some time. He occupied himself with *farā'iḍ*, arithmetic, *mīqāt*, *handasa* and other (topics) until he excelled and distinguished himself in some of them.[78]

23 Muḥammad b. ʿAlī [. . .] al-Ḥaṣkafī, then al-Maqdisī, [. . .] known in his place as Ibn al-Ḥimṣī and in these regions by his *kunya* (i.e., Abū l-Luṭf). He was born in 819 in Ḥiṣn Kayfā from the region Bakr and grew up there. He read the Qur'ān [. . .], grammar, inflection, [. . .] logic, [. . .] metrics, [. . .]

75 al-Sakhāwī (see footnote 2), vol. 7, pp. 5–6.
76 al-Sakhāwī (see footnote 2), vol. 7, pp. 115–116. The date mentioned in this extract means 865 h.
77 al-Sakhāwī (see footnote 2), vol. 7, pp. 259–261.
78 al-Sakhāwī (see footnote 2), vol. 8, pp. 220–221.

logic with Sirāj al-Rūmī in Jerusalem and al-Kāfiyājī in Cairo together with listening to an appropriate part of al-ʿAḍud's commentary on *al-Mukhtaṣar*. But he (also) read with him two chapters from al-Sayyid's commentary on *al-Mawāqif* and (he took) *ʿilm al-hayʾa* and *handasa* and arithmetic and *al-ḥarf* on the polymath Qiwām al-Dīn al-Shīrāzī and music on al-Ḥājj Qalandar in Ḥiṣn Kayfā and al-Ḥājj Zayn al-Dīn Ṭāhir b. Qāḍī al-Mawṣilī, reading with him *al-Adwār* by al-Ṣafī ʿAbd al-Muʾmin al-Urmawī in an ascertained reading and rhetoric and prosody with al-ʿAlāʾ ʿAlī al-Kurdī, teacher at the Safāḥiyya in Aleppo and other things and *fiqh* with ʿUbayd al-Bābī, imām at the great Friday mosque in Aleppo, and al-Zayn Māhir in Jerusalem; from him he took *farāʾiḍ* and arithmetic; [. . .].[79]

24 Muḥammad b. Muḥammad [. . .] al-Lakhmī, al-Santarāwī originally, al-Qāhirī, the son of the paternal uncle of my shaykh, one of those who read with him and heard from him *al-Minhāj al-aṣlī* and al-Bisātī; he heard from Abū l-Qāsim al-Nuwayrī in the reading by al-Muḥibb al-Ṭabarī, the imām, on the *Mukhtaṣar* of Ibn al-Ḥājib; from Ibn Imām al-Kāmiliyya he heard his commentary on al-Bayḍāwī and from Abū l-Faḍl al-Mashdāllī he heard *al-ʿAḍud;* and from him he took on logic, *handasa* and *kalām*.[80]

25 Muḥammad b. ʿAlī [. . .] al-Samannūdī originally, al-Miṣrī [. . .], he was known as Ibn al-Qattān. [. . .] He composed a book on the seven readings, which he called *al-Sahl*. I heard from him something of it. (He composed) a book on *farāʾiḍ* and arithmetic and *al-handasa* which he called *Jamʿ al-jumal*. I heard from him from it in lectures. I read with him much from the beginning of *al-Ḥāwī al-ṣaghīr*.[81]

26 Muḥammad b. Muḥammad [. . .] Abū l-Faḍl [. . .] al-Mashdāllī, [. . .] this refers to a tribe in al-Zawāwa, al-Zawāwī al-Bijāyī [. . .] and he was known in the East as Abū l-Faḍl and in the West as Ibn Abī l-Qāsim [. . .] he memorized the two (texts called) *al-Shāḥibiyya*, al-Khirāzī's *Rajaz* on *al-rasm* and *al-Kāfiya al-shāfiya* and *Lāmiyat al-afʿāl* by Ibn Mālik on grammar and inflection and *Ghālib al-tashīl* and his *Alfiya* and Ibn al-Ḥājib's *al-farʿī* and the *Risāla* and al-Tilimsānī's *Urjūza* on *farāʾḍ* and about a quarter of Saḥnūn's *Mudawwana* and *Ṭawālīʿ al-anwār* on *uṣūl al-dīn* by al-Bayḍāwī and Ibn al-Ḥājib's *al-aṣlī* and al-Khūnjī's *Jumal* and *al-Khazrajiyya* on metrics and *Talkhīṣ* by Ibn al-Bannāʾ on arithmetic and *Talkhīṣ al-Miftāḥ* and the *Diwān* of Imruʾ al-Qays, al-Nābigha al-Dhubyānī, Zuhayr b. Abī Sulmā, ʿAlqama al-Faḍl and Ṭarafa b. al-ʿAbd. Then he turned towards comprehension and thus studied inflection and metric with Abū Yaʿqūb Yūsuf al-Rīfī, Arabic, logic, *al-uṣūl* and *al-mīqāt* with Abū Bakr al-Tilimsānī and he took *al-mīqāt* also with Abū Bakr b. ʿĪsā al-Wānsharīsī. Then he studied grammar with Yaʿqūb al-Tīrūnī,

79 al-Sakhāwī (see footnote 2), vol. 9, p. 8.
80 al-Sakhāwī (see footnote 2), vol. 9, pp. 9–10.
81 al-Sakhāwī (see footnote 2), vol. pp. 9, 180–182.

grammar and logic with Abū Isḥāq Ibrāhīm b. Aḥmad b. Abī Bakr. Then he researched arithmetic with Mūsā b. Ibrāhīm al-Ḥasnāwī and again arithmetic together with inflection, grammar, the two *aṣl*, rhetoric and the revealed sciences of *tafsīr* and *ḥadīth* and *fiqh* with his father. Then he studied the two *asl* with Abū l-Ḥasan ʿAlī b. Ibrāhīm al-Ḥasnāwī, who is the brother of Mūsā, I believe. At the beginning of the year 40 he travelled to Tlemcen and studied there with the famous scholar Muḥammad b. Marzūq b. Ḥafīd and with Abū l-Qāsim b. Saʿīd al-ʿUqbānī, Abū l-Faḍl b. al-Imām, Abū l-ʿAbbās Aḥmad b. Zāghū, Abū ʿAbdallāh Muḥammad b. al-Najjār, known for the strength of his knowledge in *qiyās* through the butcher's knife of concluding by analogy. (He also studied) with Abū Rabīʿ Sulaymān al-Būzīdī and Abū Yaʿqūb Yūsuf b. Ismāʿīl and Abū l-Ḥasan ʿAlī b. Qāsim and Abū ʿAbdallāh Muḥammad al-Būrī and Ibn Afshūsh. He (studied) with the first *tafsīr* and *ḥadīth* and *fiqh* and the two *aṣl* and *adab* in its branches and logic and dialectic and *falsafiyyāt* (the philosophical sciences) and medicine and *handasa*. (He studied) with the second *fiqh* and *uṣūl al-dīn* and with the third *tafsīr* and *ḥadīth* and medicine and the ancient sciences (*al-ʿulūm al-qadīma*) and *taṣawwuf*. With the fourth (he studied) rhetoric and arithmetic and *farāʾiḍ* and *handasa* and *taṣawwuf*. With the fifth (he studied) *uṣūl al-fiqh* and rhetoric. [. . .] With the sixth (he studied) *fiqh*. [. . .] With the seventh (he studied) arithmetic and *farāʾiḍ*. With the eighth (he studied) arithmetic and algebra and other (things) from its branches and *hayʾa* and the movement of weights. And with the eighth (he studied) ephemerides and *mīqāt* in its branches of the arts of the astrolabes and the plates and the sines and *hayʾa* and number theory and music and talismans and what resembled them of the science of burning mirrors and optics and the science of the horizons and with the tenth (he studied) medicine. Then he returned to Bijāya.[82]

82 al-Muḥibbī (see footnote 27), vol. 1, p. 178.

APPENDIX 3

Al-Muḥibbī's linking of mathematics to the occult or the rational sciences

1 Aḥmad b. Tāj al-Dīn al-Dimashqī, originally from Medina, *muwaqqit* of *al-Ḥarām al-nabawī* and *kātib al-inshāʾ* of al-Sharīf Saʿd b. al-Sharīf Zayd al-Aʿlam. He was one of his era in the knowledge of the occult sciences such as mathematics and astrology and *simiyā* and all the others. He was very gifted in making astronomical instruments.[83]
2 Muḥammad b. ʿAbd al-Malik al-Baghdādī, [. . .], inhabitant of Damascus, al-Shaykh, al-Imām, al-Muḥaqqiq. He was from the great scholars, in particular in the rational (sciences) such as metaphysics, natural philosophy and mathematics. He belonged to the group of the arch-scholar of the time, Munlā Muṣliḥ al-Dīn al-Lārī.[84]

83 al-Muḥibbī (see footnote 27), vol. 4, p. 31.
84 al-Muḥibbī (see footnote 27), vol. 1, p. 31.

APPENDIX 4

Texts and authors on theoretical or practical geometry in al-Muḥibbī's dictionary

1 Aḥmad b. Muḥammad [. . .] al-Ḥaṣkafī [. . .], known as Ibn al-Munlā. [. . .] He worked very much with al-Rāḍī b. al-Ḥanbalī, the author of *Taʾrīkh Ḥalab*. He took from him his treatise *Sarḥ al-muqlatayn fī masḥ al-qullatayn* (in the form of) *dirāya* (transmission through reading and explanation) and he was a companion in the hearing of his work *Makhāyil al-malāḥa fī masāʾil al-misāḥa*. And he (also had knowledge) in algebra.[85]

2 Khān Aḥmad al-Kīlānī al-Sharīf al-Ḥusaynī, sultān of Gilan, from a ruling house (on the side of his) father and grandfather. In addition to coming from kings, he was one of the rare (people) of the world in the mathematical and philosophical sciences. He acquired *hayʾa*, *handasa* and *falak*. He was taught by al-Qūshjī in *hayʾa*. In him was the end of music and Persian poetry. [. . .] He died in the year nine after one thousand.[86]

3 ʿAlī b. Mubammad, called by the name of ʿAlāʾ al-Dīn Nāṣir al-Dīn, al-Ṭarābulsī originally, al-Dimashqī, [. . .] the shaykh of reciting in Damascus and the imām of the Umayyad Mosque. He was an arch-scholar in reciting, *al-farāʾiḍ*, arithmetic, *fiqh* and other (things). [. . .] and he read *farāʾiḍ* with the shaykh Muḥammad al-Najdī al-Ḥanbalī al-Faraḍī and with al-Shihāb al-ʿAlmawī, who was called by the name of Shikāra, and arithmetic and algebra together with *handasa* with the shaykh ʿAbd al-Laṭīf b. al-Kayyāl, the *muwaqqit* of the Umayyad Mosque. And he took from him much from ʿ*ilm*

85 al-Muḥibbī (see footnote 27), vol. 1, p. 277.

86 al-Muḥibbī (see footnote 27), vol. 1, pp. 373–374. It is highly unlikely that he studied with ʿAlāʾ al-Dīn ʿAlī b. Muḥammad al-Qūshjī since the latter died in Istanbul in 878/1474. Rosenfeld, Ihsanoğlu (see footnote 4), pp. 285–286. Even if the teaching of *hayʾa* was handled as the teaching of *ḥadīth*, for which I have no evidence, it remains implausible that he could have met Qūshjī. Since at least the seventh/thirteenth century, babies and toddlers were brought to lectures on *ḥadīth* in order to be able to claim later to have listened to a famous transmitter. In order to be presented to Qūshjī as a baby, Khān Aḥmad must have been born several years before 1474. Since he died in 1009/1599, this seems to be highly unlikely.

al-falak. He took the fundamentals of this science, until he excelled in it, with the shaykh Abū Bakr Taqī al-Dīn al-Saḥyūnī.[87]

4 Muḥammad b. Ḥusayn [. . .], called by the name of Bahā' al-Dīn 'Izz al-Dīn, al-Ḥārithī, al-'Āmilī, al-Hamadānī, [. . .]. Then he came to Jerusalem. Al-Rāḍī b. Abī l-Luṭf al-Maqdisī reported: [. . .] I asked him then to read about some of the disciplines of knowledge. He said: under the condition that this remains secret. I read with him things from *hay'a* and *handasa*. Then he travelled to al-Shām in order to return to Bilād al-'Ajam.[88]

5 Muḥammad b. 'Abd al-Qādir [. . .] al-Yamanī [. . .], he wrote a number of books in many disciplines, among them [. . .] a treatise on surveying called *al-Mashamma al-nafāḥa bi-taḥqīq al-misāḥa*, in which he brings together much (that he) collected from the books in this discipline according to the most direct way and the shortest method.[89]

6 Muḥammad b. Muḥammad b. Sulaymān al-Fāsī [. . .], [. . .] it was said that he [. . .] was a master in philosophy, logic, natural philosophy and metaphysics who did not reach his rank through corruption. He was well versed in Euclid's mathematical disciplines, *hay'a*, conics, the *Middle Books* and the *Almagest*. He knew the branches of arithmetic, algebra, number theory, the method of two false positions, music and surveying in a knowledge that nobody else shared with him except for the obvious (parts) of these disciplines without their particularities and the understanding of their facts. [. . .] And he was skilled in the occult sciences such as geomancy, magic squares, letter magic, *sīmiyā* and alchemy.[90]

7 Maḥmūd al-Baṣīr al-Sāliḥī al-Dimashqī [. . .], our excellent shaykh [. . .]. He read in Damascus with the most important of the teachers, among them our shaykh, the arch-scholar Ibrāhīm al-Fattāl, and by him he was trained and (educated) in (various) disciplines. He read with him Arabic, rhetoric and logic. He took the mathematical (sciences) from the Shaykh Rajab b. Ḥusayn and metaphysics from al-Munlā Sharīf al-Kurdī. [. . .] And I took from him logic, *handasa* and *kalām*. While I took *handasa*, he used tricks for determining their figures (or: theorems) well by giving examples from wax to me. His professor, the mentioned shaykh Rajab, had imitated them for him. He made them in a strong precision. While I read *handasa* with him I was astonished by his figurative representation (*taṣwīr*) as he had taken them from his professor. It was said: if the figure appears which he fabricated, then it corresponds with the figure which is in the book.[91]

8 Muṣṭafā b. Aḥmad [. . .] al-Būlawī, Muftī l-salṭana, [. . .] he taught in his house [. . .] he wrote among other works [. . .] a super-commentary on the

87 al-Muḥibbī (see footnote 27), vol. 3, pp. 186–187.
88 al-Muḥibbī (see footnote 27), vol. 3, pp. 440. 442–443.
89 al-Muḥibbī (see footnote 27), vol. 4, p. 11.
90 al-Muḥibbī (see footnote 27), vol. 4, pp. 304, 306.
91 al-Muḥibbī (see footnote 27), vol. 4, pp. 330–331.

commentary of *Ashkāl al-taʾsīs* and other (things) from the recensions (*al-taḥrīrāt*).[92]

9 Yaḥyā b. Taqī al-Dīn [. . .] al-Ḥalabī al-Dimashqī, known as al-Faraḍī. [. . .] he used to read in the school (*maktaba*) of the Mosque al-Darwīshiyya and he was the leader (*raʾīs*) in *handasa*, *hayʾa*, arithmetic, and *farāʾiḍ*.[93]

Bibliography

Brockelmann, C.: "al-Muḥibbī". *Encydopaedia of Islam*. Edited by: P. Bearman, Th. Bianquis, C. E. Bosworth, E. van Donzel and W. P. Heinrichs. Brill, 2007. Brill Online. 02 February 2007 www.brillonline.nl/subscriber/entry?entry=islam_SIM-5431

Høyrup, Jens: Practitioners – school teachers – 'mathematicians': the division of pre-modern mathematics and its actors. Talk given at Les Treilles 2003.

King, David A.: *A Survey of the Scientific Manuscripts in the Egyptian National Library*. Winona Lake, IN: Eisenbrauns, 1986.

al-Muḥibbī, Muḥammad Amīn b. Faḍl Allāh: *Khulāṣat al-athar fī aʿyān al-qarn al-ḥādī ʿashar*. Bayrūt, s.d., 2 vols.

Petry, C. F.: "al-Sakhāwī". *Encyclopaedia of Islam*. Edited by: P. Bearman, Th. Bianquis, C. E. Bosworth, E. van Donzel and W. P. Heinrichs. Brill, 2007. Brill Online. 23 January 2007 www.brillonline.n1/subseriber/entry?entry=islam_SIM-6503

Rosenfeld, Boris A.; Ihsanoğlu, Ekmeleddin: *Mathematicians, Astronomers & Other Scholars of Islamic Civilisation and their Work (7th–19th c.)*, Istanbul: IRCICA, 2003.

al-Sakhāwī, Shams al-Dīn: *al-Ḍawʾ al-lāmiʿ li-ahl al-qarn al-tāsiʿ* Bayrūt, s.d., 12 vols.

92 al-Muḥibbī (see footnote 27), vol. 4, p. 371.

93 Al-Muḥibbī (see footnote 27), vol. 4, p. 466.

8

ON FOUR SCIENCES AND THEIR AUDIENCES IN AYYUBID AND MAMLUK SOCIETIES

Having recently published a discussion of why decline is an unsuitable historiographical category for studying the history of the sciences in Egypt and Syria under the Ayyubids and Mamluks, I wish to present in this paper some examples that testify to the various interests that scholars and students had in the various non-religious sciences.[1] Whether any of these examples also speaks of decline depends, of course, on the yardstick we apply to evaluate a situation. The standard term of comparison, albeit never precisely formulated, is the quality of works produced during the ninth, tenth and eleventh centuries. These are compared to works of ancient Greek authors. When the scholars of the ninth to eleventh centuries achieved new results, developed new methods or built a new theory, they are appreciated as having been innovative or as having advanced knowledge. Works produced after 1100 are often considered to lack such features. Hence, they were labelled as representing decline. Such a procedure is beset, of course, with several serious problems. The question of whether it is appropriate to compare these texts is not raised. By now it is well known that in some areas later scholars indeed wrote texts that contain new methods or advanced theoretical contributions. Thus, if there was a decline, it was not an all-encompassing phenomenon. The procedure also ignores the fact that the alleged decline is diagnosed on the basis of teaching texts, while those that are seen as high-standard products were the outcome of research. It thus fully disregards the purpose of the compared texts and hence the most elementary context that should be taken into consideration when undertaking such a comparison. The existence of advanced texts or elementary works with new elements, after 1100, points to two other methodological shortcomings of the paradigm of decline. One is the glaring mistake of judging – on the basis

1 Brentjes, "The Prison of Categories – Decline and Its Company".

DOI: 10.4324/9781003372455-9

of a very limited knowledge – a whole range of disciplines without considering possible differences in the practices of these disciplines. The research of the last forty years has left no doubt that the situation in astronomy, logic and philosophy differed greatly from arithmetic or geometry, for example, with regard to theoretical advances after 1100. Thus, even if one were to adhere to this way of evaluating the intellectual activities in Islamic societies after 1100, decline is by no means a homogenous phenomenon that encompasses all disciplines, periods and regions equally. The other mistake committed by those attesting decline to all and everything after 1100 is that regional or even local as well as temporal differences that shaped the production of scientific texts, instruments and images are overlooked. Again, research on the various mathematical sciences, logic, philosophy and medicine in the last decades has amply demonstrated that the various disciplines and their representatives flourished in certain areas and periods, while not in others. Neither innovation nor decline were homogenously spread over the entire Islamic world. Decline as a type of grand narrative was and is a profound methodological aberration because of these different manners of disregarding the historicity of all human activities and achievements. It is like comparing apples and oranges after removing them from their trees, orchards, villages and large regions, without any concern for the people caring for them and their means to do so as well as the different kinds of use of the fruits of their labour. Hence, in this paper I want to illustrate what we have already learned by going beyond the limitations of the paradigm of decline and what else we can learn from studying the extant sources in a much narrower time frame, with a more precise spatial focus and with an open eye for the themes discussed in the sources.

1 Novelties in the Mamluk period

One of the disciplines that are usually subsumed under the label 'ancient sciences', 'Arabic sciences', 'sciences in Islam' or similar terms is *ʿilm al-mīqāt*, the science of timekeeping. For this discipline, David A. King and his students have successfully demonstrated – in the course of the last thirty years – the innovative achievements of its practitioners, who invented new instruments and developed new methods for handling big amounts of astronomical data as well as new mathematical solutions for all the standard problems of spherical astronomy.[2] The innovative character of this field that emerged under the Mamluk dynasty was not limited to innovations in content and instrumentation. The field constituted itself via two cultural activities, as King and Charette have argued. King sees the emergence of a new type of professional with a specialization in *ʿilm al-mīqāt*, the *muwaqqit*.[3] This new professional was socially recognized by being included in the madrasa system via *wuqūf*. As King has shown, this process is documented

2 King, *In Synchrony with the Heavens*.
3 King, "On the Role of the Muezzin and the Muwaqqit in Medieval Islamic Societies".

by the new apposition *al-muwaqqit* and similar terms that such experts began to add to their names from about the late thirteenth century onwards and by several *waqf* documents that stipulated a salaried position for such a person in addition to the *mudarris* and other already well-established offices. He defines them as 'mosque officials.'[4] Charette has, in addition, pointed to the formation of a new astronomical discipline, *ʿilm al-mīqāt*, which was fused from spherical astronomy, Euclidean geometry, arithmetic and instrumentation.[5] The immensely successful integration of this new discipline into the educational networks of Mamluk scholars has been emphasized by Charette and by myself.[6] *Muwaqqits* were clearly not only new experts in determining prayer times, the *qibla* or the beginning and end of Ramadhan or any other month. Many of them were first and foremost *mudarrisūn*.[7] Teaching *ʿilm al-mīqāt* was only one and not even the primary one of their many teaching obligations. They taught, not surprisingly, many hours of *fiqh* and its sub-discipline *ʿilm al-farāʾiḍ*. Furthermore, they gave classes in Arabic, philosophy (*falsafa* and *ḥikma*) and Sufism. Other mathematical disciplines, such as arithmetic, algebra and geometry, formed another important component of their teaching schedules. Teaching was the primary means of spreading the knowledge of *ʿilm al-mīqāt* among the *ʿulamāʾ* in major Mamluk cities like Cairo and Damascus, but also in provincial towns and beyond the Mamluk borders. Thus, in addition to innovation in content and instrumentation, professional specialization and the foundation of a new discipline, *ʿilm al-mīqāt* and its representatives achieved their social and cultural success by using the standard tool kit of Mamluk *ʿulamāʾ*, such as the diversification of their teaching profiles, the acquisition of several teaching and other positions and travelling for the sake of education or pilgrimage. This means that the scholarly and societal success of *ʿilm al-mīqāt* decisively depended on the vivacity of the educational and scholarly system at large. It was the result of the willingness of the religious scholars to engage with these themes and the respective methods, either directly by taking or giving classes, or indirectly by including those who studied and taught such topics in their local and long-distance networks and their narratives.

2 Mamluk attitudes to astrology

The study of the texts composed by Mamluk *muwaqqit*s together with the lack of a courtly office for a head astrologer has induced historians of science to assume that, with a few exceptions, Mamluks were not interested in astrology. King remarks that "the Mamluk astronomers (. . .) wrote surprisingly little on

4 King, "On the Role of the Muezzin and the Muwaqqit in Medieval Islamic Societies", 631–632, 642–647.

5 Charette, *Mathematical Instrumentation in Fourteenth-Century Egypt and Syria*.

6 Charette, "The Locales of Islamic Astronomical Instrumentation", see in particular: Brentjes, "Shams al-Dīn al-Sakhāwī on *Muwaqqits*", 129.

7 Brentjes, "Shams al-Dīn al-Sakhāwī on *Muwaqqits*", 134, 137–139, 143–144.

astrology".[8] He points out, however, that a few authors indeed wrote on theoretical aspects of astrology and that the astronomical handbooks that Mamluk astronomers compiled need to be examined for their astrological content.[9] With reference to one of the leading Cairene *muwaqqit*s of the second half of the fourteenth and the first half of the fifteenth centuries, Ibn al-Majdī (736–850/1336–1447), Charette comments that he "was something of an exception: A noted religious scholar, he nevertheless treated the topic of mathematical astrology in his *al-Jāmiʿ al-mufīd* and even cast a horoscope for a Mamluk amīr".[10] Both think that the relative lack of astrological texts compiled by Mamluk *muwaqqit*s resulted from their connection with religious institutions, since they were employed "either as *muwaqqit*s in mosques or as teachers in *madrasas* or Sufi convents".[11] King considers it "safe to assume that the Mamluk sultans had considerable interest in astrology, even though there are so few surviving texts to prove it".[12] Yet he also believes that "religious scholars, of course, frowned upon faith in such prognostications", pointing to Ibn Qayyim al-Jawziyya's (d. 749/1349) polemics against the discipline.[13] Ibn Qayyim al-Jawziyya pointed the finger at the *muwaqqit*s, who in his view did not take science, i.e., astronomy, seriously, but corrupted the faith of their contemporaries by practising astrology.[14] A similar lack of copies of and commentaries on Euclid's *Elements* can be attested. However, based on Mamluk biographical literature I could argue that this lack did not signify a lack of occupation with the book and the field on the part of Mamluk scholars.[15] Hence, this might also be the case with astrology. Moreover, astrology might have been considered merely as one of several means of counselling and prognostication, as was already the case among some of the Ayyubid princes. As in the case of geometry, Mamluk biographical literature provides important additional information that helps to achieve a more adequate understanding of astrology's place/s in Mamluk society and to determine who practised it and who consulted its practitioners.

The first observation to be made when studying Mamluk biographical dictionaries concerns the regular and normal usage of terms like *ʿilm an-nujūm* and *munajjimūn*, which in most cases refer to astrology and its practitioners. Despite occasional critical comments, references to authors who wrote against the discipline either as a science or as an acceptable belief for a Muslim and cases of false astrological predictions, the biographers largely regarded astrology as a legitimate

8 King, "The Astronomy of the Mamluks", 550.

9 King, "The Astronomy of the Mamluks", 550.

10 Charette, "Ibn al-Majdī"; see also http://islamsci.mcgill.ca/RASI/BEA/Ibn_al-Majdi_ BEA.htm (21 August 2011).

11 Charette, "Ibn al-Majdī", 562; King, "The Astronomy of the Mamluks", 550.

12 King, "The Astronomy of the Mamluks", 551.

13 King, "The Astronomy of the Mamluks", 551.

14 Livingston, "Ibn Qayyim al-Jawziyyah", 96–103.

15 Brentjes, "The Study of Geometry According to al-Sakhāwī (Cairo, 15th c) and al-Muḥibbī (Damascus, 17th c)".

element of scholarly education and praised those who excelled in it. The rhetoric of the biographers is not one of ostracism. The second observation indicates that this positive and at times laudatory depiction of astrological knowledge and its scholars is by no means limited to people of the past or to previous dynasties that had patronized astrologers at their courts. Mamluk biographers report on astrological knowledge also as part of the scholarly cultures of their neighbours like the Ilkhanids or the Rum-Seljuqs. These two observations suggest that not all Mamluk scholars and readers of historical literature rejected astrology. They may not have favoured it, but at least they considered it as a standard and even praiseworthy element of their narratives about themselves, their neighbours and their predecessors. Yet how did they depict astrology in their own environment? The first appearance of an astrologer under Mamluk rule is narrated for the end of the transition from Ayyubid to Mamluk rule in Egypt. Aḥmad ibn ʿAlī al-Maqrīzī (765–845/1364–1442) reports that the new ruler of Egypt, the Mamluk ʿIzz ad-Dīn Aybak (d. 655/1257), was warned by an unnamed astrologer of his death at the hand of a woman. Al-Maqrīzī specifies that this woman was Shajarat ad-Durr (d. 655/1257). As a result, "the frostiness between" the two increased in that year, although according to the story Aybak had already decided to kill his wife. The astrologer's prediction did not improve the relationship between the couple. Rather, as al-Maqrīzī believed, his newly acquired knowledge of fate changed Aybak's attitude towards Shajarat ad-Durr and made him offer for a daughter of the Atabeg of Mosul, Badr ad-Dīn Luʾluʾ (d. 659/1259).[16] While it is unclear whether there was indeed such an astrological prediction of ʿIzz ad-Dīn Aybak's death at the hand of his wife, al-Maqrīzī's story illustrates that he did not consider it unlikely. Moreover, the astrologer's presence provides the story about the violent death of the two unequal rulers of Egypt with a dramatic beginning. It serves as a justification for Aybak's attempt to form an alliance with the ruler of Mosul and a legitimization of his murder as predicted by the stars and thus as fate.

Other stories about activities of astrologers connected with Mamluk *umarāʾ* also and not at all surprisingly focus on the capacity of astrologers to influence human fate. In one such story, the Damascene astrologer of Amīr Almās al-Ḥājib, who had served him already while he was a member of the household of the sultan, i.e., an-Nāṣir Muḥammad (r. 693–694/1293–1294, 698–708/1299–1309, 709–741/1310–1340), was imprisoned and killed, when in the year 735/1334 his master had been accused of joining forces with Amīr al-Jamāl ad-Dīn Ārqūsh, the governor of al-Karak. No explanation is given why the astrologer received this harsh sentence. His master was only beaten with cudgels and canes while maintaining his innocence after his fate had apparently been determined by various *umarāʾ*, who exercised geomancy in a session outside the courtyard.[17] While there is no hint at any astrological counselling, it may well have taken place,

16 Al-Maqrīzī, *as-Sulūk*, vol. 1, 132.
17 Al-Maqrīzī, *as-Sulūk*, vol. 1, 491.

because Ārqūsh had voiced his plan to become the next sultan.[18] Being a Manṣūrī Mamluk, some twenty years earlier such an attempt to commit treason would have undoubtedly led to the conspirators' execution, given that the beginning of the sultan's third reign was marked by removing his father's and his brother's loyal supporters from major offices and speeding up their demise.[19] Yet six years before his own death an-Nāṣir obviously felt secure in his position as the head of the state and the military, so that he could give lower ranks a free hand in extorting hundreds of thousands of dirham from his deputies in ash-Shām rather than killing them.[20]

The pre-eminence of geomancy over astrology when it came to determining an amīr's fate points to the importance of *ʿilm ar-raml* as a divinatory practice in the Mamluk period. Often both disciplines are mentioned together, or people who practised astrology were also perceived as masters of geomancy, as is the case in the story about Qurqmās ash-Shaʿbānī, amīr of the arms, and his ambition to become sultan. This story presents the practitioners of astrology and geomancy as one group of experts, besides some "whom the ignorant among the people fancied as friends of God" and some interpreters of dreams who all perceived of the amīr as the next sultan due to their specific skills.[21] After the amīr's ambitions had become known, they talked to him about their knowledge of the hidden. In addition to the two groups already mentioned, others approached him to inform him about his fate on the basis of astrology and geomancy. Yet Qurqmās, so al-Maqrīzī claims, buried this prognostication in his mind and did not count on it. Moreover, he did not act on it as long as Sultan al-Ashraf Barsbay (r. 825–841/1422–1438) was alive.[22] In this case, loyalty or fear prevailed over the belief in prognostica-

18 Al-Maqrīzī, *as-Sulūk*, vol. 1, 491.
19 Holt, "The Sultanate of al-Manṣūr Lāchīn (696–698/1296–1299)", 525; Reuben, "The Remaking of the Military Elite".
20 Al-Maqrīzī, *as-Sulūk*, vol. 1, 491.
21 Al-Maqrīzī, *as-Sulūk*, vol. 3, 420.
22 Al-Maqrīzī, *as-Sulūk*, vol. 3, 420; Ibn Taghrī Birdī (d. 5 Dhū l-Ḥijja 874/5 June 1470) repeats somehow the story told by al-Maqrīzī, but clearly changed its thrust by turning the relationship between the amīr and the diviners into a case of ambition coming first and prognostication coming second, as a consequence of a widespread knowledge about the ambition. Ibn Taghrī Birdī, *History of Egypt*, part V, 19. In agreement with this basic outlook, he repeatedly stressed that it was Qurqmās' ambition and lust for power that led him to rebel against Jaqmaq after the latter had become Sultan aẓ-Ẓāhir Sayf ad-Dīn (r. 842–857/1438–1453). Ibn Taghrī Birdī, *History of Egypt*, part V, 18–20, 29–41. This difference between the two stories supports the methodological point that I will make in the following section with regard to Sayf ad-Dīn al-Āmidī's fate in 631/1233. Stories told by medieval historians and biographers cannot be taken at face value, but need to be studied against the background of the complete set of reports and their various narratives, if we want to understand their meaning. I do not pursue this approach in the case of Qurqmās, because I am not interested here in the question as to which of the lines of narration carries greater weight and is more likely to correspond to what actually occurred. My goal here is to present different kinds of presentation of the relationship between Mamluk sultans and *umarāʾ* on the one hand, and astrologers and occasionally other diviners on the other.

tion. However, al-Maqrīzī's choice of words leaves little doubt that he considered these experts of the "science of the concealed," i.e., the Sufis, dream interpreters, geomancers and astrologers as little more than charlatans and irresponsible stirrers of social unrest.

Another story emphasizes the notoriously imprecise nature of predictions concerning human affairs. A few years before Qurqmās' end, on 17 Rabīʿa II 839/8 November 1435, the heads of six men were delivered by an escort of the Qara Qoyunlu ruler Iskandar ibn Qarā Yūsuf (r. 823–841/1420–1438) to al-Ashraf Barsbay. They were displayed for three days at the Gate az-Zuwayla, and the entire population of Cairo rejoiced.[23] The reason for the men's misfortune was an astrological prediction given to their leader ʿUthmān Qarā Yūluk, saying that he would enter Cairo, which made him believe that one day he would rule Egypt. Al-Maqrīzī scolds the amīr, calling for his head to be impaled on a cudgel and exhibited publicly as "an exemplary warning from God".[24] Yet whether the historian's scorn is directed towards the man's gullibility, ambition or some other previous misbehaviour remains unclear.

A different type of story concerns the impact that astrologers had upon Mamluk society when they predicted natural events. In Jumāda II 834/Mid-February–Mid-March 1431, Cairene astrologers warned of an impending solar eclipse. According to al-Maqrīzī, the authorities in Cairo proclaimed a fasting, which everyone obeyed to his or her best.[25] However, the eclipse was not visible and people were upset with the astrologers.[26] This surprising result of the fast's ostensible efficacy throws a light on what people expected from the experts and what they believed about the purpose of religious deeds. They were not relieved that an event they considered to be dangerous apparently did not take place, but instead blamed the astrologers for getting it wrong. Neither did they fast in order to appeal to God Almighty to prevent the event from happening in their city. Then, what did they fast for or against? Al-Maqrīzī does not comment on this point. He reports that the eclipse actually did take place and was sighted in al-Andalus in the afternoon of 28 Jumāda II 834/13 March 1431. The information that arrived from the Peninsula was very specific. The eclipse was almost total except for an eighth of the sun that did not enter the earth's shadow.[27] Hence, astrologers were not merely expected

23 According to Ibn Taghrī Birdī, the ruler of Amida had died more than two months earlier, on 5 Ṣafar 839/29 August 1435. Moreover, Iskandar ibn Qarā Yūsuf did not send his enemy's head to the Mamluk sultan in Cairo, but to the Ayyubid ruler of Ḥisn Kayfa, al-Malik al-Ashraf (?, ruled at some point between 1378 and 1433). Hence, Ibn Taghrī Birdī's story conflicts with the dates for the rulers of Ḥisn Kayfa, since al-Malik aṣ-Ṣāliḥ Ṣalāḥ ad-Dīn began to rule in 1433. Ibn Taghrī Birdī, *History of Egypt*, part IV, 205–206; *Islamic Desk Reference*, compiled from the *Encyclopaedia of Islam* by E. van Donzel, Leiden, New York and Köln: Brill, 1994, 45.

24 Al-Maqrīzī, *as-Sulūk*, vol. 3, 381.

25 Al-Maqrīzī, *as-Sulūk*, vol. 3, 346.

26 Al-Maqrīzī, *as-Sulūk*, vol. 3, 346: فلم يظهر الكسوف، ووقع الإنكار على من أندر به.

27 Al-Maqrīzī, *as-Sulūk*, vol. 3, 346.

to be right, but also precise, and they indeed lived up to this expectation, at least sometimes in some places. Whether this report convinced the authorities and the population of Cairo that their own astrologers had got it right somehow and thus deserved an apology remains unclear. At least, al-Maqrīzī did not bother to make any further comment.

Another solar eclipse took place in Cairo two years later on 28 Shawwāl 836/7 June 1433 at the end of the fourth hour. This time the astrologers had apparently refrained from warning the city. Since the occultation was rather weak and the light changed only slightly, very few people, so al-Maqrīzī, took notice of the event.

> Almost nobody was performing the traditional prayer or going to one of the city's Friday mosques. Then the eclipse dissolved quickly. But there was somebody who claimed to have some understanding of astrology by some reading and much boldness. He had spread lies and rumours days before the event. He disgraced the affair of the eclipse and all that pointed to it until his calumnies and defamations became widely known and entered some peoples' erroneous imagination. After the eclipse had not been something big, the sultan summoned a group of the *people of the ephemerides* (*ahl at-taqwīm*) who practised this discipline and blamed them and intimidated them.[28]

This story is not only interesting because it blames the experts for the eclipse not turning out to be strong and clearly visible. It also indicates that people without knowledge could present themselves as experts and arouse peoples' anxieties in such a manner that the sultan felt it necessary to interfere. He did not attempt to protect the professional space of the city's experts, but threatened them because of their alleged shortcomings. Al-Maqrīzī's choice of words suggests that he disapproved of both the charlatan's incompetence and provocation of ill feelings and the sultan's inappropriate treatment of a group of scholars Another remarkable point in the author's rhetoric is his continued avoidance of the term *muwaqqit*, while referring to his colleagues as the people who produce ephemerides. It is possible that he considered *muwaqqit* too narrow a term for describing the scope of responsibilities of those who observed and described eclipses. His terminological deviation from the standard apposition *al-muwaqqit* and the sultan's interference and displeasure also indicate that in the first half of the fifteenth century scholars in Cairo (and probably elsewhere in the Mamluk dominion) were not only teaching spherical astronomy and related topics at madrasas, constructing instruments and calculating multi-entry tables for determining prayer times and prayer directions. They were also responsible for a continuous and careful observation and recording of major events in the sky.

28 Al-Maqrīzī, *as-Sulūk*, vol. 3, 366.

Other themes related to astrology in Mamluk times are rare in al-Maqrīzī's history. He hardly ever paid attention to the astrological or other scientific education of high-ranking Mamluks and scholars. An exception among the former was his own patron Sayf ad-Dīn Bazlār al-ʿUmarī (d. 791/1389) and the latter's sons. The father had been governor of various Syrian cities and finally of Damascus, before he fell into disgrace and ended his life in prison. He possessed a good education in literature, excelled in the art of writing and had some knowledge in the sciences (religious and others), in particular in astronomy and astrology.[29] A scholar to whom al-Maqrīzī ascribes some knowledge of astrology is the *muwaqqit* Shihāb ad-Dīn Aḥmad ibn Ghulām Allāh al-Kawmrīshī, who died 26 Ṣafar 836/22 October 1432, although al-Maqrīzī did not call him by this title. Al-Maqrīzī describes him as very good in extracting an ephemerid (*taqwīm*) from an astronomical handbook (*zīj*) and as someone who had no successor like himself.[30]

Other Mamluk and later Ottoman sources draw a slightly different picture and add information beyond that found in al-Maqrīzī's history. They present more scholars with astrological interests and report on astrological predictions other than only death and rebellion. They confirm al-Qayyim ibn al-Jawziyya's claim that *muwaqqit*s also worked as astrologers by giving scholars both appellations. Like al-Maqrīzī they join astrology with geomancy and speak of the fear of eclipses and the possible negative effects that people in those days believed they could cause. Surprisingly, I could find no clear indication that any of the Mamluk sultans used the services of an astrologer for personal or political goals except for the one by al-Maqrīzī referred to above. All stories of Mamluk sultans and their counselling by an astrologer refer to a period prior to their ascendance to the throne. In none of the cases the name of the respective astrologer is mentioned. It is as if these people who spoke of future chances to rule Egypt did not belong to the same group of people who studied or taught spherical astronomy and related topics at a *madrasa*. Hence it remains difficult to judge who those astrologers were and what kind of service the *muwaqqit*s provided when they acted as astrologers. The same applies to the experts who predicted, observed and described solar and lunar eclipses. The stories told by Mamluk historians about these natural events seem to imply that predicting, observing and describing such events was part of a regular service of astronomically and astrologically educated experts and that this service was somehow supervised by the Mamluk state. Yet no names are given, nor are the places of observation and the employed instruments discussed. Finally, the historians' rhetoric suggests that most of them took an ambivalent stance towards astrology, considering it to be unreliable, potentially dangerous and socially undesirable. However, they also recognized it as a teachable discipline – with a subject matter, methods and instruments – that could enhance the reputation of a scholar if he mastered it.

29 Al-Maqrīzī, *as-Sulūk*, vol. 2, 423.
30 Al-Maqrīzī, *as-Sulūk*, vol. 3, 360.

Mūsā ibn Muḥammad al-Yūnīnī (d. 726/1326) tells a second story of an astrologer's involvement in the early years of the Mamluk dynasty. As al-Maqrīzī, al-Yūnīnī presents astrology and geomancy as being practised jointly by one and the same professional as a normal matter in his days. In contrast to al-Maqrīzī, he not only delivers a story with a positive outcome, but one of a client inviting an expert to find a proper context for revealing his origins and his future plans for revenge that included the take-over of the fragile Mamluk state. It appears that the story's narrator was Amīr al-Fāris Aqṭa'ī al-Musta'rib, who at the time was in the service of the Mu'izzī Mamluk Sayf ad-Dīn Qutuz and later Sultan al-Malik al-Muẓaffar (r. 1259–1260). One day an astrologer from North Africa or al-Andalus came to Qutuz. He was very skilled and knowledgeable in *'ilm ar-raml* and *'ilm al-falak*, i.e., in geomancy and the art of compiling astronomical handbooks or in astronomy at large. Qutuz ordered most of his retainers to leave, and only the most senior of his Mamluks, among them Aqṭa'ī, stayed on. When only those whom he trusted were left in the room, Qutuz told the astrologer to perform a prognostication.[31] The astrologer did so and told Qutuz what he had divined. Then a further divination was ordered to find out who would rule Egypt after Aybak, who would defeat the Mongols and who was doomed to perdition by Aqṭa'ī's hand. The astrologer remained silent for some time, pondering over his figures. Then he spoke about their technicalities, counting points and describing their features and some of their implications, apparently unwilling to reveal too much. Qutuz asked him why he did not speak of Maḥmūd ibn Mamdūd. The astrologer confirmed that this was the only name that he could read in the figures. Qutuz came forward and revealed himself to be Maḥmūd ibn Mamdūd, vowing that he would crush the Mongols and avenge his maternal uncle, the Khwārazmshāh. Aqṭa'ī continues his story by describing how all the others in the room were caught by surprise and their oath to maintain silent about what had happened. The astrologer who had performed the two geomantic sessions received a generous remuneration of 300 dirham and was dismissed.[32] Al-Yūnīnī ends the story piously: "God Almighty decided to make him king and let him crush the Tatars, defeating them".[33]

Al-Yūnīnī tells further stories in which Qutuz speaks of his conviction that one day he would rule Egypt and defeat the Mongols. One of them is about a dream in which the Prophet Muḥammad appears to him and declares: "You will rule Egypt and crush the Tatars!"[34] In another story, astrologers again play a role. The narrator of this story was Atabeg Badr ad-Dīn Baktūt, who claims that he, Qutuz and Baybars Bunduqdārī grew up together. One day they agreed to visit an astrologer on one of the streets of Cairo. Qutuz demanded: "Watch the stars,

31 The formula is simply "beat!", meaning that the astrologer should create the necessary sixteen lines of randomly distributed dots.

32 Al-Yūnīnī, *Dhayl Mir'āt az-zamān*, vol. 1, 140.

33 Al-Yūnīnī, *Dhayl Mir'āt az-zamān*, vol. 1, 140.

34 Al-Yūnīnī, *Dhayl Mir'āt az-zamān*, vol. 1, 145.

(perform geomancy) and calculate".[35] The astrologer informed him: "You will rule Egypt and crush the Tatars!"[36] Then, seizing the opportunity, Baybars inquired about his destiny. The astrologer told him that he too would one day rule Egypt, adding other things the narrator does not disclose.[37] Then the astrologer invited Badr ad-Dīn to look into the stars, and he spoke the apparently ritualistic formula: "Watch for me, (perform geomancy) and calculate!" He was told that one day he would become the amīr of one hundred riders and that Baybars would be the one to appoint him to this position. Badr ad-Dīn concludes story with what seems to be an equally ritualistic exclamation: "What a miraculous agreement (between the predictions and the later events)!"[38]

According to al-Yūnīnī, astrologers were also involved in the fate of Qutuz after he had become sultan. This time, his informant was one of fifteen astrologers who allegedly met in 685/1286 in the Fortress al-Jabal in Cairo after al-Malik al-Muẓaffar had announced his decision to leave Cairo and fight the Mongols. Referred to as an-Najm an-Naḥḥās, the astrologer, he is the first expert in such a relationship with a Mamluk sultan to be given a name. He is also the first one whose place of birth is mentioned, namely Damascus. Since the time and place of the meeting are specified, it may well be that the story is fairly reliable despite its suggestion of a conflict of interests between Mamluks and astrologers. An-Najm an-Naḥḥās narrates that he and his fellow astrologers stayed in the fortress for eight days. At some point in time Mamluks arrived with the salary, saying to the astrologers: "Look for a lucky horoscope for the success of the sultan and his safe return!"[39] An-Najm claims: "We all agreed that we would take the horoscope (regarding) the success of the Muslims and the victory over the defeated enemy. We did not think of al-Malik al-Muẓaffar's success and his return!"[40] Thus things took their course.

Apart from these highly significant stories about the role of astrologers in the early years of the Mamluk dynasty, al-Yūnīnī also talked about the relationships between scholars and the functions that astrologers had in social and cultural settings, just like any other scholar. One such occasion was the death of the physician Abū l-Faraj ibn Yaʿqūb Amīn ad-Dawla in 685/1286. The astrologer Sayf ad-Dīn Abū Bakr delivered the funeral oration for which he composed a poem of two verses.[41] Another story that demonstrates the inclusion of scholars with an interest in astrology and astronomy in the cultural fabric of Mamluk society despite the ambivalent attitude towards this profession tells of a person who was knowledgeable about the stars and *ʿilm al-hayʾa* (mathematical cosmography) by

35 Al-Yūnīnī, *Dhayl Mirʾāt az-zamān*, vol. 1, 146.
36 Al-Yūnīnī, *Dhayl Mirʾāt az-zamān*, vol. 1, 146.
37 He reigned as al-Malik aẓ-Ẓāhir from 1260 to 1277.
38 Al-Yūnīnī, *Dhayl Mirʾāt az-zamān*, vol. 1, 146.
39 Al-Yūnīnī, *Dhayl Mirʾāt az-zamān*, vol. 1, 144.
40 Al-Yūnīnī, *Dhayl Mirʾāt az-zamān*, vol. 1, 144.
41 Al-Yūnīnī, *Dhayl Mirʾāt az-zamān*, vol. 2, 126.

the name of ʿAbd ar-Raḥmān ibn ʿAbdallāh al-Jazarī (d. 672/1274). He is praised for his study of the Qur'an and he conducted the Friday prayer at the Shrine of ʿAlī in Baalbek.[42]

Ibn al-Ḥajar al-ʿAsqalānī (773–852/1371–1448) on his part emphasized the descent of some astrologers from well-positioned families of religious scholars and holders of high civil offices. An example is the uncle of a certain Kamāl ad-Dīn, who was *muḥtasib* of Cairo under the reign of an-Nāṣir. The uncle's name was Faraj ʿAlī ibn Muḥammad, Nūr ad-Dīn ibn ash-Shāhid. Ibn Ḥajar calls him *al-mīqātī* and *al-munajjim*, i.e., a person who dealt with *ʿilm al-mīqāt* and astrology. He was the head of those who compiled astronomical handbooks and wrote ephemerides. He also had knowledge of geomancy and other things. At the end of his life (d. 801/1398) he was close to Sultan aẓ-Ẓāhir Barqūq (r. 783–791/1382–1389, 792–801/1390–1399) and the head of a Sufi order.[43] Ibn al-Ḥajar also presents other areas of astrological counselling, such as journeys. One of Cairo's head judges, for instance, avoided sailing on the Nile whenever possible, because he had either dreamt or had been told by an astrologer that travelling on a boat was a danger to his life. His name was Ṣadr ad-Dīn Abū Maʿālī Muḥammad ibn Ibrāhīm al-Munāwī (d. 803/1400), who was the son of Bint Zayn ad-Dīn ʿUmar al-Bisṭāmī (d. after 748/1347), the daughter of another head judge of Cairo, and a student of ʿIzz ad-Dīn ibn Jamāʿa (d. 766/1367), a famous teacher of *ʿilm al-uṣūl*, medicine and philosophical subjects in Mamluk Cairo.[44]

The Ayyubid prince and scholar Abū l-Fidāʾ (672–732/1273–1331) reports that people were of different opinions as to the cause of Sultan al-Ashraf Barsbay's death in 841/1438. Some believed that the total lunar eclipse that took place while he was in Damascus had caused his demise.[45] In his history he included some comments by Sibṭ al-Abharī from Hama, who was very talented in the mathematical sciences, the determination of the times and the construction of sundials. He might also have offered astrological advice.[46] However, a later reader seems to have added the information that he had been promoted by al-Malik al-Muʾayyad, i.e., Abū l-Fidāʾ, but fell into disgrace afterwards and moved to Aleppo, where he died. Abū l-Fidāʾ or this later reader added: "I say: the people of Hama contested his faith."[47]

In his *Shadharāt adh-dhahab fī akhbār man dhahab* the Ottoman historian Ibn al-ʿImād (1032–1089/1622–1679) finally exemplifies the ambivalent stance towards astrology. His text contains several critical or even derogatory remarks.[48] Yet, contrary to al-Maqrīzī, he also includes positive stories about astrological

42 Al-Yūnīnī, *Dhayl Mirʾāt az-zamān*, vol. 1, 360.
43 Ibn Ḥajar al-ʿAsqalānī, *Inbāʾ al-ghumr*, vol. 1, 228.
44 Ibn Ḥajar al-ʿAsqalānī, *Inbāʾ al-ghumr*, vol. 1, 263.
45 Abū l-Fidāʾ, *Tārīkh Abī l-Fidāʾ*, vol. 3, 68.
46 Abū l-Fidāʾ, *Tārīkh Abī l-Fidāʾ*, vol. 3, 224.
47 Abū l-Fidāʾ, *Tārīkh Abī l-Fidāʾ*, vol. 3, 223.
48 Ibn al-ʿImād, *Shadharāt adh-dhahab*, see for instance vol. 5, 339.

predictions of human fate, for instance, the case of al-Ashraf Barsbay, who was later to become sultan, as foretold by unnamed astrologers. Ibn al-ʿImād's informant exclaimed: "This was one of the miraculous agreements."[49] He also highlights the rejection that scholars with astrological knowledge could experience in fifteenth-century Cairo, while on the other hand someone was appointed as a madrasa teacher, although he was known to be an astrologer.[50]

The stories told by Mamluk and Ottoman historians show without doubt that astrologers were active in Mamluk society in the thirteenth, fourteenth, and fifteenth centuries. They served high-ranking Mamluks as well as the general populace. Their recommendations and predictions were listened to by the authorities, individual clients of the uppermost echelon of Mamluk military leadership and by the commoner. Despite religious disapproval of astrology and contemporary criticism of their practitioners by well-known religious scholars each of the examples suggests that individuals as well as groups were prepared to place their trust in astrological counselling. Only when things turned out differently, they remembered the precarious nature of astrological predictions. Then they heaped scorn upon the counsellors and doubted their expertise. In the case of planned treason an astrologer could well lose his head, while his client could get away with a less severe punishment. Fear of fate did not necessarily include fear of the messenger. As regards impending natural events, astrologers were expected to be reliable and precise observers. If the event fell short of expectations, the experts were the ones to be blamed and chastised. Concerning future human deeds, they were considered to be unreliable and imprecise. In such cases the clients were often the ones ridiculed. All stories on human destinies in al-Maqrīzī's history include the astrologer's predictions of crimes and dangers, quite often at the highest echelon of society and involving high treason. No other matter is considered to be worthy of mention, such as selecting auspicious days for a certain kind of social activity or casting horoscopes. This seems to reflect al-Maqrīzī's attitude regarding the possibility of knowing the hidden and of predicting the future of men and women. By choosing sinister plots and deadly dangers as the focus of the activities of astrologers and other experts from similar fields of knowledge al-Maqrīzī enhances the dramatic impact of his storytelling. Yet he also delivers a warning and a critique. The likelihood that al-Maqrīzī's choice of stories was intentional and that it reflects not merely his ambivalent attitude towards the discipline, its practitioners and their practices, but also a morally edifying purpose is strengthened by the presence of a richer array of themes presented by other contemporary and later writers. Mamluk astrologers were not always eager to predict death or stir rebellion by insinuating the possibility of rulership to lower-ranking Mamluks. They also gave advice on more mundane and private affairs like travelling on a boat. The relative frequency of stories of death and future ascent to the throne indicate, however, a perception

49 Ibn al-ʿImād, *Shadharāt adh-dhahab*, vol. 5, 292.

50 Ibn al-ʿImād, *Shadharāt adh-dhahab*, vol. 5, 366; vol. 7, 9.

of the astrologers' substantial impact on affairs of public interest and political relevance. Maybe this role assumed by perhaps too many practitioners of the field is one of the reasons why, once having become sultan, a Mamluk would abstain from or at least conceal any connection with such people. The stories about eclipses, religious activities linked to them, views about their impact on human life and reactions when eclipses did not take place to their full extent throw a new light on what was expected from astrologers and the pressures they were exposed to. They imply a much lesser dissemination of basic knowledge about eclipses than the many extant manuscripts about elementary astronomy and cosmology or the reports about classes on such topics at *madrasa*s suggest. Despite the undisputable fact that *madrasa*s and the teaching system connected with them led to a greater spread of elementary mathematical and astronomical knowledge than previously available through courtly patronage and private tutoring, the reactions of the ordinary inhabitants of Cairo and Damascus, scholars and sultans to eclipse reports, charlatans and experts show that the understanding of the causes and properties as well as the visibility of eclipses was very limited. Scholars like al-Maqrīzī or Ibn Qayyim al-Jawziyya felt compelled to defend faith against superstition through their narrative choices or polemical treatises. They argue or insinuate that good science, i.e., observations and calculations, in combination with the traditional prayers and the fulfilment of other obligations was the only protection against straying from the truth and proper behaviour. The name- and facelessness of most astrologers in the stories told by the historians may point, at least partly, in the same direction. Thus, a serious investigation of the complexity of astrological practice and its evaluation by different members of Mamluk society is desirable, not only for a better understanding of the role that this specific set of knowledge and skills played, but for uncovering the multitude of expectations, reactions, qualifications and interests behind the highly contradictory attitudes and actions highlighted in the stories selected for this paper.

3 Philosophy at Ayyubid courts and in Mamluk Madrasas

Since Henry Corbin's studies of Safavid philosophical writers it is a well-known fact that philosophical studies in Iran continued after the death of Ibn Sīnā (d. 428/1037). By now it has become clear that many more scholars were interested in philosophical themes and their relation to theological questions between the early eleventh century and the sixteenth century than previously assumed. In many cases they followed the perspectives opened up by Ibn Sīnā and his students. Although much more needs to be done in order to obtain a clear picture of philosophical studies in the Iranian world during these five hundred years, it is already recognized that in that period philosophy had a home in this region. But what about Egypt and Syria? These two parts of the Arab world are often ignored in such discussions. The interest in philosophical topics among scholars patronized by the Ayyubid dynasty has been emphasized only very recently, for instance in papers by Arif and Endress.[51] Dealing with the study of Ibn Sīnā's texts at *madrasa*s Endress' paper includes a section on Fakhr ad-Dīn ar-Rāzī (d. 605/1209) and the role of his students in dispersing Ibn Sīnā's and Fakhr ad-Dīn ar-Rāzī's works in northern Iraq, Syria and Anatolia.[52] The scholars involved in this dissemination were Zayn ad-Dīn al-Kashshī and Quṭb ad-Dīn al-Miṣrī (d. 617/1221) in Khurasan, Afḍal ad-Dīn al-Khūnjī (d. 646/1249) in Egypt, Shams ad-Dīn al-Khusrawshāhī (d. 652/1254) in Damascus, Athīr ad-Dīn al-Abharī (d. 663/1265 ?) in Irbil and Baghdad (?), Tāj ad-Dīn al-Urmawī (d. ?) in Anatolia and Sirāj ad-Dīn al-Urmawī (d. 681/1283) in Konya.[53] The texts they spread included Ibn Sīnā's *Ishārāt wa-tanbihāt*, *'Uyūn al-ḥikma* and *Kitāb ash-Shifā'* as well as Fakhr ad-Dīn's *al-Muḥaṣṣal* and *al-Mulakhkhaṣ fī l-ḥikma wa-l-manṭiq*.[54]

In his survey – which is not always correct – Endress names two of the features that characterized the teaching of philosophy in Ayyubid Syria and Egypt. Physicians played an important role in the spread and teaching of Ibn Sīnā's philosophy. They studied the theoretical part of the *Qānūn fī l-ṭibb* and parts of Ibn Sīnā's encyclopaedias in order to acquire the needed theoretical foundations of their profession. These works from Iran reached Cairo via Baghdad or via Damascus.[55] A second important practice of disseminating knowledge in the ancient sciences was the *riḥla*. Visits by Iranian scholars and students in Damascus, Aleppo, Cairo or Mosul for studying or teaching and by natives from those and other cities in the Arab world were an important factor in the dissemination of scientific knowledge in the sixth/twelfth and seventh/thirteenth centuries.[56] Endress demonstrates this very lively dispersion of Ibn Sīnā's and Fakhr ad-Dīn's philosophical texts through student–teacher networks that include Ayyubid Cairo and Damascus.[57]

A closer look into major biographical dictionaries and historical chronicles of the Ayyubid and Mamluk period shows that Endress's concept of a dispersion of philosophy from Iran to Egypt, Syria and Anatolia by physicians in the sixth/twelfth and early seventh/thirteenth centuries and by students of Fakhr ad-Dīn ar-Rāzī needs to be expanded to include the role that courts supporting these scholars played in these two centuries, the possible contribution of book trade and itinerant physicians and pharmacists and the planned visits to philosophers and physicians in Iran by scholars

51 Arif, "Al-Āmidī's Reception of Ibn Sīnā: Reading al-Nūr al-Bāhir fī al-Ḥikam al-Zawāhir"; Endress, "Reading Avicenna in the Madrasa".

52 Endress, "Reading Avicenna in the Madrasa".

53 Endress, "Reading Avicenna in the Madrasa", 404.

54 Endress, "Reading Avicenna in the Madrasa", 407. Regarding the character of the two texts by Fakhr ad-Dīn and his development as a philosopher and his intellectual relationship to Ibn Sīnā, see, for instance, the new evaluation by Shihadeh, "From al-Ghazālī to al-Rāzī".

55 Endress, "Reading Avicenna in the Madrasa", 390–391.

56 An example in case is Sharaf ad-Dīn aṭ-Ṭūsī (d. 1186). He taught in Aleppo, Damascus and Mosul. One of his most famous students was Kamāl ad-Dīn ibn Yūnus (d. 1242), who in the later twelfth and first four decades of the thirteenth centuries taught many important scholars. See Endress, "Reading Avicenna in the Madrasa", 394–395.

57 Endress, "Reading Avicenna in the Madrasa", 404–408.

from Syria or Egypt.[58] Kamāl ad-Dīn ibn Yūnus – in addition to his pivotal role in Mosul regarding the dissemination of the ancient sciences including philosophy and logic by teaching and the link he provided between Khurasan, Azerbaijan, Baghdad, Damascus and Cairo – also wrote a book on philosophy entitled *al-Lughz* that in his view no one understood, except for himself and two student visitors from Syria.[59] Badr ad-Dīn Lu'lu', the lord of Mosul, patronized Kamāl ad-Dīn for his excellence in the ancient sciences including philosophy. He also gave his support to the session with Frederic II's ambassador in Mosul, which Ibn Abī Uṣaybiʿa portrayed as a major cultural and political event that included many of the town's scholars and involved dressing in accordance with the rank of the visitor.[60] Shams ad-Dīn al-Lubūdī (d. 621/1224), "the most excellent of his time in the philosophical sciences and the art of medicine", travelled to Iran in order to study philosophy (*ḥikma*) with Najm ad-Dīn Asʿad al-Hamadānī and medicine with the country's greatest doctors.[61] The close contacts with Iran in philosophy and medicine continued well into the Mamluk period. A third element that characterizes the spread of philosophy under Ayyubid and Mamluk rule is the presence of scholars interested in it in smaller towns of Syria and Egypt, such as Safad, Qus or ar-Raqqa.[62] A certain Ḥasan b. Muḥammad al-Iṣfūnī al-ʿUthmānī, who had been born in al-Karak and had studied in Syria and Egypt, spent the rest of his life in Safad, leading a pious life and being obsessed with the books by Abū Naṣr al-Fārābī and Ibn Sīnā.[63] Thus, Abū Naṣr al-Fārābī continued to be available and occasionally read long after ʿAbd al-Laṭīf al-Baghdādī (558–628/1163–1231). Other cases of scholars with a particular interest in philosophical texts can be found elsewhere. Aṣ-Ṣafadī, for instance, reports on ʿAbd al-Qādir b. Muhadhdhab (d. 725 or 6/1325 or 6), a nephew of Kamāl ad-Dīn Jaʿfar al-Adfawī, that he was a philosopher (*faylasūf*) and spent much time reading the Pseudo-Aristotelian *Uthulūjiyā* and *Tufāḥa* and other philosophical books.[64] The main focus in the teaching of philosophical texts in the early Mamluk period was, however, on Ibn Sīnā. Shams ad-Dīn al-Iṣfahānī (d. 749/1348), ʿIzz ad-Dīn b. Jamāʿa, Shams ad-Dīn Shirwānī (ca. 782–834/1380–1430) and Muḥyī ad-Dīn al-Kāfiyajī (d. 878/1474) are the teachers of philosophy at madrasas in Mamluk Cairo most frequently mentioned. The works by Ibn Sīnā used in teaching, in particular by Shams ad-Dīn al-Iṣfahānī, were the *Kitāb ash-Shifā'*, *ʿUyūn al-ḥikma*, *Aqsam al-ʿulūm al-ʿaqliyya* and *Ishārāt wa-Tanbihāt* plus commentaries on the last work. Students of philosophy were either physicians or religious scholars. Some of them were seriously interested, while others only looked for some generalities. As

58 Ibn Abī Uṣaybiʿa, *ʿUyūn al-anbā'*, vol. 1, 342.
59 Endress, "Reading Avicenna in the Madrasa", 394–397, 407; Ibn Abī Uṣaybiʿa, *ʿUyūn al-anbā'*, vol. 1, 272.
60 Ibn Abī Uṣaybiʿa, *ʿUyūn al-anbā'*, vol. 1, 272.
61 Ibn Abī Uṣaybiʿa, *ʿUyūn al-anbā'*, vol. 1, 435.
62 Ibn Abī Uṣaybiʿa, *ʿUyūn al-anbā'*, vol. 1, 489.
63 Ibn Ḥajar al-ʿAsqalānī, *ad-Durar al-kāmina*, vol. 1, 191.
64 Aṣ-Ṣafadī, *al-Wāfī bi-l-wafayāt*, vol. 6, 210.

is the case with any student population, not everyone had the talent for the chosen topic. The physician Muḥammad b. Ibrāhīm Ṣalāḥ ad-Dīn, known as Ibn al-Burhān (d. 743/1341–2), and Sheikh Rukn ad-Dīn b. Quba' are good examples of the more average philosophy student in fourteenth-century Cairo and demonstrate that studying some philosophical writings had by then become quite common.

Ibn al-Burhān, one of the personal doctors of Sultan an-Nāṣir and a client of two high-ranking Mamluk amīrs, wealthy in his own right due to commerce, began to study philosophy at an advanced age with Shams ad-Dīn al-Iṣfahānī. He learned parts of Ibn Sīnā's texts by heart, among them the *Ishārāt*, wrote a commentary on them and carried Ṭūsī's commentary on the *Ishārāt* around with him all the time.[65] His teacher, when asked about his student's abilities, did not think much of them. He is said to have replied: "His study surpasses his intelligence".[66] Rukn ad-Dīn b. Quba' desired to learn some philosophy and hence began to meet with Ibn al-Burhān regularly in the shop of the witnesses near the Ṣāliḥiyya gate. The physician would talk about something Ibn Sīnā had written, either from the *Ishārāt* or some other book, and would comment on it in a parallel commentary. However, Rukn ad-Dīn was incapable of grasping the meaning of either of the texts and mumbled: "God gracious, someone with a mind like this one should study philosophy (*falsafa*)! The meaning of these words is such and such, i.e., he is in this camp and you are in that camp".[67] Ibn Khallikān sighed that his words indicated the conditions among the general populace.[68]

Given that this and further material has been available for more than a century, it is surprising that among modern researchers the Ayyubids could ever acquire the reputation of having been hostile towards the 'ancient sciences' in general, and philosophy in particular. This evaluation results from a combination of assumptions, prejudices, limited familiarity with the many stories told in the sources and methodological simplicity. The assumptions and prejudices concern issues like what was taught at madrasas by whom, when and where, and what different scholars as well as princes or courtiers thought about the different types of knowledge, their purposes and their working modes. Two cases are usually presented as prime examples of the oppression of philosophy and logic by particular Ayyubid rulers: the execution of Shihāb ad-Dīn as-Suhrawardī (548–586/1154–1191) and the house arrest of Sayf ad-Dīn al-Āmidī (d. 631/1233). I will focus on the latter case and its interpretation over the last one hundred years. The scholar who more than anyone else shaped the view regarding the attitude of Muslim religious scholars towards the so-called ancient sciences was Ignaz Goldziher. In his article of 1915, he presented the case as an example of the attitude expressed in *fatwās* by the *ḥadīth* scholar Taqī ad-Dīn Ibn aṣ-Ṣalāḥ (d. 643/1245). He had no doubt that the reason for al-Āmidī's house

65 Al-Maqrīzī, *as-Sulūk*, vol. 2, 234–235.
66 Al-Maqrīzī, *as-Sulūk*, vol. 2, 234–235.
67 Al-Maqrīzī, *as-Sulūk*, vol. 2, 234–235.
68 Al-Maqrīzī, *as-Sulūk*, vol. 2, 234–235.

arrest, which was purportedly ordered by al-Malik al-Ashraf (r. in Damascus 629–634/1229–1237), was the ruler's strong dislike for the scholar's alleged teaching of philosophy and logic, similar to the antagonism aggressively expressed by *fuqahā'* in Cairo more than a decade earlier. However, Goldziher believed that al-Āmidī had taught neither of the two subjects in Cairo,[69] ignoring Ibn al-Qifṭī's claims to the contrary.[70] Goldziher abstained from studying all the available narratives of the case before formulating a judgment, focusing only on a very small number of sources. Among the latter, he completely ignored the stories told by Ibn al-Qifṭī (567–645/1172–1248), despite mentioning his work in the footnotes. The possibly modified version of Ibn al-Qifṭī's narrative of al-Āmidī's loss of his teaching position as transmitted in the epitome of his book written in 748/1348 goes as follows:

> In 631 h [1233], al-Malik al-Kāmil [r. 615–635/1218–1237], the ruler of Egypt and head of the Ayyubid family, took possession of the town of Āmid; then it was reported that her ruler whom she was taken from had corresponded secretly with as-Sayf to come to him and to supervise the judges of Āmid. He (al-Kāmil) was angry with him (al-Āmidī) because of that and the fact that a message had been sent without being transmitted to him (al-Kāmil). Thus, his hand (al-Āmidī's) was taken away from the madrasa and he was without a position.[71]

Nor did Goldziher investigate Ibn Abī Uṣaybi'a's (590–668/1194–1270) silence about this affair despite pointing it out.[72] Even if these two contradictory approaches by contemporaries of Sayf ad-Dīn did not suffice to arouse Goldziher's suspicion regarding the all too easy ascription of the Ayyubid decision to philosophy and logic, he should have hesitated in his judgement after analysing the stories by Ibn Khallikān (608–680/1212–1282), on whom he relied first and foremost. Ibn Khallikān described the events in Damascus in a conspicuously imprecise manner, relating al-Āmidī's fall from grace to a certain "matter that one suspected him of".[73] Yet, like numerous later scholars, Goldziher bases his judgement on previous events in Cairo, ignoring Ibn Khallikān's way of presenting them and commenting on them. Ibn Khallikān described this affair as follows:

> Then a group of *fuqahā'* of the town envied him and they formed an alliance against him. They accused him of corruption of faith, laxity of moral principles, *ta'ṭīl* and of (adhering to) the doctrine of the philosophers and

69 Goldziher, *Die Stellung der alten islamischen Orthodoxie zu den antiken Wissenschaften*, 38–39.

70 Ibn al-Qifṭī, *Tārīkh al-ḥukamā'*, 240, reported: ". . . in Egypt, he carried out disputations and presented his works on the sciences of the ancients. These were transmitted and read under his direction by those who had a strong desire for such things."

71 Ibn al-Qifṭī, *Tārīkh al-ḥukamā'*, 240–241.

72 Goldziher, *Die Stellung der alten islamischen Orthodoxie zu den antiken Wissenschaften*, 39.

73 Ibn Khallikān, *Wafayāt al-a'yān*, vol. 3, 294.

> the sages (or: physicians). They wrote a protocol which (contained these accusations). They put on it their signatures so that it may be considered lawful (to shed) his blood. I was told about a man among them, who had intelligence and knowledge, that, after he saw their attacks (*taḥāmulahum*) against him and the excess of obstinacy written on the document, he wrote, when they brought it to him to write on it according to what they wrote: "They envied the man because they could not equal him in merit."[74]

This text leaves no doubt that its author did not believe in the legitimacy of the accusations, but, in accordance with his witness, regarded them as a pretext for getting rid of a superior teacher and scholar. Moreover, al-Āmidī's interest in philosophy was only the last of four accusations raised against him. The third one was the potentially dangerous one. The other three merely served to bolster the case. While the story makes it clear that studying or teaching philosophy could enrage colleagues and cause trouble, it also clarifies that in al-Āmidī's case at least some of his contemporaries regarded it as nothing more than a pretext. Academic honesty demands to differentiate between these two different conditions of persecution.

Sourdel's entry on al-Āmidī in the *Encyclopaedia of Islam*, second edition, makes it clear, however, that this simple difference was not acknowledged in the twentieth century. He categorically declared: "(H)e was dismissed from his post after 629/1229 (sic) by al-Malik al-Ashraf for having taught philosophy".[75] In addition to Ibn al-Qifṭī and Ibn Khallikān, Sourdel used two later sources, namely Tāj ad-Dīn as-Subkī's (d. 1368/70) *Ṭabaqāt ash-shāfiʿiyya al-kubrā* and ʿAbd al-Qādir an-Nuʿaymī's (d. 927/1521) *Ad-Dāris fī tārīkh al-madāris*.[76] Tāj ad-Dīn as-Subkī remained silent about any specific case of confrontation. Hence, Sourdel's only new source beyond those already relied upon by Goldziher was an-Nuʿaymī's much later report. This indiscriminate use of contemporary and later sources is one of the methodological issues that have marred academic interpretations of al-Āmidī's fate and continue to do so.

Since Sourdel, interpretations have not changed much. This even applies to cases in which parts of other stories – including the patronage conflict referred to already by Ibn al-Qifṭī – were taken into account. Humphreys, for instance, was well aware of this patronage conflict, although he worked with a later version by the historian of the Ayyubid dynasty, Ibn Wāṣil (604–697/1208–1298), who in addition also pointed to intellectual conflicts between the protagonist and his colleagues and patrons in Damascus:

> One day, al-Malik an-Nāṣir held a *majlis*, in which he united a group of the best scholars in Damascus – before she was taken from him. In this

74 Ibn Khallikān, *Wafayāt al-aʿyān*, vol. 3, 293–294.
75 Sourdel, "al-Āmidī".
76 Sourdel, "al-Āmidī".

> *majlis* participated Shaykh Shams ad-Dīn Khusrawshāhī, Shaykh Sayf ad-Dīn [al-Āmidī], Shaykh Tāj ad-Dīn al-Urmawī, and he was one of the greatest and best authors, and the judge Shams ad-Dīn al-Khūy'ī [d. 637/1239], the judge of Damascus. They were the best in the disciplines. They debated, and the word of those "Persians" was united against Sayf ad-Dīn. They were against him as if they were a single hand . . . [When an-Nāṣir left Damascus for al-Karak and took various scholars with him, al-Āmidī remained in town. – S.B.] Al-Malik al-Ashraf opposed him and loathed him. Then al-Malik al-Maṣʿūd, the lord of Āmid, sent and asked for him. Then Āmid was taken from him and al-Malik al-Kāmil said to the captured lord of Āmid: "In your country, you have no one of excellence." But he erred and the lord answered: "I had sent to the Shaykh Sayf ad-Dīn to demand of him (to come to me), and I was promised that he would come to me." This was so painful for al-Malik al-Ashraf and al-Malik al-Kāmil that the two of them grew angry against Sayf ad-Dīn. As a consequence, al-Malik al-Ashraf dismissed him from the teaching (position) at the ʿAzīziyya Madrasa. He withdrew to his garden and stayed there as an oppressed until he died in this year, being half a year older than 80 years.[77]

Humphreys dismisses this story as "almost certainly only a pretext".[78] The reason for al-Āmidī's loss is more likely to have been "his attachment to the rational sciences", as claimed by Sibṭ b. al-Jawzī (d. 654/1256).[79] Ignoring the patronage story completely despite his familiarity with Ibn Wāṣil's chronicle, Endress also gives preference to Sibṭ b. al-Jawzī's narrative:

> Al-Malik al-Ashraf made al-Āmidī leave the ʿAzīziyya and proclaimed: "he who teaches something else than *tafsīr* and *fiqh* and turns to the doctrine of the philosophers I will expel." Al-Āmidī stayed anonymously (?) in his house after the light of his happiness had been extinguished, until he died in Ṣafar and was buried at (mount) Qāsiyūn in his mausoleum.[80]

In contrast to Humphreys though, Endress acknowledges the narrative about the intellectual tensions in Ayyubid Damascus between al-Āmidī and the students of Fakhr ad-Dīn ar-Rāzī. These tensions influenced the scholars' relations with the Ayyubid rulers of Damascus in the 1220s, al-Muʿaẓẓam ʿĪsā (r. 615–624/1218–1227) and his son an-Nāṣir Dāwūd (r. 624–626/1227–1229 sultan of Damascus; 626–646/1229–1248 amīr of al-Karak after al-Kāmil and al-Ashraf had taken

77 Ibn al-Wāṣil, *Mufarrij al-kurūb*, vol. 5, 39–41.
78 Humphreys, *From Saladin to the Mongols*, 209.
79 Humphreys, *From Saladin to the Mongols*, 209.
80 Sibṭ b. al-Jawzī, *Mirʾāt az-zamān*, vol. 8, 691.

Damascus). The narration consists of al-Āmidī's rejection of Fakhr ad-Dīn's interpretation of Ibn Sīnā's philosophical works and of positions he took in *uṣūl ad-dīn*, *kalām* and *tafsīr*. In addition, eloquent and sharp as al-Āmidī was, his comments on positions he rejected could be very biting.

> He was eager to refute Fakhr ad-Dīn ar-Rāzī exploring his words and weaknesses and exaggerating his libels. He obviously envied ar-Rāzī and believed that he was more knowledgeable or at least equal to him. He saw, however, that Fakhr ad-Dīn was more famous, that the latter's works were more accepted and that he was more honoured and praised.[81]

Reflecting about the reason of the rift between al-Āmidī and al-Muʿaẓẓam, Ibn Wāṣil pointed out that during a visit the Damascene poet Sharaf ad-Dīn b. ʿUnayn discovered that al-Āmidī had an inimical attitude towards Fakhr ad-Dīn ar-Rāzī. He became very upset and went to inform al-Malik al-Muʿaẓẓam, whose vizier he was.[82] As a result, al-Āmidī lost the ruler's devotion and admiration.[83]

Humphreys has already suggested a connection between al-Āmidī's fate and the persecution of a group of Sufis by al-Ashraf. He believes that it was the prince's soldierly narrow-mindedness that compelled him to commit these acts of repression.[84] Pouzet takes this idea of a link between al-Ashraf's different acts of violence against inhabitants of Damascus one step further. He adds further measures, i.e., the closing of establishments and the banning of certain people, ascribing them all to al-Ashraf's inclination towards Hanbalism, although he admits that except for claims on the part of individual historians regarding al-Ashraf's preference for *ḥadīth* there is no proof that he indeed adhered to this particular *madhhab*.[85]

Pouzet refers to the various cases, including the one of our scholar, to argue in favour of al-Ashraf's Ḥanbalī positions. However, he does not explain why al-Āmidī's philosophical activities could have been the only reason for his persecution. He simply takes this cause for granted.[86] Neither do any of the other mentioned modern interpretations of al-Āmidī's fall from grace offer arguments that would support this choice of story line from among the competing positions offered by medieval authors. They believe what fits in with their perception of the contested position of philosophy in post-classical societies. However, is it really plausible that al-Ashraf removed al-Āmidī from his teaching position at the *ʿAzīziyya* Madrasa because of the latter's involvement with philosophy and logic? I do not believe so. There are three major arguments in favour of the story about the loss of face of the two Ayyubid princes in front of Āmida and hence of

81 Ibn Wāṣil, *Mufarrij al-kurūb*, vol. 5, 36.
82 Pouzet, *Damas au VIIe/XIIIe siècle*, 80–96.
83 Ibn Wāṣil, *Mufarrij al-kurūb*, vol. 5, 38.
84 Humphreys, *From Saladin to the Mongols*, 209.
85 Pouzet, *Damas au VIIe/XIIIe siècle*, 36, 91, 378.
86 Pouzet, *Damas au VIIe/XIIIe siècle*, 36, 91, 378.

a violation of patronage rules. The first is the date of his dismissal. The second concerns methodological shortcomings. The third, finally, consists in the analysis of the scholarly members of al-Ashraf's entourage and the intellectual activities in Damascus in the 1220 sand early 1230s.

Ibn al-Qifṭī dates al-Āmidī's loss of his teaching position to the year 631/1233 and after the Ayyubid conquest of Āmida. Ibn Wāṣil specifies the year of the dismissal as the year of his death, i.e., 631, indicating that the event took place in the immediate aftermath of this conquest.[87] Sibṭ b. al-Jawzī, the author of the second oldest report on al-Āmidī's loss of position, avoids giving a precise date, but claims that the scholar "remained anonymously" in his house until he died. Since he linked this event to the alleged claim of al-Malik al-Ashraf that he would expel everybody who taught philosophy and since he wrote specifically that al-Āmidī kept his position at the *ʿAzīziyya* Madrasa until al-Muʿaẓẓam's death, he seems to imply that al-Āmidī lost it with al-Ashraf's conquest of Damascus in 626/1229.[88] Ibn Khallikān is surprisingly unspecific, not only regarding the cause for al-Āmidī's loss of position, but also the time.[89] The closeness between the dates of the two events given by Ibn al-Qifṭī and Ibn Wāṣil makes their version seem more plausible than Sibṭ b. al-Jawzī's, which would imply that al-Āmidī hid in his house for more than four years.

In terms of methodology, taking historical accounts at face value and giving preference to one account over the other suggests a lack of critical reflection on the part of modern historians. There can be no doubt that all four scholars (and others not mentioned here) believed strongly that many religious scholars and Muslim princes were profoundly hostile towards the ancient sciences in general, and philosophy and logic in particular. They also explicitly or implicitly accepted the idea that these domains of knowledge were banned from public teaching and that there was a fundamental difference between public and private teaching, i.e., between teaching in the precincts of a *madrasa* or any other teaching institute and teaching at the place where either the teacher or the student lived. While many arguments have been brought forward against such a rigid separation of the public and the private in Ayyubid and Mamluk educational practices, these positions continue to be held until today, as made clear by Endress, for instance.[90] As my admittedly limited discussion of the various reports by al-Āmidī's contemporaries indicates, it is difficult to determine which bits and pieces of such narratives can or should be trusted and why. The subsequent transmission of these thirteenth-century narratives by Mamluk historians led to further streamlining, cutting and adding. As a result, the patronage story is almost completely suppressed, the supposed hostility to the ancient sciences is pushed to the forefront, and Ibn Khallikān's report on

87 Pouzet, *Damas au VIIe/XIIIe siècle*, 41.
88 Sibṭ b. al-Jawzī, *Mirʾāt az-zamān*, vol. 8, 691.
89 Ibn Khallikān, *Wafayāt al-aʿyān*, vol. 3, 294.
90 Endress, "Reading Avicenna", 382, 393, 395, 400.

the events in Cairo is used in a contradictory way, with some historians marking them as part of an intrigue and others emphasizing that al-Āmidī was the one at fault.[91] Isolating a particular narrative strand instead of analysing the entire extant representation of such a conflict or series of conflicts can only lead to erroneously privilege certain elements, which are either more to the liking of the modern interpreter or which are emphasized in the intentionally or accidentally chosen sources. This does not only apply to the general preference given by today's scholars to the story about the ancient sciences and al-Āmidī as culprits. It also applies to the dismissal of the patronage story and a further story closely linked to this set of narrative elements skilfully manipulated by the Ayyubid and Mamluk writers. The additional, closely related story is that of Taqī ad-Dīn Ibn aṣ-Ṣalāḥ's effort to study logic with Kamāl ad-Dīn b. Yūnus. Goldziher had already realized that the famous *ḥadīth*-scholar did not have the brains for it as Ibn Khallikān was not shy to point out about his teacher, based on the authority of a legal scholar in Mosul.[92] However, this fact did not induce him to read Kamāl ad-Dīn's soft refusal to continue wasting his time with the young man as a kind effort to spare the latter's pride. Humphreys ignores this point altogether, Endress understands it as meaning that Kamāl ad-Dīn never taught logic to the young Kurd and others, after embellishing the story with details not found in the original Arabic text, surprisingly consider it as an element of al-Ashraf's supposedly stern control of the intellectual life in Damascus.[93] Some medieval Muslim readers of Ibn Khallikān's remarks, for instance Abū l-Fidā', left no doubt that they understood the story to mean that the famous scholar simply did not know what he was talking about when it came to these two ancient disciplines.[94] Hence, I am inclined to look at Ibn aṣ-Ṣalāḥ's later rejection of logic and philosophy as grounded in this experience and as reminiscent of Aesop's fox and the sour grapes.

The dismissal of the patronage story rests on the assumption that certain Ayyubid rulers were intelligent, generous and tolerant, while others were narrow-minded,

91 Brentjes, *"Orthodoxy": Ancient Sciences, Power, and the madrasa ("college") in Ayyubid and early Mamluk Damascus*, 28–33.

92 Goldziher, *Die Stellung der alten islamischen Orthodoxie zu den antiken Wissenschaften*, 39.

93 Humphreys, *From Saladin to the Mongols*, 208; Endress, "Reading Avicenna", 396; http://sonic.net/~tallen/palmtree/ayyarch/ch9.htm#log.dig (19 August 2011): "What is fascinating in the context of al-Malik al-Ashraf's strictures upon intellectual life is an anecdote Ibn Khallikân gives elsewhere: While I was at Mosul, a certain jurisconsult related to me that Ibn al-Salâh had obtained permission from him [Kamâl al-Dîn ibn Mûsâ] to read secretly under his direction a part of the Logic or Organum of Aristotle. He went to him regularly for some time but was unable to understand anything of it; so at length he [Kamâl al-Dîn] said to him: 'My opinion is this, doctor! that you had better renounce the study of this science.' The other asked him for what reason, and received this answer: 'The public looks upon you as a good and pious man, and considers those who apply to this branch of knowledge as holding pernicious opinions on religious matters; you risk, therefore, losing their esteem, without even acquiring any knowledge of the science.' [Ibn al-Salâh] took his advice and gave up the study."

94 Abū l-Fidā', *Tārīkh*, vol. 2, 476.

coarse soldiers. This, at least, is Humphreys's categorization of al-Muʿaẓẓam (first type) and al-Ashraf (second type).[95] Al-Muʿaẓẓam and al-Kāmil, however, were not that different from their brother al-Ashraf. As I will show, the latter, like the two other brothers, surrounded himself with numerous scholars highly educated in the so-called ancient sciences, including philosophy, while the former severely punished their clients for real or supposed breaches of patronage etiquette and undertook equally ruthless conquests like al-Ashraf. Al-Kāmil, for instance, actively supported al-Ashraf's campaign against al-Muʿaẓẓam's son an-Nāṣir and pushed into the territories beyond Ayyubid borders, as al-ʿĀdil and Ṣalāḥ ad-Dīn had done before him.[96] While often praised for offering his patronage to a variety of scholars, al-Muʿaẓẓam reacted harshly when crossed. In 616/1219, for instance, he clubbed the judge Zakī ad-Dīn b. az-Zakī with a cudgel, because he had gone to his aunt Sitt ash-Shām with witnesses to formalize her testament and the conversion of her house into a *madrasa* without his approval. The judge fell ill and died the following year.[97] He dismissed Ismāʿīl b. Ibrāhīm al-Māridīnī, known as Ibn Fallūs (590–630/1194–1232), from his position at the *Tarkhāniyya* Madrasa and apparently forced a house arrest upon him, after this teacher of *fiqh*, medicine and the mathematical sciences had refused to provide him with a *Ḥanafī fatwā*, allowing the Ayyubid ruler to drink date wine and other alcoholic beverages.[98]

My third argument against interpreting al-Āmidī's fate as an expression of al-Ashraf's hostility to philosophy and logic rests on data about the scholarly clients of the Ayyubid prince. This group included two famous and some less well-known physicians. The two famous doctors were Ṣadaqa as-Sāmarī and Muhadhdhab ad-Dīn ʿAbd ar-Raḥīm b. ʿAlī, known as ad-Dakhwār, founder of a medical *madrasa* and one of Ibn Abī Uṣaybiʿa's teachers. Among the less well-known clients was Abū ʿAbdallāh Muḥammad b. Ibrāhīm abī l-Maḥāsin, another student of Muhadhdhab ad-Dīn. Ibn Abī Uṣaybiʿa describes Ṣadaqa as-Sāmarī as strong in *falsafa* and good in its study. He served the Ayyubid prince, who held him in great respect and over many years bestowed on him honours of all kinds until his death in 625/1228.[99] Muḥammad b. Ibrāhīm studied philosophical texts with Muhadhdhab ad-Dīn and learned all those by heart that students of medicine would customarily memorize.[100] Muhadhdhab ad-Dīn studied philosophy (*ʿulūm ḥikmiyya*) with Sayf ad-Dīn al-Āmidī. He read numerous of his works with him, among them al-Āmidī's commentary on Ibn Sīnā's *Ishārāt wa-tanbīhāt*, and partly memorized some of them. With Abū l-Faḍl al-Isrāʾīlī he studied astronomy and astrology,

95 Humphreys, *From Saladin to the Mongols*, 208.

96 See, for instance, Ibn Khallikān, *Wafayāt al-aʿyān*, vol. 5, 81; Abū l-Fidāʾ, *Tārīkh*, vol. 2, 382, 410, 417, 490 et al; al-Maqrīzī, *as-Sulūk*, vol. 1, 65; Ibn ʿAdīm, *Bughyat aṭ-ṭalab fī tārīkh Ḥalab*, vol. 3, 411.

97 Ibn Kathīr, *al-Bidāya wa-n-nihāya*, vol. 13, 84.

98 Ibn Kathīr, *al-Bidāya wa-n-nihāya*, vol. 13, 136.

99 Ibn Abī Uṣaybiʿa, *ʿUyūn al-anbāʾ*, vol. 1, 467.

100 Ibn Abī Uṣaybiʿa, *ʿUyūn al-anbāʾ*, vol. 1, 492.

and he owned many books on these two fields as well as copper instruments. He possessed sixteen different texts alone on the construction of the astrolabe. This specific knowledge earned him al-Ashraf's patronage offer and invitation in 622/1225. He spent the enormous sum of 20,000 Dirham on the needed instruments for al-Ashraf, who praised him for his efforts and paid him a stipend of 1,500 Dinar per year. After al-Ashraf had conquered Damascus, he made Muhadhdhab ad-Dīn the head of the physicians and arranged for him a *majlis* for teaching medicine.[101] Only one of al-Ashraf's doctors, Muwaffaq ad-Dīn al-Minfākh (d. 642/1244), is not mentioned for his study of philosophical works.[102] Given that al-Ashraf patronized these scholars and showed this kind of interest in at least some of the ancient sciences, among them one that was also contested, it does not seem plausible that he punished al-Āmidī for analogous interests in 631/1233, i.e., four years after he had taken over the rule of Damascus, without any further reason than his old animosity towards al-Āmidī.

Philosophical activities in Ayyubid Damascus seem to have taken place primarily in two domains, i.e., teaching and sessions headed by the respective ruler. A third, temporary engagement with philosophy and other ancient sciences was the exchange of questions and answers between Frederic II and his court scholars and al-Kāmil and various scholars in Syria and the Jazīra, among them ʿAlam ad-Dīn Qayṣar (574–649/1178–1251) and Kamāl ad-Dīn b. Yūnus. ʿAlam ad-Dīn Qayṣar was one of Kamāl ad-Dīn b. Yūnus brightest students, equally skilled in the mathematical sciences, including music, architecture and engineering. As is well known, he corresponded with Naṣīr ad-Dīn Ṭūsī about how to prove Euclid's parallel postulate, and he built water lifting devices, walls, a palace and astronomical instruments in Damascus, Hama and elsewhere. A celestial globe, which he designed in 623/1225–6 for al-Malik al-Kāmil, is extant today at Naples in the Museo Nazionale di Capodimonte.[103] Having participated in the correspondence with Frederic II, he also seems to have had some philosophical knowledge. He certainly was one of the most versatile scholars of the ancient sciences in the Ayyubid realm.

According to Ibn Abī Uṣaybīʿa, most of the young men who studied philosophy in the 1220s did so because they intended to acquire competence in medicine in general, or ophthalmology in particular. He himself undoubtedly took some classes with Sayf ad-Dīn al-Āmidī, in which they also discussed the life and œuvre of al-Fārābī.[104] As the quotes on Muhadhdhab ad-Dīn and Muḥammad b. Ibrāhīm given above suggest, a corpus of philosophical texts had been acknowledged as providing the fundaments of medical training. In addition to Ibn Sīnā's *Kulliyāt*,

101 Ibn Abī Uṣaybiʿa, *ʿUyūn al-anbāʾ*, vol. 1, 477.

102 Ibn Abī Uṣaybiʿa, *ʿUyūn al-anbāʾ*, vol. 1, 492.

103 Savage-Smith, *Islamicate Celestial Globes: Their History, Construction, and Use*, 25–26, 218–219; http://brunelleschi.imss.fi.it/galileopalazzostrozzi/object/AlamAlDinQaysarCelestialGlobe.html (19 August 2011).

104 Ibn Abī Uṣaybiʿa, *ʿUyūn al-anbāʾ*, vol. 1, 398.

books by Aristotle were possibly part of this corpus, if the occasional references to them can be interpreted in this sense.[105] A second group of students taking classes on philosophical matters or teachers giving them were those interested in *kalām*, *fiqh* and *uṣūl ad-dīn*. One member of this group was al-Afḍal Khūnjī, who taught Ibn Abī Uṣaybiʿa Ibn Sīnā's *Kulliyāt* and other parts of the *Qānūn* in Cairo in 632/1234. His work *al-Jumal* on logic, arithmetic, *fiqh* and *uṣūl ad-dīn* became a very successful textbook at madrasas in the Mamluk state and northern Africa. At the end of his life, Khūnjī was the head of the judges in Cairo. Shams ad-Dīn Khūyʾī achieved the same position in Damascus. He had studied with Fakhr ad-Dīn ar-Rāzī and was knowledgeable in the philosophical parts of medicine and excellent in the religious sciences. Having been honoured already by al-Malik al-ʿĀdil (r. 603–614/1207–1218), he was endowed with a stipend by al-Muʿaẓẓam, who engaged him as a particularly close client. Shams ad-Dīn Khūyʾī participated in al-Muʿaẓẓam's intellectual sessions and compiled a book on philosophical matters (*rumūz ḥikmiyya*) for his Ayyubid patron.[106]

A further religious scholar and judge, Rafīʿ ad-Dīn Jīlī (d. 641/1244), had come to the Ayyubids from Iran like the two former men. He excelled in philosophy, natural philosophy and medicine, in addition to the religious sciences. He lived in the *ʿAdhrawiyya* madrasa, where he had a *majlis* for teaching all kinds of sciences. Ibn Abī Uṣaybiʿa studied some philosophical topics with him. He served as a judge in Baalbek and was appointed head judge of Damascus in 638/1240–1, some months after Shams ad-Dīn Khūyʾī had passed away. Rafīʿ ad-Dīn did not exercise this legal position well, which in combination with his break with the vizier Amīn ad-Dawla and his stinging criticism of the current Ayyubid ruler of Damascus led to a cruel death near Baalbek, a death that had nothing to do with his philosophical interests. Like Shams ad-Dīn Khūyʾī, Rafīʿ ad-Dīn Jīlī wrote a philosophical text, a commentary on Ibn Sīnā's *Ishārāt wa-tanbīhāt*, for an Ayyubid prince, al-Muẓaffar Taqī ad-Dīn ʿUmar b. al-Malik al-Amjad.[107]

A fourth prominent teacher of philosophy in the 1220s in Damascus was the already mentioned Shams ad-Dīn Khusrawshāhī. Hence, al-Āmidī was not alone in his interests and activities. Any concerted effort by al-Ashraf to ban the teaching and study of philosophy and logic in Damascus or even only at the city's *madrasa*s would surely also have included these scholars, except for Khusrawshāhī, who left town with his patron an-Nāṣir Dāwūd. There is no evidence however, that al-Ashraf undertook any measures to limit their freedom of teaching. The medical clients of al-Ashraf and their careers confirm that al-Ashraf did not interfere with their philosophical studies either. Perhaps there were minor scholars who were told not to

105 Ibn Abī Uṣaybiʿa's uncle Rashīd ad-Dīn ʿAlī b. Khalīfah (579–616/1184–1220), for instance, studied Aristotelian works in Cairo with ʿAbd al-Laṭīf al-Baghdādī, a friend of Ibn Abī Uṣaybiʿa's grandfather; see Ibn Abī Uṣaybiʿa, *ʿUyūn al-anbāʾ*, vol. 1, 479.

106 Ibn Abī Uṣaybiʿa, *ʿUyūn al-anbāʾ*, vol. 1, 124.

107 Ibn Abī Uṣaybiʿa, *ʿUyūn al-anbāʾ*, vol. 1, 425.

read books beyond those on *fiqh*, *tafsīr* or *ḥadīth*? Yet even this does not seem to have been the case. At least so far, I have not found a single biographical entry that could serve as evidence. Hence, at the moment there is no indubitable evidence of al-Ashraf restricting the philosophical education of either medical or juridical students, whether bright or mediocre. As long as no such evidence can be offered, the case of al-Āmidī presented as a rejection of philosophy and/or logic remains isolated and unique. However, if interpreted as a disturbed patron–client relationship, it loses much of its peculiarity and appears as one of a series of dismissals, punishments and even death sentences that Ayyubid patrons meted out to hapless clients.

4 Another suspect intellectual endeavour: alchemy

In his study of Ibn Qayyim al-Jawziyya's *Miftāḥ dār as-sa'āda*, Livingston states:

> But the position of the great scientists of earlier centuries in Islamdom, back when royal patronage of an Abbāsid, Sāmānid, or Fāṭimid court offered generous support to science, had in Mamlūk-ruled Arabic-speaking lands given way to practitioners of the occult; patronage of the philosophical and exact sciences was not an outstanding feature in the courtly culture of the Burjī or Baḥrī sultans, and men pursuing the scientific tradition were then forced to make a living as best as they could.[108]

Having argued in section 1 and 3 that neither the Ayyubids nor the Mamluks actually neglected the mathematical or the philosophical sciences, and that hence the issue of the perceived or genuine loss in substance in these disciplines needs to be approached from a different perspective, which in my view should primarily focus on teaching practices in scholarly networks, I wish to explore in this final section some of the material available on practitioners of alchemy in the Mamluk era. According to biographical dictionaries, alchemy seems to have been excluded from the official teaching activities of Mamluk scholars. Nonetheless, as in the case of philosophy, it is possible to find physicians and religious scholars who spent time and money on acquiring alchemical knowledge. Occasionally, also craftsmen turned into scholars are to be found among the adepts in this field.

Ibn Qayyim al-Jawziyya, a scholar of the early Mamluk period, had apparently acquired a fairly well-rounded education in philosophy and the sciences. According to Livingston, he had read astronomical and astrological works by Ibn al-Haytham, al-Bīrūnī and Abū 'Alī 'Īsā b. 'Alī, understood the basics of Ptolemaic astronomy and optics, had learned directly or indirectly the basics of Aristotelian, Neoplatonic, and Islamic philosophical doctrines and may have studied texts by al-Fārābī and Ibn Sīnā.[109] He knew, however, much less about the sciences and

108 Livingston, "Science and the Occult", 599.
109 Livingston, "Science and the Occult", 600, 602–607.

their adherents of his own time.[110] While this familiarity with past scientific and philosophical writings as opposed to contemporary debates among scientists and philosophers posits a fascinating topic for further study, it is of no relevance for the topic of this section, except for the point that Ibn Qayyim limited his polemics against what Livingston translates as 'the occult sciences' to two disciplines only, astrology and augury.[111] Alchemy, apparently, was not an issue that caused the Ḥanbalī scholar sleepless nights. But was he right not to care? Berkey, following Haarmann, argues that Mamluk elite society was saturated with magic, alchemy, and divination. The rulers, who continued to hold shamanistic beliefs from Central Asia, met in these three fields of medieval knowledge with the local Egyptian and more broadly Near Eastern traditions of the ruled, which led to the construction of a mixed set of Islamized religious practices.[112] Based on al-Jaʿbartī's chronicle, Hanna confirms for Mamluk households of the eighteenth century that their libraries still included works on alchemy, understood as science.[113] Then what did Ibn Ḥajar al-ʿAsqalānī complain about when he criticized the Mamluk Alṭunbughā al-Jāwulī (d. 748/1348) for his 'addiction to alchemy' – religious practices or scientific beliefs or both?[114] This cannot be answered for this particular Mamluk, since Berkey does not give the needed reference. However, in the two biographical works by Ibn Ḥajar that I have access to the spectrum of activities of the men he connected with alchemy and the terms he used to talk about them do not address alchemy as situated between specific mixed forms of religiosity and scientific bibliophily, the two poles marked by Berkey and Hanna. Ibn Ḥajar rather focused on alchemy as science versus alchemy as a waste of time and goods. In the all in all very small number of cases his rhetoric is mostly neutral, but at times he leaves no doubt that he really considered alchemy to be a useless art, unless it was linked to practical matters like working with precious stones. In *Inbāʾ al-ghumr* one of the neutral statements is that ʿAlī b. Ibrāhīm al-Anṣārī (d. 774/1372) copied many books, in particular on alchemy, by his own hand.[115] Another one concerns Ibrāhīm b. ʿAbdallāh al-Khilāṭī ash-Sharīf, who had grown up in *Bi-lād al-ʿAjam* (Iran or Anatolia?). There he had specialized in the art of lapis lazuli and exercised it professionally. He finally settled in Cairo where he may have practised alchemy.[116] In both cases, Ibn Ḥajar's report leaves no doubt, however, that Cairene society in general and Mamluk elite in particular highly appreciated the two men. ʿAlī b. Ibrāhīm al-Anṣārī had good relations with the people affiliated with the dynasty, in particular the Copts.[117] Ibrāhīm b. ʿAbdallāh al-Khilāṭī allegedly

110 Livingston, "Science and the Occult", 603, fn 17.
111 Livingston, "Science and the Occult", 600, 606 et al.
112 Berkey, "The Mamluks as Muslims", 163–173.
113 Hanna, "Cultural Life in Mamluk Households (late Ottoman period)", 196–207.
114 Berkey, "The Mamluks as Muslims", 170.
115 Ibn Ḥajar, *Inbāʾ al-ghumr*, vol. 1, 14.
116 Ibn Ḥajar, *Inbāʾ al-ghumr*, vol. 1, 200.
117 Ibn Ḥajar, *Inbāʾ al-ghumr*, vol. 1, 14.

never left his house, but people often came to visit him. One day, even the sultan came and had a conversation with him.[118] Reporting what he had read in another history book, Ibn Ḥajar told his readers that this man lived like the kings, would not visit anyone and was suspected of apostasy (*rafḍ*), because he never went to the mosque for prayer and preached to some of his followers that he was the *mahdī*.[119] His alleged alchemical interests may not have been exclusively practical, since he also wrote notes on philosophy, astrology, and geomancy. Nonetheless, no one molested him and he died a natural death in 799/1396. One year later, Muḥammad b. ʿAlī b. ʿAbdallāh aṭ-Ṭaybarsī died. He had lived in the Friday mosque of the same name and was one of those upon whom Ibn Ḥajar heaped his scorn, claiming that "he was infatuated with the art of alchemy and wasted all his life and time with it to no avail".[120] Nine years later, the sultan's imam Nāṣir ad-Dīn aṭ-Ṭanāḥī died after he "had wasted all his wealth" on his interest in alchemy.[121] According to Ibn Ḥajar's dictionary, the only person with alchemical interests that had died an unnatural death had died a year earlier. Yet the historian did not link the man's death to alchemy, but rather to his own murderous activities in Yemen. Muḥammad b. ʿAbdallāh al-Ḥuḍarī, born in Egypt, was a physician who had also acquired knowledge in alchemy, talismans, and astrology. When on pilgrimage, he remained in Mecca for some time and then moved on to Yemen, where he was introduced to the sultan whose doctor he promptly poisoned. Moreover, he was suspected of having poisoned a merchant. These two murders led to his own death.[122] As these examples indicate, people did not only take an interest in alchemy for professional reasons, like physicians and craftsmen. Religious scholars of high and low status, including a student of Ibn Ḥajar himself, who also took classes with ʿIzz ad-Dīn ibn Jamāʿa, squandered their time and money and thus obviously did more than merely buy and read books.[123] In all likelihood they practised the art of transmutation, hoping to produce gold.

Such a low profile of alchemy was typical of many Mamluk histories. If their authors write at all about it they talk of the prophetic origin of the art claimed by some of its adepts, Ibn Sīnā's refutation of the possibility of transmutation, ʿAbd al-Laṭīf al-Baghdādī's expression of relief that he had not fallen for it thanks to his study of that treatise, Mu'ayyad ad-Dīn aṭ-Ṭughrā'ī's opposition to Ibn Sīnā and his death, the costly nature of experimenting, Ibn ʿArabī's licit interest in alchemy, since he did not practise it to earn money, or they vaguely hint at alchemical studies carried out by various religious scholars in the Mamluk period, among them the famous

118 Ibn Ḥajar, *Inbā' al-ghumr*, vol. 1, 200.
119 Ibn Ḥajar, *Inbā' al-ghumr*, vol. 1, 200.
120 Ibn Ḥajar, *Inbā' al-ghumr*, vol. 1, 214.
121 Ibn Ḥajar, *Inbā' al-ghumr*, vol. 1, 326.
122 Ibn Ḥajar, *Inbā' al-ghumr*, vol. 1, 325.
123 Ibn Ḥajar, *Inbā' al-ghumr*, vol. 1, 478.

ḥadīth-scholar and historian Taqī ad-Dīn Ibn al-ʿĪd (d. 702/1302).[124] Some stories emphasize the waste of money on experiments with unsatisfying results. Others raise the issue of poisoning as a danger in connection with physicians or surgeons with alchemical leanings. But most often, the remarks that a contemporary scholar had shown some inclination towards this art are vague, stating either that people said he had submitted to alchemy or that he was linked to alchemy. This lack of information and the hesitation to talk about alchemy clearly sets it apart from astrology and philosophy, including logic. It indicates that these three disciplines had acquired a very different status in the eyes of Mamluk historians and probably also of their readers. Philosophy and logic were only rarely presented as objectionable, and if labelled as unbelief, the accusation appeared most often in connection with events of a long-gone past. The names of highly regarded scholars apparently vouched for the respectability of their study. Numerous students did not hesitate to read the textbooks written by Athīr ad-Dīn al-Abharī and Najm ad-Dīn al-Kātibī in the thirteenth century and the later commentaries by ash-Sharīf al-Jurjānī and other scholars. As pointed out in connection with the stories above, Ibn Sīnā's works as well as Fakhr ad-Dīn ar-Rāzī's and Naṣīr ad-Dīn aṭ-Ṭūsī's commentaries on them were appreciated as school texts and used for commentaries not only in the Ayyubid period, but at least until the fifteenth century. Philosophy and logic, as portrayed in Mamluk biographical dictionaries, thus appear to have been the least contested disciplines. Astrology obviously ranked somewhere between philosophy and alchemy. It was ridiculed and rejected, but simultaneously desired and regarded with high expectations based on beliefs in the reliability of observations and predictions. The maxim transmitted by aṣ-Ṣafadī sums up nicely the more pragmatic reasons for rejecting first and foremost alchemy, then magic and finally astrology: "If alchemy were true, why would we need taxes? If the science of talismans were true, why would we need the army? If astrology were true, why would we need the messenger and the post?"[125]

Bibliography

Abū l-Fidāʾ, *Tārīkh Abī l-Fidāʾ al-musammā Mukhtaṣar fī akhbār al-bashar*. 2 vols. Beirut: Dār al-Kutub al-ʿIlmiyya, 1997.

http://brunelleschi.imss.fi.it/galileopalazzostrozzi/object/AlamAlDinQaysarCelestialGlobe.html (19 August 2011).

Arif, Syamsudin, "Al-Āmidī's Reception of Ibn Sīnā: Reading *Al-Nūr al-Bāhir Fī al-Ḥikam al-Zawāhir*," in ed. Tzvi Langerman, *Avicenna and His Legacy: A Golden Age of Science and Philosophy*. Turnhout: Brepols, 2010, 205–219.

124 Ibn Khallikān, *Wafayāt al-aʿyān*, vol. 1, 327; vol. 2, 224; vol. 4, 74; vol. 5, 160; vol. 6, 157, 216; Ibn Kathīr, *al-Bidāya wa-n-nihāya*, vol. 1, 310; vol. 8, 236; vol. 9, 80; vol. 10, 294; vol. 12, 175, 185; vol. 13, 156; al-Maqrīzī, *al-Khiṭaṭ wa-l-athār*, vol. 1, 34, 44, 46, 155, 166, 171, 175, 221, 300; vol. 2, 20; vol. 3, 85, 258, 290; Abū l-Fidāʾ, *Tārīkh*, vol. 1, 78, 119, 340; vol. 2, 199; aṣ-Ṣafadī, *al-Wāfī bi-l-wafayāt*, vol. 2, 11.

125 Aṣ-Ṣafadī, *al-Wāfī bi-l-wafayāt*, vol. 6, 278.

Berkey, Jonathan P., "The Mamluks as Muslims: The Military Elite and the Construction of Islam in Medieval Egypt", in eds. Thomas Philipp, Ulrich Haarmann, *The Mamluks in Egyptian Politics and Society*. Cambridge and New York: Cambridge University Press, 1998.

Brentjes, Sonja, *"Orthodoxy": Ancient Sciences, Power, and the madrasa* ("*college*") *in Ayyubid and early Mamluk Damascus*. Berlin: Max Planck Institute for the History of Science, preprint 77, 1997.

Brentjes, Sonja, "The Prison of Categories – Decline and Its Company", in eds. Felicitas Opwis and David Reisman, *Islamic Philosophy, Science, Culture, and Religion*. Leiden and Boston: Brill, 2012, 131–156.

Brentjes, Sonja, "Shams al-Dīn al-Sakhāwī on *Muwaqqits*, *Mu'adhdhins*, and the Teachers of Various Astronomical Disciplines in Mamluk Cities in the Fifteenth Century", in eds. Emilia Calvo, Mercè Comes, Roser Puig and Mònica Rius, *A Shared Legacy: Islamic Science East and West: Homage to professor J.M. Millàs Vallicrosa*. Barcelona: Universitat de Barcelona, Publicacions i Edicions, 2008, 129–150.

Brentjes, Sonja, "The Study of Geometry According to al-Sakhāwī (Cairo, 15th c) and al-Muḥibbī (Damascus, 17th c)", in eds. Joseph W. Dauben, Stefan Kir-schner, Andreas Kühne, Paul Kunitzsch and Richard Lorch, *Mathematics Celestial and Terrestrial: Festschrift for Menso Folkerts zum 65. Geburtstag*. Halle, Saale: Deutsche Akademie der Naturforscher Leopoldina, 2008, 323–341.

Charette, François, "Ibn al-Majdī: Shihāb al-Dīn Abū al-ʿAbbās Aḥmad ibn Rajab ibn Ṭaybughā al-Majdī al-Shāfiʿī", in eds. Thomas Hockey et al., *The Biographical Encyclopedia of Astronomers*. New York: Springer, 2007, 561–562.

Charette, François, "The Locales of Islamic Astronomical Instrumentation", *History of Science* 44 (2006), 123–136.

Charette, François, *Mathematical Instrumentation in Fourteenth-Century Egypt and Syria: The Illustrated Treatise of Najm al-Dīn al-Mīṣrī*. Leiden: Brill, 2003.

Endress, Gerhard, "Reading Avicenna in the Madrasa: Intellectual Genealogies and Chains of Transmission of Philosophy and the Sciences in the Islamic East", in ed. James E. Montgomery, *Arabic Theology, Arabic Philosophy: From the Many to the One: Essays in Celebration of Richard M. Frank*. Leuven and Louvain: Peeters, 2006, 371–422.

Goldziher, Ignaz, *Die Stellung der alten islamischen Orthodoxie zu den antiken Wissenschaften*. Abhandlungen der königlich-preussischen Akademie der Wissenschaften, Phil.-Hist. Klasse 8, 1915.

Hanna, Nelly, "Cultural Life in Mamluk Households (Late Ottoman Period)", in eds. Thomas Philipp and Ulrich Haarmann, *The Mamluks in Egyptian Politics and Society*. Cambridge and New York: Cambridge University Press, 1998.

Holt, Peter M., "The Sultanate of al-Manṣūr Lāchīn (696–8/1296–9)", *Bulletin of the School of Oriental and African Studies* 36 (1973), 521–532.

Humphreys, R. Steven, *From Saladin to the Mongols: The Ayyubids of Damascus (1193–1260)*. Albany: SUNY Press, 1977.

Ibn Abī Uṣaybiʿa, Aḥmad ibn al-Qāṣim, *ʿUyūn al-anbāʾ fī ṭabaqāt al-aṭibbāʾ*, ed. Nizār Riḍā. Beirut: Dār Maktabat al-Ḥayāt, 1965.

Ibn al-ʿAdīm, *Bughyat aṭ-ṭalab fī tārīkh Ḥalab*, ed. Suhayl Zakkār. 12 vols. Damascus, n. p., 1988–1989.

Ibn al-ʿImād, ʿAbd al-Ḥayy, *Shadharāt adh-dhahab fī akhbār man dhahab*. 8 vols. Cairo: Maktabat al-Qudsī, 1931–1933.

Ibn al-Qifṭī, *Tārīkh al-ḥukamāʾ*, ed. Justus Lippert. Leipzig, 1903.

Ibn al-Wāṣil, *Mufarrij al-kurūb fī akhbār Banī Ayyūb*, eds. J. Shayyāl, S. ʿAshūr, H. Rabī. 5 vols. Cairo: Maṭbaʿat Jāmiʿat Fuʾād al-Awwal, 1953–1977.

Ibn Ḥajar al-ʿAsqalānī, *Inbāʾ al-ghumr bi-anbāʾ al-ʿumr*. 3 vols. Cairo: Lajnat Iḥyāʾ at-Turāth al-Islāmī, 1969.

Ibn Ḥajar al-ʿAsqalānī, *ad-Durar al-kāmina fī aʿyān al-miʾa ath-thāmina*, ed. Muḥammad Sayyid Jād al-Ḥaqq. 5 vols. Beirut: Dār al-Kutub al-Ḥadītha, 1966.

Ibn Khallikān, *Kitāb wafayāt al-aʿyān wa-anbāʾ abnāʾ az-zamān*. 7 vols. Beirut: Dār aṣ-Ṣādir, 1882.

Ibn Taghrī Birdī, Abū l-Maḥāsin, *History of Egypt 1382–1469 A.D.* Part IV, 1422–1438 A.D. and Part V, 1438–1453 A.D. Translated . . . by William Popper. Berkeley and Los Angeles: University of California Press, 1958 and 1960.

Islamic desk reference, compiled from *The Encyclopaedia of Islam* by E. van Donzel, Leiden, New York and Köln: Brill, 1994.

King, David A., "The Astronomy of the Mamluks", in *Isis* 73 (1983), 531–555; reprint in David A. King, *Islamic Mathematical Astronomy*. London: Variorum Reprints, 1986, III.

King, David A., "On the Role of the Muezzin and the Muwaqqit in Medieval Islamic Societies", in David A. King, *In Synchrony with the Heavens: Studies in Astronomical Timekeeping and Instrumentation in Medieval Islamic Civilization*, vol. 1. Leiden: Brill, 2004, 623–677.

King, David A., *In Synchrony with the Heavens: Studies in Astronomical Timekeeping and Instrumentation in Medieval Islamic Civilization*. 2 vols. Leiden: Brill, 2004–2005.

Livingston, John W., "Ibn Qayyim al-Jawziyyah: A Fourteenth Century Defense against Astrological Divination and Alchemical Transmutation", *The Journal of the American Oriental Society* 91 (1971), 96–103.

Livingston, John W., "Science and the Occult in the Thinking of Ibn Qayyim Al-Jawziyya", *The Journal of the American Oriental Society* 112 (1992), 598–610.

al-Maqrīzī, Aḥmad ibn ʿAlī, *as-Sulūk li-maʿrifat duwal al-mulūk*. 8 vols. Beirut: Dār al-Kutub al-ʿIlmiyya, 1997.

al-Maqrīzī, Taqī ad-Dīn Aḥmad ibn ʿAlī: *Kitāb al-mawāʿiẓ wa-l-iʿtibār fī dhikr al-khiṭaṭ wa-l-athār*, ed. Gaston Wiet. Reprint: Frankfurt a.M.: Johann Wolfgang-Goethe-Universität, 1995.

Pouzet, Louis, *Damas au VIIe/XIIIe siècle: Vie et structures religieuses d'une métropole islamique*. 2nd edition. Beirut: Dar-el-Machreq Sarl Éditeurs, 1991.

Reuben, Amitai, "The Remaking of the Military Elite of Mamlūk Egypt by Al-Nāṣir Muḥammad b. Qalāwūn", *Studia Islamica* 72 (1990), 145–163.

aṣ-Ṣafadī, Aybak Khalīl, *al-Wāfī bi-l-wafayāt*, eds. Aḥmad Arnāʾūṭ and Turkī Muṣṭafā. 25 vols. Beirut: Dār Iḥyāʾ at-Turāth al-ʿArabī, 2000.

Savage-Smith, Emilie, *Islamicate Celestial Globes: Their History, Construction, and Use*. Washington, DC: Smithsonian Institution Press, 1985.

Shihadeh, Ayman, "From al-Ghazālī to al-Rāzī: 6TH/12TH Century Developments in Muslim Philosophical Theology", *Arabic Sciences and Philosophy* 15 (2005), 141–179.

Sibṭ b. al-Jawzī, *Mirʾāt az-zamān fī taʾrīkh al-aʿyān*. 1 vol. in 2 parts. Hyderabad: Dāʾirat al-ʿUthmāniyya, 1951.

Sourdel, Dominique, "al-Āmidī, ʿAlī b. Abī ʿAlī", *Encyclopaedia of Islam*, vol. 1, New Edition. Leiden: Brill, 1986, 434.

al-Yūnīnī, Mūsā ibn Muḥammad, *Dhayl Mirʾāt az-zamān*. 4 vols. Hyderabad, 1954–1961.

http://islamsci.mcgill.ca/RASI/BEA/Ibn_al-Majdi_BEA.htm (21 August 2011). http://sonic.net/~tallen/palmtree/ayyarch/ch9.htm#log.dig (19 August 2011).

9

NARRATIVES OF KNOWLEDGE IN ISLAMIC SOCIETIES

What do they tell us about scholars and their contexts?

Abstract

Current history of science, including cartography and geography, in Islamic societies has lost its mainstream status that it occupied some decades ago. The major reason for this unfortunate development is the change of what constitutes mainstream today in comparison to the past. Mainstream history of science in other than Islamic societies is much more focused on the study of the sciences in culture than on the study of the content of scientific texts or instruments alone. Moreover, numerous of its representatives apply various theory-based approaches and methods taken from other humanities and other fields of the historical sciences. Historians of science in Islamic societies have refused to engage in a productive manner with these newer developments and hence have found themselves marginalized with high costs for subsequent generations in the field. Some newer developments in the US indicate that younger scholars have taken up the challenge. It is, however, too early to judge whether this will lead to a renewed recovery of mainstream positions. On the other hand, mainstream history of science in other than Islamic societies is plagued by problems caused by its shift to new grazing grounds. In many cases, the specialized training as historians of science has caused a loss of qualification characteristic for previous generations in the field – a more than fleeting familiarity with at least one of the sciences and a knowledge of more than their native language/s. Some of this equally unfortunate development has caught on in history of science in Islamic societies with the worrisome result of an empty rhetoric.

This rhetorical kind of history of science holds little promise. I am more attracted by the challenge of the first generation of colleagues who made context, culture and theory mainstream in history of science. I think history of science in Islamic societies has still to go a very long way to reach their level of skills and

 DOI: 10.4324/9781003372455-10

sophisticated insights. The subject of my talk to the workshop on how to become mainstream was chosen to discuss one of the many possibilities to advance on that road. Hence my paper will offer some observations and reflections on narratives of knowledge in Islamic societies and possibilities for uncovering what they meant for their narrators and their audiences.

Introduction

In my view, five types of narratives of knowledge can be discerned in Arabic and Persian biographical and historical literature, which I call "origin narratives", "achievement narratives", "career narratives", "authority narratives" and "wisdom narratives". Origin narratives speak of the origin/s of a field of knowledge, the motivation for "inventing" or "establishing" the kind of knowledge and the field's subsequent development. They often consist of at least two, if not three layers: coming into being; first practitioner; dissemination. These layers can be connected among each other in weak or strong forms. Origin narratives are found primarily in biographical and bio-bibliographical dictionaries. Achievement narratives focus primarily on the fields of knowledge a scholar was engaged in, his works and some of his colleagues, students or patrons. Career narratives are more or less complete and consistent presentations of the path of a scholar from his, and in the religious disciplines also her, childhood education to his/her death or the year of the composition of the biography. Achievement and career narratives present different ideals of scholarly personas and appear mainly in biographical dictionaries. Some specimens can also be found in some historical chronicles and geographical accounts. This is in particular the case in post-Mongol Persian works. Authority narratives are stories about a person or a group of people who are invested by the narrator with the authority to decide on the truth-value of any given scholarly result. Wisdom narratives are collections of individual entries that combine biographical snippets with selected sayings of the person of the entry.

In a somewhat simplifying manner, these narratives have been studied so far from four perspectives: as carriers of older narrative patterns, mostly introduced in Antiquity and modified by Christian or Jewish groups appropriating "pagan lore"; as reports about "real" people and events, i.e., as data providers; as representatives of agendas by the author of any given work; as indicators of attitudes towards knowledge and its disciplines in various Islamic contexts. Depending on the methodological positions of an interpreter of these narratives the choice of one perspective strictly excluded any other position or at least focused the view on one approach only. This selectivity also included a preference for one type of narrative neglecting the others.

The older generation of historians of science privileged routinely the use of those narratives that provided possibilities for data collection, i.e., primarily narratives about scholarly achievements and careers and for the classical period also some of the wisdom narratives.[1] Two of the three authority narratives known currently to me have also found attention among historians of science, while origin narratives did not manage to attract them. This also applies to the other three kinds of narratives when taken as a whole, not merely as resources of factual information. The two authority narratives that historians of science are familiar with are the representation of the ʿAbbasid caliph al-Maʾmūn (r. 813–833) as the first sponsor and advocate of Greek sciences and scientific expeditions and the description of the Timurid prince and later sultan Ulugh Beg (r. 1411–1449) as the leading astronomer and mathematician of his time and the allegedly last patron of a program of astronomical observation and tabulation. In particular, reports about the expeditions sponsored by al-Maʾmūn have been analyzed for the scientific content and veracity more than once.[2] This focus on content and veracity, as necessary as such analyses are, has led to overlooking the narrative structure of these reports and hence the question of what it means that these reports were told in the manner they were formulated, modified and transmitted by astronomers, jurists and historians. In the case of the aggrandizement of Ulugh Beg, his participation in the Timurid network of inter-familial exchange of objects of knowledge, Ilkhanid emulation and astrological guidance of policy and decision-making has been ignored to the detriment of other princes in the network, princes whom art historians showed to have been eager patrons of the sciences and central figures of authority narratives of their own.[3]

Wisdom narratives were studied by numerous historians of medicine, philosophy and culture of whom only a few can be mentioned here – Max Meyerhof, D.M. Dunlop, ʿAbd al-Rahman al-Badawi, Wadad al-Qadi, Dimitri Gutas and Joel L. Kraemer. They edited, translated and analyzed a group of texts that seems to have taken its origin in the 10th century in a circle of students of philosophy in Baghdad, but later spread across Iran well into the 14th century as the very least. Wisdom narratives share a limited number of common points with achievement narratives in so far as they offer information about the fields a scholar contributed to and name some of his works, peers, students and patrons. Achievement narratives occasionally use wisdom sayings or quotes from scholarly works compiled in wisdom narratives. Hence, the boundaries between these two types of narratives are fluid and one can count several biographies among both of these two types.

The first who paid attention to origin narratives is Keren Abbou who analyzed some of them as part of a history of medicine in her doctoral thesis.[4] Her approach focuses on the narrative aspects of the stories. She asks which purposes the narrators pursued when presenting them in the specific form and place found in their

1 Examples abound. See, for instance, Matvievskaya 1967; Saliba 1994; Charette 2003 and King 2004.

2 Mercier 1992, 176–181; King 2000, 229–231.

3 Lentz and Lowry 1989.

4 Abbou 2008.

biographical dictionaries and which attitudes these features express towards medicine in the environments of the narrators.

As Arabic and Persian biographical dictionaries and historical chronicles were often understood by older generations of Orientalists as exercises in copying and plagiarizing with little novelty and rare pieces of truthfulness, modernist historians prefer to emphasize the ancient nuclei of origin stories and their adaptation from a pluralist theistic world view to monotheism in Christian and Jewish communities of Late Antiquity.[5] It needed the break with this mainstream under the impact of postmodernist perspectives and theories to understand that such "old" stories were not merely copied mechanically, but woven into a new text in order to present the author's own views of events, identities, goals and values to his contemporaries. The present in such compilations is not told independently, as a mere follow up of the past, but is questioned, criticized and lauded through stories about the past.

In this paper I will limit myself to the discussion of one example of an origin narrative, because the issues that need to be addressed are too numerous and complex for a thorough discussion of more than one example.

Origin narratives

Origin narratives appear in three principal forms. Form 1 speaks of divine origin through creation, revelation, inspiration through dreams and other manners or prophetic gift. Examples are stories about the ancient Greek physician Aesculapius turned into a God or the Jewish prophets Daniel and Enoch. Form 2 ascribes a field's origin to mythical cultural heroes, i.e., to superhuman gifts or talents. Examples are Hellenistic (Graeco-Egyptian) stories of Hermes and their Sasanian variants plus Iranian stories of Jamshīd. Form 3, finally, talks of ordinary human beings as inventors of skills or knowledge forms.

Due to the complexity of such stories, the problems with disentangling their layers and the difficulties with determining the web of contemporary interlocutors and envisaged audiences, I decided to present only one case here. The narrator of this complex origin story was Ibn Abī Uṣaybiʿa (d. 1269/70), a physician from a family of physicians who worked in various cities of Syria and Egypt for members of the Ayyubid family (r. 1172–1260). His history of physicians called *ʿUyūn al-anbāʾ fī ṭabaqāt al-aṭibbāʾ* (*Sources of News on the Classes of Physicians*) and written in Damascus in 1242 is structured chronologically and geographically:

(1) on how the art of medicine came into being and the beginning of its existence; (2) on the classes of physicians to whom parts of the art of medicine appeared and who were the beginners in it; (3) on the classes of the Greek physicians in the lineage of Aesculapius; (4) on the classes of physicians to whom Hippocrates revealed medicine; (5) on the classes of physicians since Galen's time or close to it; (6) on the classes of the Alexandrian physicians and the Christian physicians

5 See, for instance, Rosenthal 1956; Strohmaier 1996, 2003.

and others who lived in their times; (7) on the classes of the physicians of the Arabs and others who existed at the beginning of the appearance of Islam; (8) on the classes of the Syriac physicians at the beginning of the dynasty of the Banū ʿAbbās; (9) on the classes of the physicians who translated books on medicine and other (books) from Greek into Arabic, and mentioning of those for whom they translated; (10) on the classes of the Iraqi physicians and those of the Jazīra and Diyār Bakr; (11) on the classes of the physicians that appeared in the country of the ʿAjam (i.e. the non-Arabs, usually referring to Iran or its western parts); (12) on the classes of the physicians from India; (13) on the classes of the physicians who originated in the Maghrib and those who settled there; (14) on the classes of the famous physicians of Egypt; (15) on the classes of the famous physicians of Syria (Ibn Abī Uṣaybiʿa 1965, 1–2).[6] The key word *ṭabaqāt* in the book's title highlights the chronological aspect of Ibn Abī Uṣaybiʿa's presentation, including the Islamic format of this division. It even may be the case that the choice to start with origin stories was triggered by this emulation of Islamic biographical and historical periodization of the past as a sequence of generations beginning with Muḥammad and his first companions, followed by the younger companions and the generations that came afterwards. One should not forget though that the discussion of origins of knowledge and in particular of medicine has a venerable history in Islamic societies going back at the very least to the late 8th century.

Ibn Abī Uṣaybiʿa tells stories of types 1 and 2, which are framed by a typology of how the knowledge came into being. He was recognizably a committed Muslim and seems to have subscribed to the Ayyubid Sunni mainstream. Within this religious frame, he undeniably appreciated the various sciences including philosophy and medicine. Both kinds of value systems were obviously not mutually exclusive nor does his adherence to religious beliefs or the integration of commitments to the sciences, philosophy or medicine into an overarching religious framework say anything about the quality or strength of his scholarly arguments and works. Instead of making such inferences in one or the other direction, as has been done so often in discussions on the relationship between "Islam" and "science" or "philosophy", it is much more important to study the specific forms in which the relationship between these two value systems are expressed as well as the specific beliefs that are mobilized to legitimize the sciences, philosophy or medicine.

Methodological considerations

As outlined above, Ibn Abī Uṣaybiʿa begins his book with three interconnected chapters. The first of them is the chapter on the views of different authors regarding the origin of medicine. The two following chapters have each one a more limited

6 This is a very literal translation of the headings. A more polished, but at times problematic translation, adapted to current English, is that made by L. Kopf in 1971 for the National Library of Medicine, Bethesda, Maryland, www.tertullian.org/fathers/ibn_abi_usaibia_01.htm. Downloaded 23 April 2012. All translations in this paper, if not specified otherwise, are my own.

scope and discuss the first human practitioner of medicine and then his heirs and successors. One of the central methodological questions is whether these three chapters are independent narrative units or whether they are interlinked in a manner that they could not be analyzed separately. The answer to this question impacts of course my definition of "origin story". A second, subsequent methodological problem that arises is the question as to where the meaning of the narrative might be located from Ibn Abī Uṣaybiʿa's perspective. A third issue is whether such a quest is in itself meaningful or whether it is more appropriate to assume that there is no such single narrative climax and that all three chapters contribute their own, particular meanings to the overall narration of a history of medicine in general and to the desire to locate and determine the beginnings of the field in particular.

A first, simple answer to these three questions can be gleaned from the formal presentation of the stories on origin, beginning and family-based continuation as three visibly different chapters. This organization of the narrative was chosen by the narrator himself since the extant manuscript copies include these headers as the introduction to the book clarifies where the general outline is given. The outline describes the book's structure as proceeding from the most general (origin of the discipline and professional practice) to the most concrete (the distribution of medical knowledge among the generations after the "first" physician) followed subsequently by later generations of practitioners until the author's own time. It constitutes a hierarchical organization of Ibn Abī Uṣaybiʿa's narrative as a whole and of the first three chapters as the part discussed here. Thus, it is possible and reasonable to take the stance that the first three chapters constitute in themselves a narrative unit in comparison to all later chapters since it is they alone that are concerned with aspects of origin, beginning and succession within a closed social group, the first practitioner's larger family. This formal organization also makes it possible and reasonable to consider the first three chapters themselves constituting individual sub-units of the narrative construction that ties them together.

A second answer comes from the explicit intertextual relations between the three chapters and their particular functions. All three chapters are linked by the same simple rhetorical device of "this is what I will tell you later/told you already earlier". Chapter 1 ends with "Asclepius I was the first who discovered the different aspects of medicine, as will be shown in the following". (Ibn Abī Uṣaybiʿa 1965, 10) Chapter 2 repeats this device at its very beginning: "Many ancient philosophers and physicians are agreed that Asclepius, as pointed out above, was the first physician mentioned by name and the first to discuss all medical topics on the basis of practical experience". (Ibn Abī Uṣaybiʿa 1965, 10) Chapter 3 also uses this device to begin its report: "As we have already mentioned, Asclepius gained his knowledge of medicine from experience, and when he started to teach his children and relatives those aspects of it that he had acquired, he bade them transmit the art to absolutely no one but their children and other descendants of his". (Ibn abī Uṣaybiʿa 1965, 15)

Within this device, there is a second link of deeper meaning, namely the repetition of the claim that Asclepius gained and discussed his medical knowledge on the basis of experience. Since this is not the only statement in ancient Greek sources about Asclepius' access to medical knowledge and also not the only one presented within Chapter 2 about this first practitioner of the arts, the emphasis on the experiential nature of his knowledge and the omission of its divine character take up a trend also visible in Chapter 1 when Ibn Abī Uṣaybiʿa presents his additions to Galen's *Commentary on the Hippocratic Oath* and discusses the strong position of Muwaffaq al-Din Asʿad b. Ilyās b. Muṭrān (d. 1191) in favor of an exclusively human origin of medicine. A third intertextual link between all three chapters is their usage of ancient Greek texts and traditions, in particular the explicit and in case of Chapter 2 also extensive use of Galen's *Commentary on the Hippocratic Oath* (Chapters 1 and 2) and the use of Johannes Philoponnus' history of physicians (Chapters 2 and 3).

A third answer can be derived from the relationship between the narrative approaches in the first three chapters. A comparison between these three individual sub-units demonstrates that they proceed in a hierarchically ordered sequence. The sequence starts with a scholarly organized presentation of a broad variety of opinions among scholars as well as those held outside scholarly communities about the question as to whether medicine originated among men or was delivered to them by God or other divine creatures. The next unit (Chapter 2) reduces the narrative scope to the question of who was the first creature on Earth who practiced healing, a question that includes animals as well as humans. The third unit (Chapter 3) focuses exclusively on human practitioners who all belonged to the larger family of the person on whom the stories in Chapter 2 settle as the first healer of true relevance to a history of physicians, i.e., all digressions into the realm of animals and the inclusion of a prophet as a teacher of the first healer into the narrative of Chapter 2 are disregarded in Chapter 3.

A fourth answer is won when the degree of narrative complexity within each sub-unit is considered. These complexities differ significantly, but proceed again in a hierarchical sequence, with Chapter 1 being the most complex one and Chapter 3 the least. The different narrative approaches allow Ibn Abī Uṣaybiʿa to prove his command of the available heritage on the three questions in form of stories, opinions and texts. That he treats these three components of knowledge differently suggests that self-representation was not the only decisive factor for this different scale of narration. Other factors must be searched in the type of stories told within each sub-unit and their functions for each narrative thread. In Chapter 1, he sets out to survey all the stories and opinions he could find or at least wished to impart on his readers. He does not pretend, however, to have had engaged in an equally broad-scale investigation of all available literature. He rather focused on two explicitly mentioned textual sources, the Galenic commentary on the Hippocratic oath which is disputed among current historians for its authenticity, and an Arabic text on medicine by Muwaffaq al-Dīn, one of the leading court physicians of the founder of the Ayyubid dynasty. Ibn Abī Uṣaybiʿa added with clear narrative purpose texts based on orally transmitted opinions. He also used additional material the sources of which are not given. This material is presented as a list of positions held by different religious and scholarly communities that did not need further documentation. This narrative set up which will be analyzed more specifically below responds to the declared goal of Chapter 1 of surveying all relevant opinions. This goal allowed Ibn Abī Uṣaybiʿa to use the format of formal classification as a rhetorical and organizational device. Not explicitly declared, but impossible to miss due to the focal points of the chosen bits of texts borrowed from earlier books and produced (or pretended to have been produced) as oral reports is a second purpose of the narrator – the presentation of a contemporary debate pro and con the divine or human creation of medicine. This is a narrative choice, not a goal in itself, since it allows Ibn Abī Uṣaybiʿa to present his own views on the topic. The narrative, but not necessarily positional clarity and comprehensiveness of this part of Chapter 1 marks the presentation of the narrator's views on the issue as the main purpose of the entire narrative. The two explicitly mentioned texts serve as entrance into and formulation of the problem as well as a defense against possible detractors. The presumably oral reports function as proofs for specific positions. The broad-scale survey on the different beliefs held by various religious and scholarly communities translates the declared goal of Chapter 1 into "factual reality" and thus represents a continuation and particularization of the classification approach with which the chapter starts. In this sense, it does not need further anchorage in specified sources. The general classification namely was already presented as taken from Galen's *Commentary on the Hippocratic Oaths*. The later survey was announced then and there as Ibn Abī Uṣaybiʿa's comments and extensions.[7]

In Chapters 2 and 3 Ibn Abī Uṣaybiʿa offers a much more limited range of stories. In Chapter 2, they bring together ancient Greek mythical and scholarly accounts of Asclepius with the legend of three Hermeses, one of whom was identified with the prophet Enoch = Idrīs who is presented as Asclepius' teacher. Against its appearance, this legend is, according to van Bladel, not of pre-Islamic origin, but a combination of different variants of pre-Islamic legends about pre- and post-diluvian figures of Hermes who lived in Egypt and Mesopotamia. They served as chronological devices to allow a comprehensive chronology for both main epochs.[8] Chapter 3 appears exclusively as a survey of ancient Greek stories with a single textual anchorage, a work by Johannes Philoponnus.

Thus, the type of narrative told in each of the three sub-units is different and necessitates different narrative strategies. In Chapters 2 and 3, Ibn Abī Uṣaybiʿa does not declare to present a survey of all possible views on the first practitioner of medicine and his dispersion of medicine among his own children. Apparently, he did not see a need for setting up an elaborate narrative scheme in order to be able to take an explicit personal stance. This major narrative difference between the three chapters suggests that the central points of Chapters 2 and 3 were not contested in

7 For a discussion of this interpretation see below.
8 Van Bladel 2009, 121–163.

Ibn Abī Uṣaybiʿa's time or at least not as controversially debated among Ayyubid scholars as the more general issue of divine versus human creation of medicine. Moreover, the integration of ancient Greek traditions on Asclepius with the legend of the three Hermeses responds, following van Bladel's analysis, to a harmonization of chronologies. Such a harmonization of chronologies was a concern of astrologers in the 9th and 10th centuries when writing astrological histories about different dynasties or the new religion. Neither of the two types of world histories were a major concern in the Ayyubid period or Ibn Abī Uṣaybiʿa's immediate context. Ayyubid legitimacy was constructed in this harmonized chronology, but not with regard to ante- and post-diluvian times. Their points of reference were located in their immediate past with the Fatimid dynasty whom they toppled and their contemporary friends and foes, the Abbasids, Seljuqs and the Crusaders. Hence, it is not particularly surprising that all that Ibn Abī Uṣaybiʿa' does in Chapter 2 in this regard is to insert into a series of ancient Greek stories a brief survey of various elements of the legend according to Mubashshir b. Fātik (11th c.) and Abū Maʿshar (d. 886), whom he both names explicitly, and the history of physicians and philosophers *Ṭabaqāt al-aṭibbāʾ w-al-ḥukamāʾ* of the 10th-century Andalusian court physician Ibn Juljul (c. 944–c. 994), whom he does not mention here by name, while Chapter 3 has not even a need for such an insertion and can follow its more restricted pattern.[9] A closer analysis of the narrative assemblage presented in Chapter 2 is a much needed desideratum, because none of the so far published translations and interpretations study this insertion as an integral part of Chapter 2.[10] Thus, in addition to the formal and structural differentiation between the three chapters, Chapter 1 is set apart from Chapters 2 and 3 in its declared goal and style and consequently in its narrative composition.

Chapters 2 and 3 are stylistically almost identical, although their narrative composition in its main format differs. Ibn Abī Uṣaybiʿa's own voice made explicit is very muted in both chapters, in contrast to Chapter 1 where it is energetic, substantial and argumentative. Each of the Chapters 2 and 3 is organized as a series of relatively short quotes from often named sources. The main difference in this set up between the two chapters is the temporal and cultural identity of these sources and their authors. Ibn Abī Uṣaybiʿa presents the Greek material about Asclepius in Chapter 2 through several Arabic sources and one dominant Greek author – Galen. The longest contribution ascribed to Galen again comes from the *Commentary on the Hippocratic Oath*, although other Galenic writings are also exploited as is the case in the last part of Chapter 1. The Hermes legends are quoted exclusively from Arabic texts, their main named authors being Abū Maʿshar and Mubashshir b. Fātik. If van Bladel is right that the stories of three different Hermeses were an invention by Abū Maʿshar, this exclusivity is not surprising, but an indirect point in favor of this interpretation. In Chapter 3, the only explicit voices are those of

9 Ibid., 124–125 (Arabic text) and 125–127 (English translation).

10 See Van Bladel 2009, 122–132 and 125, fn 12.

Johannes Philoponnus and Ibn Abī Uṣaybiʿa. They do not, however, dominate visibly the scene, although much of the information seems to come at least indirectly from Philoponnus.

As a result of these comparisons, I conclude that meaning is not located exclusively in one of these three chapters and that it is not distributed sequentially in the sense that the last point made or the last story told reveals the author's final conviction. Ibn Abī Uṣaybiʿa's purpose did not consist in clarifying dates, names and other kinds of facts by accessing even newly available texts in a temporal sequence. Each chapter has its own specific goal according to which the stories are selected and arranged. Neither does Chapter 2 supplant Chapter 1 in delivering a message about Ibn Abī Uṣaybiʿa's "true" or "final" opinion on the origin of medicine nor does it foreclose Chapter 3 in its explanation on how the medical knowledge of the first practitioner of medicine was disseminated. If Ibn Abī Uṣaybiʿa indeed carefully planned the narrative nexus between the three chapters it is expressed in his explicitly made intertextual links described above. They indicate that divine and human activities were important in the creation of medicine, that human experience was an indispensable foundation of medicine that even gave rise to one particular approach to healing and discoursing and that Ibn Abī Uṣaybiʿa considered it appropriate and possible to criticize ancient authorities directly and indirectly.

Furthermore, I conclude that while following an overall scheme for the entire book and a particular, coherent narrative in the first three chapters about the beginnings of medicine Ibn Abī Uṣaybiʿa divided this particular narrative unit into three largely independent narrative sub-units. His sequential style of narrating origin, first practitioner and dissemination allows me to focus here exclusively on the analysis of the first unit about origin.

I recognize, however, that a complete separation of these three chapters is misleading and can cause misinterpretations. As my discussion of the different layers of similarity and difference has shown the three chapters share important elements in format, structure and style, which, as I will argue below for Chapter 1, are the intentional result of Ibn Abī Uṣaybiʿa's creative approach to the issue of origin and in all likelihood also to the two subsequent questions of first practitioner and dissemination.

Ibn Abī Uṣaybiʿa's approach to collective biography writing

Ibn Abī Uṣaybiʿa focused on narrating the conditions of his own environment in Egypt and Syria. According to his own testimony, he gave the finished text to some of his best friends and most esteemed colleagues in Damascus for a critical reading before he allowed it to circulate freely. Placing the two chapters on Egypt and Syria at the end of his long collection of biographies, poems, stories and book titles cannot have presented an insult to them. Otherwise, they would not have approved of it. By emphasizing their status as the most famous members of their profession and region, Ibn Abī Uṣaybiʿa ascertained that his addressees felt properly appreciated and singled out as the culmination of his entire work. Consequently, the stories that he told about his various contemporaries are well-measured laudatory concoctions about the most high ranking and influential members of the medical class in Ayyubid times. This does not mean that he abstained from critique. But it is clear that he tried to please his very own circles of relatives, friends, colleagues and patrons.[11]

In addition to the focus on his own context, Ibn Abī Uṣaybiʿa pursued two structural lines of narration – one of emulation of standardized Islamic biographical history and the other of propagation and praise of the virtues and values of the leading physicians of his time. His classification of origin narratives is formal and lacks emotional depth and drama. It opposes in a binary manner divine and human history and invention. Except for Chapter 2, Ibn Abī Uṣaybiʿa does not return to these elements in later biographies, in particular not in those of his own group. I am convinced, however, that Ibn Abī Uṣaybiʿa chose the formal, binary approach to narrating origin stories for medicine, because there was a debate about these positions among physicians and religious scholars in the entourage of the founder of the Ayyubid dynasty and in the Abbasid capital. This binary, formal approach provided him with distance and scholarly status, which he apparently sought in order to offer his very skilled narrative set up of this debate and come out of it unscathed, precisely because he wished to emphasize the importance of the human element in the origin of medicine as a professional and scholarly field of knowledge.

Throughout the book, Ibn Abī Uṣaybiʿa has a strong voice of his own, uses direct quotes to emphasize his main agenda and introduces witness reports and anecdotes in order to strengthen his argument.

Ibn Abī Uṣaybiʿa's narrative format in chapter 1

Ibn Abī Uṣaybiʿa's narrative format consists in Chapter 1 of five main features. First, he constructed a cohesive text that comprises five parts. Part 1 starts with a brief introduction into the difficulties to decide the issue of origin followed by Part 2, an extract from the *Commentary on Hippocrates' Oath* ascribed to Galen. Part 3 is a long extract from Muwaffaq al-Dīn's *Bustān al-aṭibbāʾ*. Part 4 is Ibn Abī Uṣaybiʿa's wrapping up the gist of the discussion of Part 3 supported by reflections of his own together with additional material collected orally among his colleagues. Part 5 is the relatively extensive survey of different opinions on the issue of origin of medicine, which he promised in Part 1. Second, he chose two texts to frame the central issue, i.e., divine versus human invention of medicine, an ancient one and a very recent one, both by leading physicians of their time. Third, in Part 2 he presents storytelling as a formal element of medical education, as part of the much-cherished mode of classifying knowledge and as an integral

11 Brentjes 2010.

component of scholarly inquiry and reasoning. As a result, storytelling appears as essential to the professional identity of a physician. Fourth, he uses Galen, his alleged beliefs and his *Commentary* as a defense against possible serious critique at the position he builds up through several steps and which privileges human before divine invention. Fifth, he structured Chapter 1 in a manner that it leads from a formal discussion of the variety of opinions in Antiquity (Part 2) to a very sharp rebuttal of the belief in divine invention by a leading Ayyubid physician (Part 3) followed by comments of Ibn Abī Uṣaybiʿa on the two preceding parts and recommendations to the readers expressed as the author's personal conclusions (Part 4). Chapter 1 ends with a fairly long survey on opinions held on the two main views among numerous religious and scholarly communities intermingled with Ibn Abī Uṣaybiʿa's own remarks (Part 5).

These five features highlight the indisputably active role that Ibn Abī Uṣaybiʿa played in the construction of the narrative of Chapter 1. Not only could he have chosen a different overall structure and different authorities to quote from, he also could have avoided to express explicitly and at length his own opinions. By choosing a leading Ayyubid physician as the central voice pro human invention and by finishing the debate in favor of this position, Ibn Abī Uṣaybiʿa emphasizes that the debate belongs to his own socio-cultural context. Several 12th-century physicians and religious scholars had offered views on the topic, among them Ibn al-Jumāyʿ (d. 1198), another one of Ṣalāḥ al-Dīn's (r. 1169–1193) physicians, and Ibn al-Jawzī (d. 1200/01), a famous Ḥanbalī scholar from Baghdad. In his presentation of the debate, Ibn abī Uṣaybiʿa focused without hesitation on Muwaffaq al-Dīn's biting critique at the otherwise unknown Abū Jābir al-Maghribī. Ibn Abī Uṣaybiʿa's main witness of the debate was clearly in a combative, outspoken mood when commenting on Abū Jābir's opinions:

> This reasoning is altogether confused and unintelligible. Even though Galen, in his *Commentary on the Covenant* says that this art is a revealed and inspired (one), and Plato in his *Book of Politics* [*Republic*] says that Asclepius was [divinely] supported and inspired, it is wrong to banish the achievement of this art from the inventions of the [human] minds and an abatement of (those) minds who invented more sublime (arts) than medicine. We may assume that the first [human being] of the world was alone and in need of medicine as the bodily needs in this world (with its) many crowds (exist) nowadays. (We may assume) that his body bothered him, his eyes reddened, and signs of an excess of blood (i.e., hyperaemia) befell him, (while) he did not know what to do. Then, due to its strength, a nosebleed befell him and that what had (afflicted) him went away. He acknowledged this. Then, when, on another time, the very same happened to him again, he hastened to his nose and scratched it, so that blood came out of it and that what had befallen him subsided This became something he remembered and taught to all of his sons and offspring whom he (encountered). The [reflections] on the art became more refined

> until finally, thanks to sophistication of mind and delicacy of sense, the vein was opened.
>
> (Ibn Abī Uṣaybiʿa 1965, 3–4)[12]

Although Ibn Abī Uṣaybiʿa did not dare or wish to fully support Muwaffaq al-Dīn's outspoken defense of medicine's human origin, but combined both positions, his choice of words and emphasis favors the human part. He achieved to establish this impression by quoting directly and indirectly long passages from the *Bustān* to which he added a few inquiries as well as some confirmations of specific forms of human medical practice by other physicians. Part 4 sums up what Ibn Abī Uṣaybiʿa wished his readers to conclude from Chapter 1:

> I (presented) these opinions, which were mentioned before, in their differences and their kinds, since at that time it was my intention to mention most of what each sect believed in. As the contradictions and differences in this are as you saw, the search for its origin became very difficult. Yet, an intelligent person, if he thinks about that according to his intellectual capacity, will find that it is not unlikely that the origins of the art of medicine resulted from these things that preceded or from most of them. I therefore say that the art of medicine is needed by the people and belongs to them wherever they are and at all times, except that it varies according to the localities, the multitude of food and the strength of discernment. The need for it may be greater among one people than among another. The reason for this is that since in some regions people have many diseases, especially if they take a great variety of food and continually eat fruit, their bodies remain ready for illness so that perhaps not a single one of them (can) escape for the rest of his life a sickness that strikes him. (People) like those are in greater need of the medical art than others who live in regions with a healthier air, eat less sorts of food and, moreover, eat less of the food that they have. Furthermore, since human beings are also different in the power of rational discernment, the most perfect of them in discernment, the strongest of them in (knowledge based on practical experience) and the best of them in personal judgment will comprehend and memorize that what occurs to them as experiential matters and other (things) in order to confront the diseases with that one of the drugs that heals them and no other. If the population of a region is frequently affected by maladies or includes a number of people of the kind I have just described, these persons will master, by their powers of perception, their genius, what they have as the experiential matters and other (things) the means of healing and will bring together, in the long run, many things from the art of medicine.
>
> (Ibn Abī Uṣaybiʿa 1965, 5)[13]

12 Kopf www.tertullian.org/fathers/ibn_abi_usaibia_01.htm, 10–11.

13 Ibid., 14.

The explicit positions taken by Ibn Abī Uṣaybiʿa in Part 5 are even more remarkable than those of Part 4. Here he presents several opinions that are clearly marked as not his own, but held by pre-Islamic and Islamic communities that specify the divine contributions to the origin of medicine. It is not medicine as a theoretical edifice and a systematic set of practices and experiences that was revealed or inspired by God and other divine entities. Divine origin of medicine rather means to reveal in dreams specific remedies for specific maladies and to inspire the sick person to try them out. Human origin of medicine, in contrast, comprises three components: 1. the systematic testing of the applicability of divinely inspired remedies and treatments; 2. the description and discussion of the results of these tests; 3. the derivation of generalizations and the formulation of theories. The human activities of 1 and 2 led to empiricism whose founder was Asclepius as Ibn Abī Uṣaybiʿa says in Chapter 2. Those of 3 yielded dogmatism whose founder was Hippocrates as stated in Chapter 3. Hence, while divine inspirations and revelations are important for medicine's coming into being, without the higher-ranking human activities medicine as a systematic body of experiences, practices and theories would not have been achieved. Ibn Abī Uṣaybiʿa's partisanship in the conflict of his time about divine and human invention of medicine is certainly presented in a careful, well-balanced array of narrative strategies. This form of narration is, moreover, not less bold and energetic than Muwaffaq al-Dīn's much stronger rejection of divine creation of medicine.

Classical interests in chapters 1 and 2 of biographical dictionary

The standard research interest in Chapters 1 and 2 concerns those parts where Ibn Abī Uṣaybiʿa quotes longer extracts from a text ascribed to Galen which is neither extant in Greek nor mentioned in Greek sources. Two authors who have written extensively about those parts are the late Franz Rosenthal and Gotthard Strohmaier. They analyzed Part 2 of Chapter 1 and the third part of Chapter 2 where Ibn Abī Uṣaybiʿa quotes or summarizes extracts from this Galenic source. Rosenthal's primary goal was to reconstruct this supposed Galenic text as much as possible including Ibn Abī Uṣaybiʿa as one of the extant Arabic witnesses. A part of this effort was Rosenthal's and Strohmaier's discussion of whether this text is genuine or a later Greek composition. Furthermore, Strohmaier studied specific aspects of the Arabic representation of Greek motives like the transformation of Greek gods into the monotheistic creator or the attributes of Asclepius.[14]

Because they did not read the fragments of the purportedly Galenic text as part of Ibn Abī Uṣaybiʿa's complete narratives in the two chapters they did not only misunderstand the function of these fragments as mere quotes from a Galenic or pseudo-Galenic text. They accepted comments made by Ibn Abī Uṣaybiʿa as being words by Ḥunayn b. Isḥāq (d. 867), one of the 9th-century translators of the

14 Strohmaier 1996, 65–66, 2003, 155–166.

text.[15] They also misunderstood the last passage of this text in Chapter 1. Strohmaier rejected to trust Ibn Abī Uṣaybiʿa's words, who had marked the passage as a direct quote from what he and other writers in Arabic believed to be a Galenic text. Strohmaier decided to consider this quote as an interpretation by the Ayyubid physician.[16] Rosenthal even failed to recognize this quote as a part of the supposed Galenic text.[17] The overall lack of interest in the complete Arabic narrative constructed by Ibn Abī Uṣaybiʿa in Chapters 1 to 3 seems to me the main reason for the misinterpretations, because once one considers all narrative elements used in them things become clearer as I will argue below.

Moreover, the two eminent scholars of Graeco-Arabica did not only abstain from analyzing Ibn Abī Uṣaybiʿa's activities as a creator of the narrative sub-units in Chapters 1 and 2. They also seem to have failed in the more traditional task of sorting out the contributions of those who rendered the Greek text into Syriac and Arabic. As a result, various partially opposing evaluations can be found in their work.

At least four scholars of the 9th century were involved in the various translations of the Greek text and the use of one of its Arabic versions. Ḥunayn b. Isḥāq translated it from Greek into Syriac. His nephew Ḥubaysh b. al-Ḥasan and his student ʿĪsā b. Yaḥyā produced two different Arabic translations of Ḥunayn's Syriac text. Ḥunayn's son Isḥāq b. Ḥunayn (d. 911) used one of these versions for his survey on who was the first physician and who followed this originator until the very days of the compiler. Of these four versions, only the last one is extant.

Rosenthal changed more than once his opinions on the character of the extant Arabic fragments of the presumed Galenic commentary. In 1954, he wrote that "(t) he identical text appears in Isḥāq b. Ḥunayn, Ta'rîkh al-aṭibbâ'"(*Isḥāq b. Ḥunayn* 1954, 73). He pointed out that the identity is not complete, because there are "some statements (that are) not found in Isḥāq's work (who) obviously used the Commentary as his source [. . .]" (*Isḥāq b. Ḥunayn* 1954, 74). In 1954, he also believed that Ibn Abī Uṣaybiʿa had used Isḥāq b. Ḥunayn's work. Two years later he had changed his mind and considered this impossible. Rosenthal's altered judgment on this last point is correct, since the difference between the two texts is rather substantive, although the content of the two versions is in principle identical. Each of the two writers followed, however, a different narrative sequence. Thus, Ibn Abī Uṣaybiʿa did not copy Isḥāq's history of the physicians nor did he appropriate its structure. The texts' vocabulary also differs. There can be no doubt whatsoever that Ibn Abī Uṣaybiʿa worked with a different textual version of the *Commentary*. Moreover, the relative closeness between the content of the two versions suggests that Ibn Abī Uṣaybiʿa did not rephrase his source, but reproduced it fairly faithfully. On the other hand, Rosenthal took Ibn Abī Uṣaybiʿa's introductory statement of Chapter 1, Part 2 about having added something to the *Commentary* as being Ḥunayn's

15 Rosenthal 1956, 55, fn 9.

16 Strohmaier 2007, 147.

17 Rosenthal 1956, 59.

words. He declared the absence of certain parts of Ibn Abī Uṣaybiʿa's extract in Isḥāq's history of physicians as "proof" for being Ḥunayn's additions or outside material. Given the narrative differences between the two versions, this judgment is not well founded. Arguments for accepting this claim as Ibn Abī Uṣaybiʿa's words will be presented in the following section.

As stated already, Rosenthal abstained from discussing the statement defined by Ibn Abī Uṣaybiʿa as a direct quote from Galen. He obviously did not understand it as belonging to the text of the *Commentary*. This is made clear in the last sentence of his translation: "This theory is the theory of Galen, and this is the text of what he mentions in his *Commentary* on the *Book of the Oaths by Hippocrates*" (Rosenthal 1956, 59). Kopf, in contrast, who translated the first five chapters of Ibn Abī Uṣaybiʿa's book into English, considered this direct quote as a part of the *Commentary*:

> This is the opinion of Galen, which he expressed as follows in his *Commentary on the Hippocratic Oath*: 'As for myself, I maintain that it would be most proper and most plausible to say that God, the Blessed and Exalted, created the art of medicine and revealed it to man, because it is unthinkable that the human mind should have been able to conceive so sublime a science. Only God, the Blessed and Exalted, is the Creator Who is truly capable of this. For we do not find that medicine is inferior to philosophy, which is generally believed to have taken its origin from God, the Blessed and Exalted, Who revealed it to mankind'.
> (Kopf www.tertullian.org/fathers/ibn_abi_usaibia_01.htm, 7)

Plessner who annotated Kopf's translation thought that Rosenthal had made an error by excluding the passage from his translation.[18] Given that the quote starts with a verb in the third singular, while Ibn Abī Uṣaybiʿa introduces his own remarks in the first person (either singular or plural), Kopf's and Plessner's reading of the Arabic text appears more plausible. Strohmaier also accepted the quote as part of Ibn Abī Uṣaybiʿa's extract from the *Commentary*. He considers the text before it as genuinely Galenic, but is convinced that these last lines are Ibn Abī Uṣaybiʿa's own words, not those of Galen. His reason for this different evaluation of these two parts is the doubt that Galen believed in an origin of medicine by divine revelation and inspiration.[19]

Ibn Abī Uṣaybiʿa's extract from Galen's Commentary of Hippocrates' Oath

Ibn Abī Uṣaybiʿa, in a sense, begins his entire work with a double claim in Part 2 of Chapter 1 that anchors himself and his understanding of the world of physicians on the one hand squarely within ancient Greek traditions and reflects on

18 Plessner in Kopf www.tertullian.org/fathers/ibn_abi_usaibia_01.htm, fn, 31.
19 Strohmaier 2007, 147.

the other his self-perception as that of an equal to Galen whom he reveres as the iconic authority of Greek medicine. In this classification of origin stories he claims first to follow the views expressed by Galen in his *Commentary on Hippocrates' Oath* and second to have taken the liberty to add to it his very own views. As mentioned before, Rosenthal believed that the person making the second claim was not the narrator Ibn Abī Uṣaybiʿa, but the translator and physician Ḥunayn b. Isḥāq. Ḥunayn's statement in his *Risāla [. . .] ilā ʿAlī b. Yaḥyā fī dhikr mā turjima min kutub Jālīnūs bi-ʿilmihi wa-baʿḍ mā lam yutarjam* (Epistle [. . .] to ʿAlī b. Yaḥyā mentioning what has been translated of Galen's books on his science (followed by) what was not translated) says, however, only: "I translated it into Syriac, adding a commentary to it which I made for its difficult parts" (Ḥunain ibn Isḥāq 1925, 32, 40). Ibn abī Uṣaybiʿa, on the other hand, declared: "Let us first begin with recording what he has mentioned together with what we attached to it with the intention of (giving an exhaustive survey) of these different opinions" (Ibn Abī Uṣaybiʿa 1965, 2).

Neither Rosenthal nor Kopf understood this passage in a substantially different manner.[20] The purposes of both additions are recognizably different. Hence, there is no truly convincing reason for assuming that all that Ibn Abī Uṣaybiʿa did here was copying the respective passage from an Arabic version of Ḥunayn b. Isḥāq's Syriac translation. This conclusion is supported by other extracts from this very same text in Chapter 2 when Ibn Abī Uṣaybiʿa talks about Asclepius. There he distinguishes carefully and explicitly between Galen's and Ḥunayn's words.[21]

But if it is Ibn Abī Uṣaybiʿa who speaks to us here, what are his additions? There is one phrase in this extract ascribed to Galen that may actually have been added by somebody else, perhaps Ibn Abī Uṣaybiʿa, because it refers to magicians from Yemen, people known as opponents of Muḥammad. Other phrases that looked like interpolations at first are insufficiently specific for accepting them necessarily as such. The entire rest of the discussion on the origin of medicine (Parts 3–5), however, is clearly an addition by the Ayyubid doctor. Before I turn to those parts, one more word on other changes found in the Galenic extract. The choice of concepts and vocabulary makes clear that this passage, while undeniably permeated by views, examples and names of Greek medical schools and some of their representatives, has undergone an adaptation to monotheistic beliefs and modes of speaking. The strongest components of this adaptation are the concepts of creation, a single god and his omnipotence. Terms like *alhama* (to inspire) and its verbal noun *ilhām* (inspiration) or *ḥudūth* rather than *khalq* for creation at the beginning of Part 2 suggest that the adaptor was in all likelihood not the

20 Rosenthal translated it as: "Let us first begin with setting down what he has mentioned, together with our additions to it, for the purpose of giving a complete survey of all those different opinions", Rosenthal 1956, 55. Kopf rendered it slightly more freely: "Let us begin, therefore, with an account of what Galen says, supplementing it with remarks of my own, with a view to comprehending all these divergent views", Kopf www.tertullian.org/fathers/ibn_abi_usaibia_01.htm, 6.

21 Rosenthal 1956, 65–73.

Muslim narrator, but the Christian translator.[22] Other translations by Ḥunayn b. Isḥāq show that he indeed transformed polytheistic parlance into monotheistic terminology.[23]

Ibn Abī Uṣaybiʿa's classification of origin stories closes with Galen's position to the question of origin, this time in form of an explicitly marked direct quote. By choosing this format, Ibn Abī Uṣaybiʿa emphasized Galen's importance as a role model and guide for Ayyubid physicians as well as his own membership in this intellectual circle. As already stated above, Strohmaier believes that the quote is not genuine. According to him, it was Ibn Abī Uṣaybiʿa who interpreted the commentary ascribed to Galen in the sense that Galen himself had believed in divine inspiration as the origin of medicine. His argument is, like the one by Rosenthal for other passages, that this part is missing in the history of physicians written by Ḥunayn's son Isḥāq.[24] Ibn Abī Uṣaybiʿa was, however, not the first who interpreted Galen's position in this sense. In Part 3 of his origin narrative, Ibn Abī Uṣaybiʿa reports that Muwaffaq al-Dīn Asʿad b. Ilyās b. Muṭrān had written in his *Bustān al-aṭibbāʾ wa-rawḍat al-alibbāʾ* (The Garden of Physicians and the Meadow of the Intelligent) that Galen had maintained in his *Commentary on Hippocrates' Oath* "that this art is revelational and inspirational" (Ibn Abī Uṣaybiʿa 1965, 3). Muwaffaq al-Dīn opposed this allegedly Galenic opinion as well as Plato's supposed position according to which Asclepius was a divinely favored human, because he believed squarely that medicine was created by humans through experience, attention and cumulative repetition. Nonetheless, the formulation that he chose to state this opposition expresses his appreciation, if not veneration of the two Greek scholars. Thus, it is highly unlikely that he invented the two beliefs. It is much more likely that he too encountered them in earlier texts.

Who ascribed when to Galen the idea that medicine came into being through revelation and inspiration is unclear. But it should be taken into account that Galen, as Strohmaier has shown, held some closely related beliefs about the transmission of suggestions for healing from divine beings to humans through dreams.[25] After this direct quote Rosenthal's and Strohmaier's investigations turn to Chapter 2 about Asclepius, ignoring the remainder of Chapter 1. The research interests of Rosenthal and Strohmaier focused here obviously on the survival of ancient texts and themes in Arabic garb. While it is certainly important to study the transmission of Greek texts in Arabic and to determine whether a text like the one ascribed to Galen in Chapter 1 of Ibn Abī Uṣaybiʿa's history of physicians is genuine, it is highly problematic to isolate such a text from its immediate narrative context. As a result, Rosenthal and Strohmaier missed the links between

22 Ibid., 84–85, although his examples refer to the later quotations about Asclepius. See for further examples also Strohmaier 1996, 216–221.

23 Strohmaier 1996, 241–242, 2003, 82–83.

24 Strohmaier 2007, 147.

25 Strohmaier 2003, 83, 2007, 164.

the Parts 2–5 of Chapter 1 and thus misunderstood the function of Part 2 and in particular its emphasis on a Galenic belief in the emergence of medicine as the result of divine revelation and inspiration. For them, Part 2 was a report about a (translated) Greek text.

Read together with Muwaffaq al-Dīn's biting critique at Abū Jābir's almost "Galenic" opinions and Ibn Abī Uṣaybiʿa's well-balanced support for Muwaffaq al-Dīn's thesis of medicine's human origins, Part 2 appears in a very different light. Presenting beliefs and opinions as elements of research and classification, its Galenic extract offers first a formal entry into a polemic about medicine's origins. Deriving it from a text believed to come from Galen and closing it with a direct quote of Galen's alleged beliefs that are at the heart of the polemic, Part 2 offers secondly an authoritative and thus safe entry into the controversy about divine versus human origin. Positing this controversy as one between the famous and influential Muwaffaq al-Dīn and some other, rather unknown writer, Ibn Abī Uṣaybiʿa creates for himself the possibility to express securely his own stance in what was a larger debate among Ayyubid physicians and religious scholars.

If secular beliefs are seen as milestones in the development of the medical as well as other sciences, as Strohmaier does, then Muwaffaq al-Dīn's outspoken avowal of medicine's human origins and Ibn Abī Uṣaybiʿa's more cautiously as well as ingenuously wrapped support for this position affirm that these two Ayyubid physicians were not living in a period of decline as Strohmaier claimed rather generally. Ibn Abī Uṣaybiʿa's very intelligent construction of Chapter 1 as the first part of the narrative about the beginnings of medicine does not merely highlight his narrative skills. The many efforts he invested into telling a well-grounded story that makes room for divine participation in the emergence of medicine, but ascribes the major contributions to human beings leaves no doubt, at least to me, that Ibn Abī Uṣaybiʿa was an active and sharp participant in a discussion relevant to his immediate predecessors as well as contemporaries.

Conclusions

The study of narratives of knowledge, in particular when done with the assumption that the authors of such narratives pursue goals that guide them in their choices of themes, arguments, sources, which authors they mention and whose names they suppress, whether they participate as speakers or pretend to appear as mere reporters and how they arrange the various bits and pieces of texts and hence show them as active creators of their narratives, not as mere copyists and compilers, can provide a number of insights in their intellectual profile, their views on debates of their times and the values they and their peers subscribed to. As I have shown for one particular narrative constructed by Ibn abī Uṣaybiʿa about the origin of medicine, his text is much more than merely the preservation of an extract of a lost ancient text by or ascribed to Galen. It is an intelligent construction of a defense of the role of humans in the development of medicine without provoking too much more conservative physicians, religious scholars and Ayyubid patrons. The acceptance of the divine role in this development is reduced to very particular and limited activities at the very beginning of medicine's emergence. The elaborate character of this construction suggests the existence of a complex, sophisticated intellectual environment in Damascus and other Ayyubid holdings in the middle of the thirteenth century, shortly before the fall of this dynasty. Analyzing other narratives of the types outlined at the beginning of my paper will yield similarly surprising results that can contribute to more than merely modifying our views of the intellectual atmosphere in later Islamic societies. Such results can bring body and depth to our views of those societies and thus attract more interests among mainstream historians of science in our work.

References

Abbou Hershkovitz, K. (2008), *The Historiography of Science between the 10th and the 14th Centuries*. PhD Thesis. Beer-Sheva: Ben-Gurion University.

Brentjes, S. (2010), "Ayyubid Princes and their Scholarly Clients from the Ancient Sciences", in Fuess, A., Hartung, J.-P. (eds), *Court Cultures in the Muslim World: Seventh to Nineteenth Centuries*, (SOAS/Routledge Studies on the Middle East). London: Routledge, 326–356.

Charette, F. (2003), *Mathematical Instrumentation in Fourteenth-Century Egypt and Syria: The Illustrated Treatise of Najm al-Dīn al-Miṣrī*. Leiden, Boston: Brill.

Ibn Abī Uṣaybiʿa (1965), *ʿUyūn al-anbāʾ fī ṭabaqāt al-aṭibbāʾ* (The Sources of Information about the Ranks of Physicians). Bayrūt: Dār Maktabat al-hayāt.

Isḥāq b. Ḥunayn (1954), "Ta'rîkh al-aṭibbâ'" ["History of Physicians"], *Oriens* 7: 55–80.

Ḥunain ibn Isḥāq (1925), "Über die syrischen und arabischen Galen-Übersetzungen", Zum ersten Mal herausgegeben und übersetzt von G. Bergsträsser, *Abhandlungen für die Kunde des Morgenlandes* XVII.2. Liechtenstein: Kraus Reprint Ltd. (Reprint Nendeln 1966).

King, D.A. (2004), *In Synchrony with the Heavens. Studies in Astronomical Timekeeping and Instrumentation in Medieval Islamic Civilization*, vol. 1, The Call of the Muezzin. Leiden, Boston: E.J. Brill.

King, D.A. (2000), "Too Many Cooks . . . A New Account of the Earliest Muslim Geodetic Measurements", *Suhayl* 1: 207–242.

Lentz, Th.W., Lowry, G.D. (eds.) (1989), *Timur and the Princely Vision: Persian Art and Culture in the Fifteenth Century*. Washington, DC: Smithsonian Institution Press.

Matvievskaya, G.P. (1967), *Uchenie o chisle na Srednevekovom Blizhnem i Srednem Vostoke* (Doctrine of numbers in the medieval Near and Middle East). Tashkent: FAN.

Mercier, R. (1992), "Geodesy", in. Harley, J.B., Woodward, D. (eds), *History of Cartography*, vol. 2, Book 1: *Cartography of Traditional Islamic and South Asian Societies*. Chicago, IL & London: University of Chicago Press, 175–188.

Rosenthal, F. (1956), "An Ancient Commentary on the Hippocratic Oath", *Bulletin of the History of Medicine* xxx, 52–87.

Saliba, G. (1994), *A History of Arabic Astronomy – Planetary Theories during the Golden Age of Islam*, New York University Studies in Near Eastern Civilization XIX. New York: New York University Press.

Strohmaier, G. (2007), *Antike Naturwissenschaft in orientalischem Gewand*, AKAN-Einzelschriften, vol. 6. Trier: WVT.

Strohmaier, G. (2003), *Hellas im Islam, Interdisziplinäre Studien zur Ikonographie, Wissenscha und Religionsgeschichte*. Wiesbaden: Harrassowitz.

Strohmaier, G. (1996), *Von Demokrit bis Dante, Die Bewahrung antiken Erbes in der arabischen Kultur*. Hildesheim, Zürich, New York: Georg Olms Verlag.

Van Bladel, K. (2009), *The Arabic Hermes, From Pagan Sage to Prophet of Science*. Oxford: Oxford University Press.

10

SANCTIONING KNOWLEDGE*

In this paper I wish to look at a few examples of forms of controlling knowledge that do not consist first and foremost of acts of censorship or material destruction. Rather, they proceed through the evaluation of results, public performances, and the establishment of norms and boundaries. I suggest considering not merely the contentious and the adversarial, but also the innocuous and the seemingly 'normal' or 'natural' as results of controlling knowledge.

The simple thesis I am pursuing is well-known from sociology, psychology and today's academic institution of peer-upheld norms and boundaries for scholars, and the knowledge they produce and distribute. Positive forms of sanctioning establish these norms and boundaries with long-term efficacy.[1] Such positive forms of sanctioning knowledge are primarily rules that prescribe how scholars may speak, write, argue, verify, and in short exercise their profession, and the practices for learning the behavior that conforms to these rules. Positive forms of sanctioning include rules for how to dress or interact with other members of society. Stories are a further positive form of sanctioning knowledge, because they link knowledge to other appreciated or condemned socio-cultural practices, thus attributing values to different forms of knowledge. Other forms encompass practices of valorizing and ranking knowledge and its particular group/s of practitioners, such as honors, prizes, monuments, laudatory speech acts or titles. Due to various limitations of time, page number, and the process of reviewing, I cannot offer a survey on the broad range of positive forms of sanctioning knowledge documented in different written, instrumental, visual and other material sources.[2] Hence, I have decided to focus on discussing stories about how scholars and princes in different historical circumstances thought about the issue of authority

* A larger version of this paper was read at the panel "Knowledge under control. Religious and political censorship in Islamic societies" in WOCMES 2010 (Barcelona 19–24 July), sponsored by the ARG-ERC KOHEPOCU and the *Acción Especial* FFI2010-09040-e (subprograma FILO), Spanish Ministry of Science and Innovation. The person responsible for the English revision is Nicholas Callaway.

1 <http://plato.stanford.edu/entries/social-norms/>.

2 Examples are structural prescriptions for scholarly texts, among them titles, introduction, chapters, subchapters, conclusions or epilogues, deductive or narrative styles of writing, modes of verification and proof, modes of explanation such as examples, recipes or descriptions, the permissible roles

DOI: 10.4324/9781003372455-11

in relationship to knowledge. I found this information in stories told in texts about astronomy/astrology and geography from the ninth, twelfth and fifteenth centuries composed in Abbasid Baghdad, Norman Palermo and Timurid Shiraz or Isfahan.

I accepted the invitation to contribute reflections on how knowledge was controlled or sanctioned in specific medieval societies within the domain of the mathematical sciences, because this is an important topic for understanding how those societies perceived such scientific knowledge, and what status and functions they accorded to it and its representatives. Historians and historians of science rarely address such questions when Islamicate societies are concerned; in fact, historians of Islamicate societies often avoid the study of scientific sources altogether. If they have to take a stance, they prefer to rely on secondary literature, which they cannot evaluate with regard to its reliability or the nature of its claims.[3] Historians of science in Islamicate societies usually focus on the technical content of the texts or instruments they study, and treat historical sources as if they were containers of facts.[4] Thus, the stories I discuss in this article are either unknown to the one or the other, or have been dealt with incompletely. Elements other than technical data like numbers, types and places of observations, or historians and titles are rarely, if at all, taken into consideration by authors of science. Historians often ignore such technical details or modernize them to fit today's knowledge and values. Depending on their methodological commitments, they tend either to take non-technical features at face value, or to consider them as rhetorical devices, propagandistic elements or forms of self-aggrandizement.[5] A dialogue between the technical content and context of such stories and non-technical elements is,

of illustrations, para-scientific elements such as dedications and matters of functionality like the relationship between page, text and image.

3 See for instance El-Hibri, *Reinterpreting Islamic Historiography*, 95–96.

4 Examples are the discussions of stories on the astronomical expeditions under caliph al-Ma'mun by Mercier, "Geodesy," in particular pp. 176–181; King, "Too many cooks . . . A new account of the earliest Muslim geodetic measurements," in particular pp. 229–231.

5 See for instance Ahmad's and Houben's modernizing and 'factual' readings of parts of al-Sharif al-Idrisi's preface to the *Nuzhat al-mushtaq* in Sayyid Maqbul Ahmad, "Cartography of al-Sharif al-Idrisi," in particular p. 156 and Houben, *Roger II of Sicily: A Ruler between East and West*, p. 104 against El-Hibri's primary emphasis on the caliph's propagandistic self-representation in the stories about al-Ma'mun and the rhetorical strategies of later historians for criticizing this propaganda in a veiled manner, and his often speculative evaluations of these stories and their elements: El-Hibri, *Reinterpreting Islamic Historiography*, 101–142. El-Hibri explicitly stated his break with previous historiographical approaches. He did not wish to construct a factual history nor "social, political or religious interpretations on the basis of the chronicle's information." He rather wished to apply methods of literary criticism and searched for "the originally intended meaning in the narratives", which he believed was not a presentation of facts, but the interpretation of political, social, religious or cultural issues through the technique of providing commentary (El-Hibri, *Reinterpreting Islamic Historiography*, 13). My own position, rather, sits between these two opposites. I consider it more plausible to assume that chroniclers from any given time period and location strove for several goals situated within their cultural and intellectual contexts. These contexts would have included the narration of certain factual items, reporting on narratives delivered by earlier and/or contemporary writers or speakers, as well as statements by the compiler or narrator of the given chronicle, whether

to the best of my knowledge, never established. One important reason for this imbalanced treatment of stories in scientific texts or texts about scientific themes is insecurity with regard to methods of analysis. This is an issue I cannot address in this paper in a satisfying manner. It will have to remain a task for further research. I will nonetheless highlight some of the questions that should be pursued in such future research, to indicate my awareness of the problems involved in my own proposed interpretations.

1 Al-Ma'mun's authority to decide on matters of astronomy

In Arabic geographical and astronomical literature there are numerous versions, some brief, others detailed, about astronomical expeditions ordered by the Abbasid caliph al-Ma'mun (r. 198/813–218/833) in order to determine various astronomical parameters, among them the *qibla* of Baghdad and the length of 1° of terrestrial latitude. The latter is reported to have served for determining either the length of the Greek *stade* or the size of the earth's circumference, which are two different aspects of the same problem. Many technical features of these stories contradict each other or are open to doubt, as Raymond Mercier and David A. King have pointed out in their analyses.[6] Because several of the texts are also corrupt and cannot be restored in a convincing manner, we do not know and perhaps cannot know what happened in the early ninth century. In order to prepare the ground for interpreting the stories that I have chosen on the basis of Mercier's and King's works, the first task is to establish the main points of agreement and disagreement between the reports. This will help in recognizing the main threads of each story and to sort them into groups. Subsequently, I will describe and discuss those elements of the stories that concern activities describing or setting norms and establishing relationships between the stories' actors.

1.1 Relations between the extant Arabic reports on al-Ma'mun's interest in astronomy

The three accounts about al-Ma'mun's interest in astronomy that I will discuss here are extant in five textual settings and differ, not surprisingly, in important details. Sanad b. ʿAli (3rd/9th c.) and Khalid b. ʿAbd al-Malik al-Marwarrudhi (*fl.* 217/832), who apparently participated in one or two of the expeditions, produced accounts as eyewitnesses; one apparently in written form (Sanad), the other submitted orally (al-Marwarrudhi). Habash al-Hasib (d. after 253/867) and Yahya b. Aktham (d. 242/857) claim to have written down what they heard from al-Marwarrudhi.

---

of an evaluative or other such nature. The qualitative and quantitative distribution of these features and their specific forms varies between different cultures and their social groups.

6 Mercier, "Geodesy," 176–181; King, "Too many cooks," 229–231.

Sanad b. ʿAli's account has been preserved by Ibn Yunus (*fl. ca.* 380/990). An extract of Habash's account has been transmitted by Ibn Yunus as well as Abu l-Rayhan al-Biruni (362/873–440/1048). In addition, the first two parts of Habash's account are found in a seventeenth-century collection of astronomical texts given by a Muslim scholar from Sanaʿa to a Rabbi before the latter left for Jerusalem.[7] Most or all of Yahya b. Aktham's report seems to be preserved in a ninth/fifteenth-century copy of a seventh/thirteenth-century Egyptian treatise by the unknown writer Siraj al-Din wa-l-Dunya.[8]

In his analysis of the different versions of these accounts, King emphasizes that the records by Habash al-Hasib, Yahya b. Aktham and Sanad b. ʿAli are often at variance. Several of the numerical results, the places of the expeditions, the number of the teams and their composition do not agree.[9] Sanad's account agrees with Habash's report in some of the numerical results and the number of the teams. In other respects, namely the location where one of the two groups operated, these two reports differ.[10] Sanad's presentation agrees here with that by Yahya b. Aktham. On the other hand, Sanad's report deviates from both accounts with regard to who participated in the expeditions and who took the decision to form the two different groups.[11]

These differences are most likely the outcome of copying, the scholarly aptitude of the various copyists, and the different levels of scientific competence of the original authors. Khalid b. ʿAbd al-Malik al-Marwarrudhi, Habash al-Hasib and Sanad b. ʿAli were professional astrologers patronized by the caliph, while Yahya b. Aktham was a judge, and for some time al-Ma'mun's chief judge.

King considers the reported values, despite their differences, as 'too good' to be true and doubts that an expedition would have carried all the heavy material either to the desert of Sinjar, 200 km north of Baghdad, or to the Syrian desert, some 500 km away, when the same measurements could have been carried out in the capital itself. He sees, however, in the survival of Sanad b. ʿAli's report a confirmation that some expedition did indeed take place under al-Ma'mun.[12]

Mercier, by contrast, suggests that the expedition to the desert north of Palmyra did not take place in the time of al-Ma'mun and may not have taken place at all, since the location is not suited for such undertakings. But at the same time, he believes that its report recalls a set of pre-Islamicate measurements executed near Tadmur.[13]

7 Langerman, "The book of bodies and distances of Ḥabash al-Ḥāsib."

8 Ibid.

9 These differences are: the desert of Sinjar versus the desert between Tadmur and perhaps al-Raqqa; two teams (Habash al-Hasib) or only one team (Yahya b. Aktham); Yahya b. Aktham's report adds Yahya b. Abi Mansur as a team member. King, "Too many cooks," 217.

10 This location is described as having been in the desert between Tadmur and perhaps al-Raqqa.

11 It adds Sanad and does not mention Yahya b. Abi Mansur (d. 215/830). It claims that al-Ma'mun took the decision about the composition of the team. King, "Too many cooks," 210.

12 King, "Too many cooks," 229–230.

13 Mercier, "Geodesy," 180–181.

It is, however, undoubted that al-Ma'mun was in al-Raqqa in 218/833 during a campaign against the Byzantine Empire, since he died there.[14] Hence, some astronomical activities could have taken place in the desert between al-Raqqa and Tadmur during that campaign.

Taking these deviations into consideration, we can make out three different story lines in the extant reports on the astronomical expeditions, based on their technical and institutional content. These differences cannot be resolved and are not the subject matter of my discussion. They suggest nonetheless that the basic story about one or more astronomical expeditions ordered by al-Ma'mun for solving specific problems should not be considered as completely lacking in reliability or, phrased differently, that this type of story was produced for the sole purpose of providing another element in the caliph's extensive and long-term engagement with narrating himself.[15] A better understanding of the relationship between these different levels of story-telling with regard to al-Ma'mun's astronomical projects will only be achieved once we have brought together all the early forms of stories about al-Ma'mun's interest in the mathematical sciences and natural philosophy, with all extant scientific texts from the period of his reign, and examined the ways in which they dialogue with one another.

1.2 Norm-setting features of the reports

For the purpose of studying attitudes towards the mathematical sciences and the relationship between rulers, scholars and judges, the issue of whether the expeditions in fact took place, in the described form/s or at all, does not truly matter. The important issues for the present purpose of studying positive forms of sanctioning knowledge are rather the following two: (1) the reports were written by or ascribed to different well-known actors of the early third/ninth century who represent the inter-human relationships in different forms; what is the meaning of these differences? (2) These reports were transmitted by leading representatives of the mathematical sciences, as well as later authors probably from the milieu of the religious scholars; both groups of transmitters present the stories' protagonists as competent and trustworthy narrators of the main (technical, institutional and relational) points; can we recognize traces of change in the reports that might explain the differences between their content and modes of narrating?

As a first result, we can see that the general role of caliph al-Ma'mun as an active participant in the determination of worthwhile tasks for astronomical research, and his leading role in the evaluation of the final results, were already accepted as socio-culturally sound one century after the events. Three centuries later, the transmitter of a version of Yahya b. Aktham's account found it

14 Cooperson, *Classical Arabic Biography: The Heirs of the Prophets in the Age of al-Ma'mun*, 32–33.

15 For these efforts and their content see El-Hibri, *Reinterpreting Islamic Historiography*, 101–108.

unproblematic to present a story about a judge who was versed in astronomy to some degree, knew the people who were experts in this field, could answer the ruler's questions and acted as a liaison between the two sides. The transmission of the stories by Ibn Yunus and al-Biruni reflects the role courts and dynasties played in the daily practice of the mathematical sciences during the fourth/tenth and the fifth/eleventh centuries, while the transmission of Yahya b. Aktham's report by an anonymous copyist of Siraj al-Din wa-l-Dunya's text bespeaks the changed socio-cultural context of these sciences, which were now taught by *madrasa* scholars.

Secondly, we discover substantial differences between the parts of the reports preserved by the later transmitters. Ibn Yunus was barely interested in the socio-cultural aspects of the astronomical activities.[16] All he takes over from the reports he claims to have read is "that al-Ma'mun ordered that one degree of a great circle on the surface of the earth should be measured."[17] Al-Biruni also summarized his sources, including Habash al-Hasib's report. Like Ibn Yunus he was primarily interested in the technical aspects of the expeditions and their conflicting reports about the final numerical results, though he paid more attention to the socio-cultural components of the reports. He stressed much more than Habash al-Hasib the religious meaning of the determination of the *qibla*, and preserved the latter's comments on why al-Ma'mun became interested in the length of a *stade* and which projects he commissioned.[18]

Thirdly, the differences between the two summaries of Habash al-Hasib's text as reported by Ibn Yunus and al-Biruni and the form found in the third version transmitted by two seventeenth-century Yemenite scholars are substantial. This seems to imply that the differences between the technical details of these versions of Habash al-Hasib's text and the report attributed to Yahya b. Aktham are perhaps the result of a modification inserted in the original of Yahya's text by the seventh/thirteenth-century transmitter.

It is not at all surprising to find that the clear differences between the textual environments of the reports are accompanied by differences with regard to their technical content, as well as their descriptions of the social behavior of the reporters and their patron. Context and content of a text condition each other, even if not always in the same substantial manner as in the case of the reports on al-Ma'mun's astronomical interests. Habash al-Hasib described two projects in the treatise he authored himself. The task of the first project was to determine the length of 1° of a great circle of the earth. When discussing it, the author speaks of the project's purpose and result, the way in which he acquired the information, and the function of the transmitters of the result. The task of the second project was to measure the azimuth of the *qibla* of Baghdad. Here, the story talks of how the caliph distrusted

16 King, "Too many cooks," 210–211.

17 Ibid., 211.

18 Ibid., 214–215.

the scholar's result, asked for a second opinion by a surveyor, and then chose between the two deviating results and evaluated their difference.[19]

The copy of Siraj al-Din wa-l-Dunya's treatise presents Yahya b. Aktham's story as part of a chapter, which deals with "some aspects of geometry (*al-handasa*) relating to the size (*dawr*) of the earth."[20] He introduces it, in opposition to the following report itself, as an explanation of the reason that led al-Ma'mun to commission observations.[21] Yahya, as cited in the copy, does not talk at all about what motivated the caliph's questions, but merely states them. He adds a third project to the two also described by Habash. In this project the caliph inquired about the maximum distance between the earth and the moon.[22]

Both reports portray al-Ma'mun as an intelligent man, interested in knowledge, eager to check its veracity and willing to go to significant lengths to obtain answers to questions of personal as well as religious relevance. This portrait is not very different from the one that Ibn Abi Tahir Tayfur (d. 280/893), al-Ma'mun's first biographer, drew of the caliph through reports and anecdotes told by his courtiers: "If the reports are any guide, the caliph's associates thought of him as a quick-witted, fair-minded and enterprising leader, temperamental but quick to forgive."[23]

The differences in detail between the two reports are substantial. This also applies to their description of the human relationships. According to Habash's report, provided below, the caliph knew how to find answers to his questions. He read the right books, discovered the proper questions, inquired about their solution with the right kind of people, criticized them for the unreliability of their knowledge, and in the end simply sent his court experts – astrologers and craftsmen – to do the work and find the correct answer. In the case of the first project, he ordered Yahya b. Aktham to produce a written testimony of al-Marwarrudhi's oral report to the judge. In the second project, Habash does not mention the judge as a mediator. He describes the caliph as receiving direct, oral information from the scholar/s and the surveyor/s who had measured and calculated the different values. Furthermore, Habash portrays the caliph as capable of evaluating the results and explaining their differences.[24]

> (. . .) (Habash) said that the Commander of the Faithful, al-Ma'mun, wanted to know the size of the earth, so he made some investigations about this and found that Ptolemy had stated in one of his books that the circumference of the earth was so many thousand *stades*. Thereupon

19 Ibid., 217–220.

20 Ibid., 223.

21 Ibid., 234.

22 Ibid., 223–224.

23 Cooperson, *Classical Arabic Biography*, 48.

24 My focus in the following quote is the description of the caliph and his inter-human relationships. Hence, I have omitted almost all technical details.

> he asked the interpreters about the meaning of *stades* and they gave different interpretations. [He said about their interpretations:] "They do not dispense with (?) what we wanted to know." {The text is corrupt here.}
>
> (Al-Ma'mun) therefore sent out Khalid ibn ʿAbd al-Malik al-Marwarrudhi, Ali ibn ʿIsa al-Asturlabi and Ahmad ibn al-Buhturi the surveyor together with a group of surveyors and craftsmen including carpenters and brass-workers, to make correctly the instruments which they would need, transporting all of them to a place which they selected in the desert of Sinjar. (. . .)
>
> I heard this (information) in this book {*sc.* in this book of mine?} from Khalid ibn ʿAbd al-Malik al-Marwarrudhi as he was conveying it to the Qadi Yahya ibn Aktham. Yahya [had been] ordered [by al-Ma'mun] {the text appears to be corrupt here} to write down for him {sc. for al-Ma'mun} all that Khalid told him, so he wrote (it) for him. I have written what I heard from Khalid himself.
>
> The Commander of the Faithful al-Ma'mun – may God be pleased with him – (also) wanted to measure the azimuth of the *qibla.* So he sent out (someone) at the time of the lunar eclipse to measure the longitude between Mecca and Baghdad. (That person) found that the meridian of Mecca was west of that of Baghdad by approximately three degrees.
>
> (. . .)
>
> When Khalid ibn ʿAbd al-Malik al-Marwarrudhi submitted (this) value of one degree of the earth('s circumference) to al-Ma'mun, (the caliph) wanted to check it and so he [sent out] {reading *wajjaha* for *wajada*} someone {singular! – compare the account of Yahya ibn Aktham} to measure the road between Baghdad and Mecca by the shortest road (. . .)
>
> (. . .) Al-Ma'mun said that this was not to be regarded as excessive because there must be (in addition to) the flat parts, inclines up and down on (the road) amounting to this (difference).[25]

Yahya's report as transmitted in the seventh/thirteenth century presents a significantly different picture. Yahya speaks of the caliph as curious, but in need of help since he lacked knowledge. He presents himself as a trustworthy person in possession of a part of the information the caliph had asked for, as a participant in all the relevant meetings between the caliph and the astrologers, and as having been entrusted with writing down the oral report of the head of the research group. Only this last point is confirmed by Habash. According to Yahya, the caliph knew, however, enough to ask pertinent questions about procedures and results. He did not trust the results that his astrologers had arrived at through observing a lunar eclipse in the two cities, measuring times and latitudes, and calculating

25 King, "Too many cooks," 218–220.

differences, sums, products and a square root, but rather preferred to have them checked by measuring the shortest road between the capital and Mecca.

> (. . .) ({Yahya b. Aktham} reported:) The Commander of the Faithful al-Ma'mun became very excited about knowing the size of the earth. So he asked me and I told him that the astronomers familiar with geometry (*al-muhandisun min ashab al-najama*) had knowledge of these matters. So he summoned Khalid ibn 'Abd al-Malik al-Marwarrudhi, Yahya b. Abi Mansur, 'Ali ibn 'Isa and Ahmad ibn al-Buhturi, and they chose a group of brass-workers and carpenters (to assist them in making instruments). Al-Ma'mun questioned them on the way to proceed and they answered him in unison that it was easy. (. . .)
>
> Then al-Ma'mun told them that he would like to know the azimuth of the *qibla* and the distance between Baghdad and Mecca, and they replied that (they could do) this at the time of a lunar eclipse. (. . .)
>
> Al-Ma'mun wanted to check the calculations made by Yahya {*sc.* Yahya ibn Abi Mansur} and his colleagues, so he sent out someone to measure the road to Mecca, and they {plural! – compare the account of Habash} found that (the number) of miles between Baghdad and Mecca by the shortest and straightest route was 700 and 10 miles, which was about 6 miles more than the calculation. Al-Ma'mun pronounced that the calculation was more accurate and that the difference was due to the depression (*istifal*) of the *wadi*s and the elevation (*irtifa'*) of (hills) on the surface of the earth. When he had become convinced of this, he became excited to know the maximum distance of the moon from the surface of the earth, so he asked them about this and they answered (. . .).[26]

The reason for the caliph's distrust, confirmed by both reporters (and their transmitters), is not explained by either of them. Habash simply states that al-Ma'mun wished to verify the quantity (*miqdar*) that al-Marwarrudhi had submitted to him as the result of the measurements and calculations. Yahya's statement that the caliph wished to check what Yahya (b. Abi Mansur) and his colleagues had calculated may point, however, to the cause of the caliph's distrust or perhaps insecurity. The methods applied by the astrologers were not at all as simple as they had claimed when asked by al-Ma'mun about their procedure, at least not for someone untrained in all the technicalities. Hence the caliph may have preferred to test the result achieved by observation and calculation against a simple measuring method, although how the craftsmen executed the measurement is not specified. Such an interpretation of Yahya b. Aktham's description seems to be supported by the beginning of the sentence he uses to introduce the third topic that the caliph wanted information about from his experts, namely "after he had

26 Ibid., 223–225.

been reassured" (or, as King translates: after he had become convinced of this) (*fa-lamma sakanat nafsuhu*).

These reports and their variants reflect and reinforce three norms or beliefs: 1) knowledge is related to social hierarchies; 2) knowledge needs scholarly and technical experts; 3) knowledge needs to be verified; substantially different methods, sources and experts can be used in this process. The two main differences in the descriptions of the specific skills of the caliph, and his roles in the verification of transmitted knowledge and the production of new knowledge, indicate two interpretations of the first norm. According to the versions of Habash's narration, the caliph is supposed to acquire the needed knowledge through his own activities and high intellect. He represents the highest human representative of knowledge and acts as an arbiter or final evaluator. The story put into the mouth of Yahya expresses a less exalted view of the relationship between the caliph and expert knowledge. Here, the worldly ruler is allowed to be a human who needs advice and support, but who learns in the process the necessary elements to successfully play the role of undisputed evaluator in the end.

1.3 Issues of interpretation

The differences regarding the display of the caliph's role in the reported astronomical activities raise three questions of interpretation and one fundamental question of methodology. The questions of interpretation are the following: (1) what was the function of these reports? (2) What is the meaning of the fact that the "scientists" appear to concede the dominant role as well as expertise to the caliph, while the jurist describes his prince as in need of help, guidance and information? (3) What do these different features tell us about the stories' potential to sanction the production, distribution and evaluation of the specific kind of knowledge discussed in the reports? The methodological question is whether it is acceptable to focus on this limited sample of stories to achieve some reliability of interpretation, or whether it is necessary to contextualize the three stories and the forms in which they were transmitted by later writers. In light of the second half of this question, it is important to ponder what necessary and/or sufficient contexts in quantity and quality might be, how we might be able to rank them, and what kind of methods has to be applied to establish relationships between the possible meanings of such contexts for the originally chosen objects of analysis. This easily leads to an infinite regress, since we would have to ask the same kind of methodological question on each contextual level. Hence it is clear that there is no ideal solution to the difficult problem of how many objects of study and comparison we need to access in order to carry out a solid and fair analysis of the three stories about the astronomical expeditions under al-Ma'mun and their subsequent transmission. This practical impossibility of reaching a satisfying degree of soundness in research and reliability in results does not mean, of course, that no effort to contextualize should be made. In the case of the three reports discussed so far, a thorough analysis of the possible ramifications of their modes of storytelling should take into account four lines of investigating context. The first line consists of a comparative analysis of the modes of storytelling within these texts, which is what I am offering in my paper. The second line would consist of comparing these stories and their various elements with extant scientific texts composed by the participants in the described activities. Such an approach to local contextualization would perhaps improve possibilities for evaluating the technical features of the stories. Since, within the framework of this paper, I am not interested in verifying the stories' technicalities, I have not studied such extant witnesses to the astronomical research undertaken at al-Ma'mun's court. The third line of contextualization would extend the study of the local conditions horizontally. It should collect references to these reports in scientific texts, historical chronicles, literature and perhaps religious texts from the third/ninth, fourth/tenth and seventh/thirteenth centuries, as the moments in time when these reports were transmitted by Ibn Yunus, al-Biruni and Siraj al-Din wa-l-Dunya. The task is to determine the additional information that these references provide on the representation of the reports' meaning and the modes of narration chosen by the transmitters. The fourth line of contextualization would extend the study of the local conditions vertically. It would collect and analyze other stories told by the reporters on caliph al-Ma'mun's scholarly skills and knowledge, with some focus, perhaps, on the mathematical sciences. I have tried to find some examples belonging to this fourth line, but was not particularly successful within the time left for revising my paper. The few statements by Yahya b. Aktham that El-Hibri presents as examples of both recognition of the caliph's scholarly stature, and of including the caliph's scholarly interests within his propagandistic self-representation, discuss al-Ma'mun's knowledge of *fiqh* or the future.[27] As such they do not lend themselves to establishing a clear, direct relationship to the stories of interest to me in this paper. The contextualization of stories about technicalities of the mathematical sciences in such chains of narrative would require greater methodological and analytical skills than those that I currently dispose of.

As for the three questions concerning the function of these three reports on the astronomical expeditions in the time of al-Ma'mun, their standard interpretation refers to the reports' factual nature, i.e., their position as the last element in the execution of specific scientific projects. In this factual function, the reports are taken to be indicative of a large and long-term research program in the mathematical sciences that was commanded, financed and evaluated by the caliph. Since reporting on research results and organizing our research as longer-term projects are features of our own scholarly practices, this function was seen for a long time as unproblematic. Al-Ma'mun was accepted as being akin to the director of a research institution. The first doubts as to this function of the reports and their interpretation are those formulated by Mercier and King, summarized above. Their reasons for doubting are geographical and scientific in nature, i.e., they imply that the reports are, after all, not serious scientific reports, since their data is

27 El-Hibri, *Reinterpreting Islamic Historiography*, 110–112 *et al.*

unrealistic in quality and content. They do not go beyond this step of expressing doubts. They do not ask whether the idealized nature of the technical data could have had a further function in addition to being technical data. Different answers to such a type of question can be imagined, and thus will have to be explored in the future.

The idealized nature of the data corresponds to the seemingly idealized nature of the caliph's technical and epistemological skills in the reports of the astrologers. This idealization, then, could speak to the narrative function of the reports, which could have been to propagate this image of the caliph's exceptional intellectual abilities. El-Hibri suggests that the caliph and his advisors did indeed include praise of the caliph's knowledge across a broad range of disciplines in their efforts to legitimize al-Ma'mun's appropriation of the caliphate and his right to continue appropriating control and power over territories, people and holders of such knowledge.[28]

We should not forget, however, that the astrologers' glorification of the caliph as superior to them in expertise and discernment may reflect nothing more than standard social forms exercised in a hierarchical relationship such as that between patron and client, ruler and subject. Before we can read the astrologers' language as more than such a standard mode of rhetoric, and indeed as an intentional element of caliphal propaganda, we need a systematic, comparative study of other depictions of al-Ma'mun's role as the leading scientific expert of the day. While such a methodological demand goes substantially beyond the scope of this paper, at least some comments can be offered thanks to El-Hibri's discussion of the caliphal propaganda efforts and their treatment by later chroniclers. El-Hibri emphasizes that the description of al-Ma'mun's religiosity, for instance, "is grounded in his learnedness in the areas of *fiqh*, hadith and the art of reasoning. (. . .) Al-Ma'mun's campaign of self-idealization no doubt provided the initial impetus for this overly favorable representation, but in the end, it leaves us wondering how the *'ulama*, reacted to this official trend of flattery. In theory, the *'ulama*, were probably willing to accept the pietistic pretensions of al-Ma'mun, so long as these merely connoted erudition or a scholarly interest in the religious sciences."[29] This interpretation seems to suggest that specific items of scholarly knowledge were seen as individual features of this particular caliph and as carrying no strong significance for how later chroniclers evaluated the caliph's idealizing self-representation. Since the reports on the astronomical projects are replete with such specific, minute details of scholarly knowledge, and since the particular scholarly knowledge discussed in two of the three projects (circumference of the earth; distance between the earth and the moon) has no direct bearing on the caliph's claims to religious leadership, it is more plausible to suspect that the narrative mode used by the astrologers was not strongly relevant to al-Ma'mun's legitimization campaigns.

28 Ibid., 101–108.
29 Ibid., 109–110.

What then does the much less glorifying and more personalizing language that we find in the report attributed to Yahya b. Aktham signify? Can it indeed be the mode of expression of the original report based on al-Marwarrudhi's oral account of the first expedition? Or do we have to consider such personalized language as very improper for a caliphal client and judge, and hence need to assume interference on the part of the transmitter/s? If the latter were the case, what might have been the aims of the transmitter/s' change to a supposedly different mode of expression than that of the original text? Would they have had an interest in divesting the caliph of extraordinary intellectual capabilities and hence of his function as a learned role model? These questions again go far beyond the scope of this paper. They are, however, important to raise and study, since they might lead us to new insights into the values upheld and contested by different groups of *madrasa* teachers and their public with regard to the mathematical sciences. Such values are, as stated in the introduction, at the heart of positive sanctions of knowledge, and maintain or undermine norms, rules and boundaries of knowledge. As I have pointed out before when discussing the modifications to Habash's text at the hands of Ibn Yunus and al-Biruni, it is more likely than not that Siraj al-Din wa-l-Dunya adapted Yahya b. Aktham's text to suit his own needs. Hence, it is impossible to decide without further inquiry whether the image of the ruler as presented in this report belongs to the early third/ninth century or to the seventh/thirteenth century, and thus whether it refers to an Abbasid caliph, an Ayyubid *malik* or a Mamluk sultan. But independently of which dynastic ideal or real ruler is addressed in this report, a closer study of the extant form of the report ascribed to Yahya b. Aktham reveals interesting details with regard to norms and rules for practicing knowledge.

The language ascribed to Yahya b. Aktham describes the three projects as individual points of personal interest, caused by the caliph's increasing zeal, ardor or, as King translates, excitement (*samat himmatuhu*). They are not portrayed as related to the realm of (religious, military or other forms of) politics. Not even the second project that Yahya's report shares with those of the astrologers and which centers on an important religious issue, namely the determination of the prayer direction between Baghdad and Mecca, is described based on more than purely technical terms and inter-human relationships.

The suggested status of these three specific projects as personal questions about technicalities corresponds well with the fact that al-Ma'mun's early biographers, like Ibn Abi Tahir Tayfur and Muhammad b. Jarir al-Tabari, do not mention any of them.[30] This could imply that the mode of storytelling presented in the extant version of Yahya's report was not impossible in the early third/ninth century. If that could be supported by other stories told by Yahya b. Aktham, this report could be read to mean that al-Ma'mun did not promote a coherent, long-term research program carried out by his expert clients. If the text is reliable in the

30 Cooperson, *Classical Arabic Biography*, 43.

sense specified here, then al-Ma'mun's astronomical interests may have been an astronomical equivalent of the problems tackled in recreational mathematics, only much more expensive.

The report contains details that support an early date for its compilation. It agrees in its technical details in principle with the reports of the astrologers, i.e., it possesses the necessary technical elements to be accepted as having been derived from al-Marwarrudhi's oral account of the first project. It agrees furthermore in principle with the astrologers' description of the second project. This strengthens its acceptability as a text which was compiled in the context of al-Ma'mun's astronomical interests and activities, and in dialogue with the reports of the astrologers. The deviations between the three reports leave no doubt, however, that the text ascribed to Yahya b. Aktham is not a simple note-taking of what al-Marwarrudhi had told the judge. The text preserved by Siraj al-Din wa-l-Dunya is a transformed literary document, which in all likelihood was composed or continued after the two astrologers had written their accounts, since it adds a further, previously unaccounted for astronomical project to al-Ma'mun's activities.

Does such a reading of Yahya b. Aktham's report as transmitted by Siraj al-Din wa-l-Dunya contradict the traditionally offered claims about the importance of astrology for the early Abbasid caliphs, including al-Ma'mun? I do not think so. None of these three projects is directly related to astrology. The language of Yahya's report reveals how its author (or its transmitter) thought he could portray the caliph and his own (or Yahya's) role in these three specific projects. A larger investigation of the stories told by al-Ma'mun's contemporaries will hopefully shed more light on the variations that can be found in such narratives. Here, El-Hibri's interpretation of fourth/tenth-century accounts about al-Ma'mun as mostly a later, subtly negative hagiography needs to be taken into account.[31] This is, however, too large a task for this paper. The overarching issue of al-Ma'mun's personal engagement with astrology proper also needs to be investigated in more detail, if we want to find out whether or not these three projects were seen as belonging to astrology or representing other kinds of knowledge.

2 Rulers as pinnacles of knowledge

As we saw in the previous stories, the final decision about the validity of conflicting results lay at the end in the hands of the caliph. The long-term impact of al-Ma'mun's attention to specific scientific problems was twofold. First, numerous sons of caliphs and their relatives received in childhood and adolescence a sound introduction to philosophy and related fields of knowledge by leading scholars of the day, such as Ya'qub b. Ishaq al-Kindi, who was Ahmad b. al-Mu'tasim's tutor. The same applies to sons of courtiers, as was the case with the sons of Muhammad, the oldest of the three Banu Musa, taught by Thabit b. Qurra, a protégé of

31 El-Hibri, *Reinterpreting Islamic Historiography*, 101–108.

the Banu Musa and a leading scholar in his own right. Some of these children, like Muhammad b. Jahm al-Barmaki (3rd/9th c.) or Ja'far b. al-Muktafi (907–987) became noteworthy scholars themselves in adulthood.[32] Secondly, al-Ma'mun and his alleged relationship to the philosophical and mathematical sciences grew into a role model. It was either approved of as something to be emulated, or rejected as having introduced alien thoughts and culture into the ideal Islamic society, hence laying the ground for corruption and digression from the right path. Later stories about two Timurid princes of Shiraz and Isfahan, Iskandar Sultan (r. 811/1409–817/1414) and Ibrahim Sultan (r. 818/1415–838/1435), and two Norman rulers of Sicily, Roger II (r. 1130–1154) and William I (r. 1154–1166), can be considered as successors in form to the same kind of narrative, since they share certain structural elements. They present the respective ruler as the person who decided which knowledge was good enough to be included in the write-up of the historical or the geographical project pursued by scholars at each of the two courts. As in the case of al-Ma'mun, the scholars did most of the work, the results of which they then presented to the ruler. However, the stories about Roger, William, Iskandar and Ibrahim also contain several new elements. Here, the ruler does not only invite the scholars to discuss their findings and opinions, but also allegedly participates in sessions at which different informants, for instance ambassadors, travelers or visiting scholars, present their particular pieces of knowledge to the scholar in charge of the project. A further difference with respect to the stories about al-Ma'mun is that the later stories about princely competence in the sciences incorporate statements about the ruler's solid education in specific fields. This enumeration encompasses, in the case of the Timurid prince Iskandar, almost all disciplines presented in other sources as standard scholarly education in the given period. Such a comprehensive description of the prince's education implies that the prince was equal to the scholarly class in his training for knowledge-based activities, including disputes and decisions. In the case of the Norman kings Roger II and William I, the descriptions name those sciences and skills that are the heart of the books which follow this laudatory passage about the princely patron. Roger II had Greek and Arab tutors with whom he studied philosophy, the mathematical sciences, medicine and political theory. William I came to the throne when he was thirteen years old, and his education was thus not yet complete, and is not specified in the material examined for the purposes of this paper. Iskandar Sultan studied a broad range of scholarly disciplines, and presents himself, as we will see below, as the author of a *summa* on astronomy and astrology. Moreover, he ordered his palace workshop to include mathematical, astronomical, astrological, medical and alchemical texts by well-known ancient, Ilkhanid and Timurid scholars in the illustrated compendia which the calligraphers and painters produced

32 Matvievskaya, and Rozenfel'd, *Matematiki i astronomy musul'manskogo srednevekov'ja i ich trudy (VIII–XVII vv.)*, vol. 2, pp. 54, 161.

for him.[33] Information about Ibrahim Sultan's education is less specific. The surviving artwork ascribed to him and his patronage proves his excellent training as a calligrapher and poet, and his intimate familiarity with religious and other literature.[34]

2.1 Norman and Timurid princely education in the sciences

Al-Sharif al-Idrisi (493/1100–560/1166?) seems to describe his first Norman patron's scientific training in the geographical work *Nuzhat al-mushtaq fi 'khtiraq al-afaq*, which he produced at the latter's court, allegedly under the guidance of Roger himself, speaking in the preface of Roger's infinite knowledge of the mathematical and practical sciences, which enabled the prince to make extraordinary inventions, full of novelty. Because these marvelous innovations are talked about in all the cities, provinces and districts of the Norman realm, there is really no need for al-Idrisi to discuss them, but because they are so wonderful, he will do so nonetheless.[35] Here the princely scholar is portrayed as superbly educated, intelligent and inventive in those sciences constituting the theoretical and technical basis for the project described in the book following this preface: universal geography and map making. Roger appears not only as the wise ruler, but as the expert qualified for this specific scholarly enterprise. Parallel to underscoring this point, however, al-Idrisi tells his readers that this Norman prince was much better than any royal before him, that he was admired for his incomparable wisdom, and that he dazzled everybody with the products of his mind and hands.

> His knowledge of the mathematical and practical sciences was countless. It was not limited by a boundary because he (learned) each of their disciplines in perfection (*al-hazz al-awfa*), shooting at it with the seventh arrow of *maysir* (i.e., being extraordinarily successful). Indeed, he himself made wonderful inventions and strange innovations, which no prince (achieved) before him, while he did it single-handedly. They indeed were manifest (when) eye-witnessed and clear in their demonstration and proof. Due to their use (*masiruha*) in the cities and (the fact that) they were talked about in all the provinces and districts, we dispensed with mentioning them in a detailed and versatile (manner) and with presenting them individually, not combined together. But if we went to describe them and wrote down the thought about their (construction), laying them

33 Soucek, "The manuscripts of Iskandar Sultan;" Roxburgh, *The Persian Album 1400–1600: From Dispersal to Collection*, 111–115.

34 Roxburgh, *The Persian Album*, 39, 40, 73, 121; <www.metmuseum.org/toah/works-of-art/13.228.1-2>.

35 Al-Idrisi, *Opvs geographicvm*, 5. For the difficulties in interpreting this passage compare Houben, *Roger II of Sicily*, 104 and Jaubert's French in Jaubert, *Géographie d'Édrisi, traduite de l'arabe en français*, XVIII.

> out one after the other, his masterpieces would dazzle us by their wonderful meanings and their powerful intentions. He who counts all the little pebbles obtains the furthest of goals![36]

This enumeration of Roger's extraordinary capabilities in the mathematical sciences and practical matters is part of a long panegyric in which Roger is described as the best of the best. He is called the most excellent of all of God's creatures. He is a prince who governs with perfect justice and impartiality. He also is an administrator who has established the best order and conditions for the greatest felicity. He is a leader of armies on land and sea, which only see the greatest success since they are under divine protection. Not surprisingly, he is a man for whom the doors of future events open, while they remain closed for others. His character and his behavior are the best imaginable.[37] Hence, Ahmad's reading of this short passage as an expression of al-Idrisi's admiration for Roger's concrete knowledge and skills, because he himself did not yet understand geography, map making and the mathematical sciences, misses its full integration into this rhetorical device of adulation of the ideal prince. Even if Roger indeed was a bright, scholarly-minded ruler and a gifted craftsman, al-Idrisi's preface does not aim at presenting a reliable biography of the Norman or at comparing their respective qualifications.[38]

Roger's son William I receives much shorter, but still lavish praise for his high intellectual level and his allegedly excellent education, in Henricus Agrippus' preface to his translation of Plato's *Phaedo* from Greek into Latin, composed for an English friend in 1160.[39] According to Henricus, William spoke as if he were a philosopher, someone who could not be outdone by others, in short, as someone who is absolutely the best with one exception – his own father. As in the case of his father, knowledge and education are clearly linked to princely power and military success. But while the description of William's intellectual achievements is pure flattery, the depiction of his military enterprises encompasses a mixture of territories both temporarily lost and recovered, as well as recently abandoned. Henricus apparently wrote this laudatory piece shortly before the outbreak of the great rebellion of the Sicilian nobles in November 1160. While not completely flattering the king with regard to his stature as a soldier, Henricus clearly did not wish to spell out the various threats to which Norman rule in Sicily was exposed during William's rule. Hence, his depiction of the king also reflects more the ideal than the real man. The point to highlight is that this ideal made explicit room for concrete, secular sciences.

> There is not such another one in the world, whose court is a school, whose retinue is a *Gymnasium*, whose own words are philosophical

36 Al-Idrisi, *Opvs geographicvm*, 5.

37 Ibid., 3–4.

38 Ahmad, "Cartography of al-Sharif al-Idrisi," 156.

39 Houben, *Roger II of Sicily*, 98–99.

> pronouncements, whose questions are unanswerable, whose solutions leave nothing to be discussed, and whose study leaves nothing untried, whose lordship is acclaimed by Sicily, Calabria, Lucania, Campania, Apulia, Libya and Africa, whose victorious hand stretches out to Dalmatia, Thessaly, Greece, Rhodes, Crete, Cyprus, Cyrene and Egypt, and whose already glorious deeds are rendered more glorious and shining by his father, that great [King] Roger.[40]

The image that Timurid sources paint of Iskandar Sulṭān's knowledge and skills goes beyond al-Idrisi's and Henricus Agrippus' rather brief panegyrics of Roger and William as the most knowledgeable and most successful among the princes. Not only is Iskandar's education in the sciences spelled out in much greater detail, but he is moreover portrayed as being directly involved in his self-representation as the best author who ever wrote about the sciences of the heavens. His knowledge enhances his qualities as a just ruler. Divine guidance lets him find the right path in all his endeavors. Constructing himself as the pinnacle of wisdom, justice and virtue undoubtedly was an important goal in Iskandar's pursuit of recognition by his relatives, with whom he competed for power. Because this legitimizing narrative is in such stark conflict with Iskandar's unwise, unjust and non-virtuous behavior, which finally cost him his rule, eyes and life, the strong reference to education, scholarly practice and intellectual superiority testifies to the cultural value of the specific kinds of knowledge enumerated and incorporated in the sources connected with his name.

Iskandar describes his education as having comprised *kalam*, *hadith* and other fields of the religious sciences, history, *'ilm al-hay'a* (planetary theory), astrology and *'ilm al-huruf* (letter magic), if the preface to a lost handbook on astronomy and astrology, allegedly authored by the prince, can be trusted.[41] He portrays himself as having mastered all parts of the available canon of the 'rational' and the 'traditional' sciences, be it major disciplines or branches thereof. Of note is the preface's emphasis on the certainty of this acquired knowledge. It may delineate one boundary that Iskandar Sultan did not wish to trespass, at least rhetorically, i.e., stepping into domains of knowledge that were contested and open to doubt and challenge. His interest in *'ilm al-huruf* in particular might have induced him to offer such a declaration.[42] Except for this one boundary, the other elements all belong to the category of positive control or supernatural protection. They

40 Ibid., 98.

41 This preface is preserved in a collection compiled by the Timurid historian and scholar of the mathematical and other sciences Sharaf al-Din 'Ali Yazdi or one of his intimates, as Aubin believes. No other part of the astronomical and astrological *summa* ascribed to Iskandar has yet been found. Aubin, "Le Mécénat Timouride à Chiraz," 80.

42 For more on this topic, see the recent study by Melvin-Koushki, *The Quest for a Universal Science: The Occult Philosophy of Sa'in al-Din Turka Iṣfahani (1369–1432) and Intellectual Millenarianism in Early Timurid Iran.*

guarantee the prince's maturity in matters such as demonstration and purpose of reflection, beyond what the best of his scholarly clients had been able to achieve.

> Thus speaks the servant of God, the Exalted and the Master who commands there his servants, Iskandar, son of ʿUmar Shaykh – may God pardon them both and be satisfied with them. (. . .) In addition to upholding the rules of law and fairness and to accomplishing the duties of justice and solicitude towards his subjects – 'because one hour of good deeds is equivalent to seventy years' – he spent the wealth of his time and the quintessence of the duration and the moment for acquiring the certain sciences and things of knowledge and for accumulating the veritable virtues and perfections that are the capital of eternal beatitude and the ornament of perpetual fortune. (. . .)
>
> Guided by the sovereign favor and privileged by divine direction (. . .) he became informed and instructed in the sum of the sciences in a short time, the rational ones as well as the revealed ones, the fundamental ones as well as the derived ones. He pushed the verification of the discussions and goals of all of (these sciences) forward to a degree that the gifted and masterful (students) of this art had not begun to achieve. In each of these sciences he discovered marvelous questions and astonishing points according to the expression, 'the masters of power are inspired', as well as delicate finesses; and all this thanks to God's grace which He dispenses on whom He wishes.[43]

Undoubtedly, diligence in childhood and youth, royal blood, virtue and divine grace had turned Iskandar into the preeminent knower.

As these excerpts demonstrate, the three princes are described or describe themselves not merely as successful soldiers and conquerors of lands and titles, but also as masters of the sciences. They have become men of the sword and the pen. This transformation is recognized by their clients, who lent legitimacy to it by telling stories about their patrons as the highest authority in matters of knowledge.

2.2 Roger's embodiment of geographical knowledge

The narrative about Roger's geographical knowledge in al-Idrisi's *Nuzha* points first to the shared reading of the available geographical literature, much of which was in Arabic. Finding many discrepancies and contradictions, Roger was sorely disappointed. First, he turned to the living sources of knowledge, the scholars of his court. He discovered sadly that they did not know more or better information than the books they read. Hence, they were of no help. He sent orders to scholars in his realm, looking for those who had traveled widely. He invited

43 Aubin, "Le Mécénat Timouride à Chiraz," 81.

them to come to court and interviewed them alone and in groups. His former experience repeated itself. There was simply no agreement between them. "He affirmed where they agreed and kept (this information), but questioned where they differed and declared (this) for invalid."[44] In order to lend the appropriate weight to this royal enterprise, al-Idrisi states that this search for information and sifting through the reports took not merely one or two years, but fifteen. There is, it seems to me, no way to corroborate whether such a lengthy period of time is indeed a reasonably fair description of the acquisition and comparison of data by the Norman king and his client. It may be more an instance of elevating praise than a correct estimate of the process of collecting and reflection, in particular because al-Idrisi emphasizes that Roger devoted all his time to the study of this art.[45] When the preparatory phase had come to its end, the king had second thoughts about his choices:

> He wished to ascertain the correctness of what these people had agreed upon with regard to the longitudinal differences between localities and their latitudes. So he had brought to him a drawing board [*lawh al-tarsim*], (on which) he traced with iron instruments (each item, while examining in the same time) the aforementioned books and (chose what) he preferred among the claims of their authors. He examined with discernment all of this in its entirety until he came to know the truth about (these values).[46]

If this is a correct description of Roger's contribution to the production of the world map and thus to the book *Nuzhat al-mushtaq*, its later medieval designation *Roger's Book* would be more appropriate, and the authorship should at least be divided between al-Idrīsī and the king. But neither did al-Idrisi consider this his duty nor did Roger insist on highlighting his royal contribution. Hence it is much more likely that this part too is a euphemistic portrayal of a patron by his client. The fascinating aspect of this passionate praise of Roger's scholarly qualities is the insistence on the methods – reading, comparing, interviewing, evaluating, selecting, constructing a draft map, comparing again and again until the truth was found satisfactory. Fairclough's analysis of the political language of New Labour highlighted a general feature of today's public language, whether political, commercial or professional.[47] Today's rhetoric of actors replaces human agency with the agency of abstract things such as the market, the corporation or the government. The language of princely adoration, however, operates differently.

44 Al-Idrisi, *Opvs geographicvm*, 6.

45 Ibid.

46 I have altered Ahmad's English translation here since I felt it was too free. Al-Idrisi, *Opvs geographicvm*, 6; Ahmad, "Cartography of al-Sharif al-Idrisi," 159.

47 Fairclough, *Critical Discourse Analysis: Papers in the Critical Study of Language*.

It collapses all human actors into the figure of the ideal ruler. The head of a dynasty is transformed into the abstraction of all the contributors to their various projects. The individual share of each one of them is thus hidden beneath the aura of an all-knowing king. Their specific knowledge has to be merged into one great whole of geographical knowledge, authorized by al-Idrisi as the only relevant other knower. As a consequence, the readers of the work were deprived of the possibility to learn the names of the other participants, to study the different types of knowledge they contributed, and to evaluate the appropriateness of Roger's and al-Idrisi's choices. Hence, they too were transformed by this story into passive recipients of royally filtered knowledge.

2.3 Ibrahim Sultan's power to decide on matters of historical knowledge

In 1425, roughly ten years after Iskandar Sultan's execution, another Timurid prince, Ibrahim Sultan b. Shahrukh, was governor of Shiraz. Like his grandfather Timur, his father Shahrukh, and his predecessor and cousin Iskandar Sultan, Ibrahim Sultan patronized the production of historical and geographical literature in order to tell and retell the coming to power of his ancestors and their dynastic legitimacy. The œuvre that Sharaf al-Din ʿAli Yazdi (d. 858/1454) composed around 827/1425, at least in its extensive introduction and first chapter, is often called *Ẓafarname* (The Book of Triumph). It is seen to emulate, rephrase and extend Nizam al-Din ʿAli Shami's (d. before 814/1411–12) work of the same name commissioned by Timur before 1404.[48] The well-known preface of ʿAli Yazdi's version of Timurid history presents the same perspective on the relationship between ruler, scholar and knowledge as al-Idrisi's preface to his *Nuzhat al-mushtaq*. Although many scholars, literati and scribes allegedly participated in compiling the book, the ruler is superior in both body and mind. He is the only person who decides as to 'truth', 'facts' and 'falsehood'. The ruler "exerted (concern and attention) from the beginning to gather and arrange this composition."[49] Copies of all previous versions of histories of Timur, whether in verse or prose, were collected and brought before him for perusal. "(T)hree classes of men, readers, witnesses and writers" served in reading each one of the manuscripts and checked whether each event was described as they reminded them as immediate participants.[50]

> After being apprised of the contents of the manuscripts and eyewitness accounts, and upon repeated examination and investigation of every jot

48 Forbes Manz, "Tamerlane and the symbolism of sovereignty;" Thackston, *A Century of Princes. Sources on Timurid History and Art*, 63. Ando takes a different position emphasizing that the work's original title was *Fath-nama-yi Sahib-Qirani*: Ando, "Die timuridische Historiographie II: Sharaf al-Din ʿAli Yazdi," 221.

49 Thackston, *A Century of Princes*, 65.

50 Ibid.

> and title, His Highness declared with his pearl-raining, jewel-dripping tongue what he decided to be correct and true, and the clerks wrote it down. Once again it was read aloud for verification and recorded. If the slightest detail remained unclear or in doubt, or if there was a discrepancy between the manuscripts and the narrators, messengers were dispatched to the ends of the realm and trustworthy witnesses, upon the veracity of whose word in that affair there was reliance, were interrogated. In this manner episode after episode was verified and penned in the royal assembly, reread several times and corrected so that the gathering, writing and ordering of this history and the introduction of each story in its proper place (. . .) were absolutely the results of His Highness's gracious concern. Then, as commanded, it was written in a clean copy in the version that had been decided upon, and once again it was listened to in the royal assembly. It was compared with the first draft and master copy, and the greatest effort was exerted to correct errors. Emendations that occurred to the royal mind were made (. . .).[51]

Care, attention, knowledge, honeyed words and numerous highly educated advisors, witnesses and scribes, while essential ingredients of the story about how the Timurid prince achieved the production of yet another history of his grandfather and the dynasty he had forged, nevertheless did not satisfy Sharaf al-Din as he sought to persuade his readers of the supreme quality of his new compilation. Before describing the ubiquitous and decisive role of Ibrahim Sultan in the genesis of his creation, Sharaf al-Din underlined that his history "shall be distinguished in three ways from all other histories of rulers and possessors of might and majesty written in prose or poetry by the ancients or moderns in Arabic or Persian."[52] The first point that set it apart in its author's mind was that the hero of its story, Timur, "himself was concerned to collect his greatest exploits."[53] These were not usual exploits, but undertakings decreed by destiny. The second point arose from the author's own machinations. Never had any author of a historical account of ancient or modern kings, he proudly claimed, explained every affair in such detail as he himself did.[54] The third point was the veracity of his account resulting from the incredibly painstaking attention that Timur had paid to ascertaining the truth of what he collected:

> (. . .) for the Sahib-Qiran, while traveling and otherwise, was constantly accompanied by great turbaned lords, sayyids, ʿulema, and jurisprudents, and by people of learning and wisdom, Uighur *bakhshī*s and Persian

51 Ibid.
52 Ibid., 63.
53 Ibid., 64.
54 Ibid.

> secretaries. As commanded, a group of them continually verified every deed and word that issued from His Majesty and everything that happened to the domain and subjects and laboriously wrote them down. It was ordered that every event be recorded exactly as it happened, without any interpolation, addition or subtraction, particularly concerning any person's bravery and courage, that there be no hypocrisy nor magniloquence, and especially in what concerned His Majesty's bravery and daring, that in that there be absolutely no exaggeration. It was also by imperial command that the writers of eloquence clothed it in phrased garments and composed it in prose and poetry with the same proviso. Many times in the royal assembly they read it for the royal hearing so that total reliability was ascertained by verification. In this manner the Turkish verse and Persian prose versions comprising His Majesty's great exploits were written and composed.[55]

As the many extant illuminated copies of Sharaf al-Din's history prove, his presentation of Timur's glorious deeds was well appreciated among his heirs. Many of his successors praised his style as eloquent and elegant. Current historians, in contrast, have found no agreement about the work's significance and meaning in the overall course of the rich Timurid historiography. Ando, for instance, considered it as a "kind of scientific history" situated "in the intellectual atmosphere" of fourteenth- and fifteenth-century Iran.[56] Woods named it "the best known representative of early Timurid historiography."[57] Quinn appreciated it as a typical representative of what constituted in his view the outstanding feature of Timurid historiography: imitation.[58] All of them, though, acknowledge that the ideological program so poetically and forcefully presented by Sharaf al-Din was the common thread that bound all Timurid histories of the fifteenth century together. This program aimed to prove Timur's legitimacy as a Chinghizid imperial ruler, and the rightfulness of Shahrukh's succession.[59] This struggle for legitimacy necessitated each family member's participation in shaping historical discourses that adapted previous narratives to new conditions, and elevated the deeds of the family leaders to the realm of the divinely destined. The most remarkable feature of these narratives is their insistence on the truth value of the new accounts. The very fact that 'truth' is guaranteed by the care, attention and labor of the family's head or one of the princes speaks to the censorial nature of the entire procedure. Thus, it is not surprising that at the center of this discourse stands ever-present control – control of information, style, presentation and people. No group that disposed

55 Ibid.

56 Ando, "Die timuridische Historiographie II: Sharaf al-Din 'Ali Yazdi," 245.

57 Woods, "The rise of Timurid historiography."

58 Quinn, "The Timurid historiographical legacy: A comparative study of persianate historical writing," 19–32.

59 Szuppe, "Historiography v. Timurid period."

of knowledge-related skills is left out of Sharaf al-Din's depiction. Alternative accounts could thus flow only from the pen of foreign visitors like Ibn ʿArabshah (791/1392–854/1450) or Ruy González de Clavijo (d. 1412).

Timurid control was, however, not the only force that gave Timurid historiography its limits. As Maria Szuppe has noted, "Timurid historiography is firmly rooted within the Persian literary tradition of official court histories of the post-Mongol period (. . .) (as well as being nourished by local traditions of regional history)."[60] While Timur and his successors may not have wished to break the hold of this tradition since it was sanctioned by major historiographical works of the Ilkhanid period, they were not completely free in the choice of their self-representation. The need for legitimation forced them to accept the narrative forms of their predecessors and the tastes of their subordinate elites. Censorship and control are thus a complex cultural enterprise with more than one dominant actor and more than one option for the lower-ranking participants to act within the limits prescribed by the censor.

2.4 Princely images of science

In addition to censorship, the stories about the Normans and the Timurids also speak to issues other than just the power to control knowledge. By focusing on the sciences as a tool for control by rulers and a marker of princely excellence, they indicate that the various sciences were held in high esteem in the middle of the twelfth and beginning of the fifteenth centuries in Norman and Timurid societies, respectively. The prince's reputation as a 'prime knower' serves to complete and exalt the ruler's persona. The excessiveness of praise, the uniqueness of each prince's scholarly prowess and the capability of the narrators to find ever more glowing words that overtrump the already fulsome tribute paid to the previous royal hero of knowledge, do more than apotheosize just the figure of the magnificently knowing ruler. Indeed, the royal glory radiates back to the sciences and to secular knowledge in general. The ruler as the arbiter of knowledge is thus the figure of triumphant science. The narratives embody the cultural success of ruler and science in Norman as well as Timurid society. Furthermore, they reflect the belief in science's fundamentally hierarchical nature. Being the head of a ruling house lends sufficient credibility to claims to superior knowledge and to the right to organize knowledge production, as one of the domains under his princely command. Control of knowledge appears here not as a vice, but as a virtue inscribed in the nature of kingship, such that science and kingship condition and complement each other.

3 Epilogue

In the few stories I have been able to offer in this paper, I have suggested that positive forms of control shape knowledge and its spaces to an even greater extent than

60 Ibid.

negative forms, because they determine behavior by setting norms and rules, and by allowing or restraining debates and compromises.

The stories told about scholars and rulers in scientific texts constituted an important medium for negotiating values and shaping beliefs and behavior. These stories indicate that positive control includes elements of exclusion, eradication and at times even institutionalized censorship. However, not all of these negative forms of control necessarily entail distorting or even discarding knowledge. Excluding an axiomatic-deductive style of thinking, speaking or writing about mathematics, for instance, can encourage a livelier atmosphere in class and a less formalistic research practice. Likewise, institutionalized control of knowledge may in fact stabilize criteria for expertise and its professionalization. Contradictory narratives often mirror contested values and ideals of knowledge, and may point to the impossibility of finding closure. The stories discussed in my paper indicate that while knowledge and its practices changed dramatically throughout these periods, knowledge continued to be highly valued among important groups of elites. Rulers and their relatives invested themselves in stories about their intellectual prowess in many different fields of knowledge. Scholars struggled ruthlessly with their colleagues over social standing, intellectual reputation and material affluence in this life using knowledge, including that of the mathematical sciences, as a tool and a weapon. While the narratives of rulers as ‘prime knowers’ speak to the triumph of science among the warrior elites, narratives of scholars as ‘warriors for *madrasa* chairs and profitable marriages’, which could not be dealt with in this paper, bear witness to the usefulness of knowledge, including mathematics and astronomy, beyond its application to practical problems of a Muslim’s daily life, such as inheritance distributions, architecture or prognostication. Hence it is not at all surprising that courts established hierarchies of knowers and included doctors and astrologers in their elaborate courtly protocols and honors. Knowledge is portrayed here as the property of elite groups, and as invested with the power to sanction access to itself.

Sources and bibliography

Sources

Al-Idrisi, *Opvs geographicvm*, Volume 1, Naples, Rome, Istituto Universitario di Napoli, Istituto Italiano per il Medio ed Estremo Oriente, 1970.

Jaubert, P. Amédée (trans.), *Géographie d’Édrisi, traduite de l’arabe en français*, Paris, L’Imprimerie royale, 1836, vol. 1.

Bibliography

Ahmad, Sayyid Maqbul, “Cartography of al-Sharif al-Idrisi,” in *History of Cartography*, Volume 2, Book 1: *Cartography of Traditional Islamic and South Asian Societies*, Chicago, IL-London, University of Chicago Press, 1992, pp. 156–174

Ando, Shiro, "Die timuridische Historiographie II: Sharaf al-Din 'Ali Yazdi," *Studia iranica*, 24 (1995), pp. 219–246.

Aubin, Jean, "Le Mécénat Timouride à Chiraz," *Studia Islamica*, 7 (1957), pp. 71–88.

Cooperson, Michael, *Classical Arabic Biography: The Heirs of the Prophets in the Age of al-Ma'mun*, Cambridge, Cambridge University Press, 2000.

El-Hibri, Tayeb, *Reinterpreting Islamic Historiography*, Cambridge, Cambridge University Press, 1999, Cambridge Studies in Islamic Civilization.

Fairclough, Norman, *Critical Discourse Analysis: Papers in the Critical Study of Language*, Harlow, Longman Group Limited, 1995.

Forbes Manz, Beatrice, "Tamerlane and the Symbolism of Sovereignty," *Iranian Studies*, 21 (1988), pp. 105–122.

Gutas, Dimitri, *Greek Thought, Arabic Culture: The Graeco-Arabic Translation Movement in Baghdad and Early 'Abbasid Society (2nd–4th/8th–10th centuries)*, London-New York, Routledge, 1998.

Houben, Hubert, *Roger II of Sicily: A Ruler between East and West*, Cambridge, Cambridge University Press, 2002.

King, David A., "Too Many Cooks . . . A New Account of the Earliest Muslim Geodetic Measurements," *Suhayl*, 1 (2000), pp. 207–242.

Langermann, Tzvi, "The Book of Bodies and Distances of Ḥabash al-Ḥāsib," *Centaurus*, 28 (1985), pp. 108–228.

Lentz, Thomas W. and Lowry, Glenn D. (eds.), *Timur and the Princely Vision, Persian Art and Culture in the Fifteenth Century*, Washington, DC, Smithsonian Institution Press, 2007.

Matvievskaya, G.P. and Rozenfel'd, B.A., *Matematiki i astronomy musul'manskogo srednevekov'ja i ich trudy (VIII–XVII vv.)*, Moscow, Nauka, 1983, 3 vols.

Melvin-Koushki, Matthew S., *The Quest for a Universal Science: The Occult Philosophy of* Sa,in *al-Din Turka Isfahani (1369–1432) and Intellectual Millenarianism in Early Timurid Iran*, Ph D Thesis, New Haven, Yale University, 2012.

Mercier, Raymond, "Geodesy," in *History of Cartography*, Volume 2, Book 1: *Cartography of Traditional Islamic and South Asian Societies*, Chicago, IL-London, University of Chicago Press, 1992, pp. 175–188.

Quinn, Sholeh A., "The Timurid Historiographical Legacy: A Comparative Study of Persianate Historical Writing," in Andrew J. Newman (ed.), *Society and Culture in the Early Modern Middle East: Studies on Iran in the Safavid Period*, Leiden, Brill, 2003, pp. 19–32.

Roxburgh, David. J., *The Persian Album 1400–1600: From Dispersal to Collection*, New Haven, Yale University, 2005.

Soucek, Priscilla P., "The Manuscripts of Iskandar Sultan," in Lisa Golombek and Maria Subtelny (eds.), *Timurid Art and Culture: Iran and Central Asia in the Fifteenth Century*. Leiden-New York-Köln, Brill, 1992, pp. 116–131.

Szuppe, Maria, "Historiography v. Timurid Period," in *Encyclopaedia Iranica*, 2003. <www.iranica.com/articles/historiography-v> (accessed 7 July 2011).

Thackston, Wheeler M., *A Century of Princes. Sources on Timurid History and Art*, Cambridge, MA, The Aga Khan Program for Islamicate Architecture, 1989.

Woods, John, "The Rise of Timurid Historiography," *Journal of Near Eastern Studies*, 46 (1987), pp. 81–107.

INDEX

ʿAbbās I 125
ʿAbbās II 125
ʿAbbās ibn Akhī al-Sharīf al-Bahāʾ 125
ʿAbdallāh Ḥāsib 130
ʿAbd al-Qādir b. Muhadhdhab 183
ʿAbd al-Raḥīm ibn ʿAlī, Muhadhdhab al-Dīn 55, 58, 60, 63
ʿAbd al-Raḥmān b. Muḥammad *see* al-Rashīdī
ʿAbd al-Sattār b. Qāsim ʿAbd al-Sattār 129
al-Abharī, Athīr al-Dīn 11, 30n9, 40, 182, 197
al-Abidī, Jalāl al-Dīn Faḍl Allāh 40
al-Abīwardī, al-Ḥussām b. Ḥasan 159n11
Abū ʿAbdallāh Muḥammad b. Ibrāhīm Abī l-Maḥāsin 191
Abū l-Faraj 67
Abū l-Faraj b. Tāj al-Dīn Mūsā, Saʿd al-Dīn 138
Abū l-Faraj ibn Yaʿqūb Amīn ad-Dawla 178
Abū l-Fidāʾ 14, 123, 125, 130, 179, 190
Abū Isḥāq 124
Abū Isḥāq Ibrāhīm b. Aḥmad b. Abī Bakr 163
Abū l-Khayr al-Ḥasan 88–89, 111
Abū Maʿshar 115, 120, 208
Abū l-Najā Muḥammad 160
Abu Saʿīd Bahādur Khān 124
Abū Saʿīd Gürägān 122
Abū Shākir ibn Abī Sulaymān
Abū ʿUmar
Abū l-ʿAbbās b. Kuḥayl 158
Abū l-Wafāʾ Būzjānī 86, 89, 91, 136, 158
Abū Yazīd Ṣahār Bakht 130
al-Adfawī, Kamāl ad-Dīn Jaʿfar 183
al-ʿĀḍīd Abū Muḥammad ʿAbdallāh 52
al-ʿĀdil 51, 51n3, 53–55, 54n22, 61, 64–65, 191, 191
al-Adkāwī, al-Shihāb Aḥmad 159
ʿAḍud al-Dawla 89–90, 100, 136, 142
al-Afḍal ʿAlī ibn Ṣalāḥ al-Dīn 51n3
Aḥmad b. al-Buḥturī 228–229
Aḥmad b. Khalaf 89, 91, 112
Aḥmad b. Shihāb al-Dīn 158
Aḥmad b. al-Muʿtasim 234
Ahmet III 128
Ahmet Fazil Köprülü 126
al-ʿAjamī, Abū Bakr 158
Akbar 89, 92, 125–127
al-Akhmīmī, Nāṣir al-Dīn 46
ʿAlam al-Dīn Qayṣar ibn Abī l-Qāsim 68n99, 69–70, 89, 93, 192
Alexander of Aphrodisias 74
ʿAlī b. ʿĪsā al-Aṣṭurlābī 112
ʿAlī Shīr Navāʾī 127
Almās al-Ḥājib 172
al-ʿAlmawī, al-Shihāb, called Shikāra 165
ʿAlqama al-Faḍl 162
al-Āmidī, Sayf al-Dīn 8, 13, 22, 48, 71–78, 71n110, 121, 127, 173n22, 184–194
Amīn ad-Dawla Abū l-Ḥasan ibn Ghazal ibn Abī Saʿīd 62
al-ʿĀmilī, Bahāʾ al-Dīn 130–131, 152, 166n4
Amīr-Shāh Muḥammad b. Ṣadr al-Saʿīd 142
al-Amjad Majd al-Dīn Bahrām Shāh 51n3, 62, 64
al-Amshatī, al-Muẓaffar 46
al-Āmulī, Muḥammad b. Maḥmūd 124
al-Anṣārī, Abū Zakariyyāʾ 13
al-Anṣārī, ʿAlī b. Ibrāhīm 195
al-Anṣārī al-Sanbakī al-Qāhirī al-Azharī al-Shāfiʿī al-Qāḍī, Zakariyyāʾ b. Muḥammad 34
Aristotle 73, 130, 190n93, 193
Asad al-Dīn 65
al-Āqsarāy'ī, Amīn 138
al-Āqsarāy'ī, Muḥammad b. ʿAbd al-Laṭīf *see* al-Mahallī 46

al-Āqsarāy'ī, Muḥammad b. 'Īsā 119n33
al-Āqsarāy'ī, Shams al-Dīn 161
al-Ashraf 174n23
al-Ashraf Barsbay 16n54, 31, 43, 173–174, 179–180
al-Ashraf Mūsā 51, 51n3, 54n22, 55, 69–70, 72, 75, 77, 89, 93, 121, 185–194, 190n93
al-'Asqalānī, Ibn Ḥajar 35, 37, 40, 45, 138, 144, 148, 179, 195
al-Asyūṭī, al-Fakhr 138
'Aṭā' Allāh Rashīd b. Aḥmad Nadīr 126
Awḥad Najm al-Dīn Ayyūb ibn al-'Ādil 51n3
Awrangzīb 'Ālamgīr 89, 95, 129
al-'Azīz 'Uthmān ibn Salāḥ al-Dīn 51n3

al-Bābī, 'Ubayd 162
Bābur Bahādūr Khān, Abū l-Qāsim 122
al-Badr 87
Badr ad-Dīn Baktūt 177
Badr ad-Dīn Lu'lu' 172, 183
al-Baghdādī, 'Abd al Laṭīf 66, 73, 183, 193n105, 196
al-Baghdādī, Ibn al-Naqqāsh 68
al-Baghdādī, Muḥammad b. 'Abd al-Malik 164
al-Baghdādī, Sayf al-Dīn 11
Bahrām Shāh ibn Farrukhshāh 70
al-Bājā, 'Alī 'Alā' al-Dīn 11, 11n22
al-Baklamishī, 'Alā' al-Dīn Ṭaybughā al-Dawādār 36
al-Baklamishī, 'Alā' al-Dīn 'Alī b. Ṭaybughā al-Dawādār 36
al-Balādurī, al-Sharns Muḥammad 158
Banū 'Abbās 204
Banū Mūsā 69, 234–5
al-Baqā'ī, Ibrāhīm b. 'Umar 138
al-Baranī, Abū l-Ḥasan b. al-'Abbās 160
al-Barinbarī, Nāṣir al-Dīn 40, 47, 158
al-Barmakī, Muḥammad b. Jahm 235
Barqūq, al-Ẓāhir 17, 138, 179
al-Baṣrī, Abū l-Ḥusayn 12
al-Bayāsī, Abū Zakariyyā' Yaḥya 68
al-Bayyāz, Abū Bakr 160
Baybars Bunduqdārī = Baybars, al-Ẓāhir Rukn al-Dīn 67, 177–178, 178n37
al-Bayḍāwī al-Makkī al-Shāfi'ī al-Faraḍī al-Ḥāsib, Ḥusayn b. 'Alī *see* al-Zamzāmī, al-Badr Ḥusayn
al-Bayḍāwī al-Makkī al-Zamzāmī, Nūr al-Dīn 'Alī b. Muḥammad *see* al-Zamzāmī, Nūr al-Dīn 'Alī b. Muḥammad
al-Bayḍāwī, Nāṣir al-Dīn 12–13, 162
al-Bayṭar, Abū Bakr b. Badr (or: Mundhir) 119n34
Bhāskara 125–126
al-Bijāyī, Muḥammad b. Muḥammad al-Zawāwī 147, 162
Bint Zayn ad-Dīn 'Umar al-Bisṭāmī 179
al-Birjandī, Niẓām al-Dīn 130
Bīrūnī, Abū Rayḥān 86, 88–89, 95, 111, 128, 194, 224, 226, 231, 233
al-Bisātī 9, 15n53, 161
al-Bisāṭī al-Qāhirī, Shams al-Dīn Abū 'Abd Allāh Muḥammad ibn Aḥmad *see* al-Bisātī
Blaeu, Joan 126
al-Bukhārī, Mubārak Shāh 130
al-Būlawī, Muṣṭafā b. Aḥmad 166
al-Bulqīnī, al-'Ilm 35
al-Burhānī al-Karakī 138
al-Būrī, Abū 'Abdallāh Muḥammad 163
al-Burullusī al-Faraḍī, al-Burhān 39, 159
al-Butījī 35, 158–159
al-Būzīdī, Abū Rabī' Sulaymān 163

al-Čaghmīnī (also Jaghmīnī), Maḥmūd 40–41, 41n81, 126, 148, 153, 155

al-Dahhān, Aḥmad b. Muḥammad 37, 44
ad-Dakhwār *see* 'Abd al-Raḥīm b. 'Alī
Daniel 203
Dārā Shikūh 125, 127
al-Davvānī, Jalāl al-Dīn 123–124
al-Dhahabī, Shams al-Dīn 20–22, 21n81, 78
al-Dimashqī al-Sāliḥī, Maḥmūd al-Baṣīr 166
al-Dimashqī al-Ṭarābulsī, 'Alī b. Mubammad, called by the name of 'Alā' al-Dīn Nāṣir al-Dīn 165
al-Dimashqī, Aḥmad b. Tāj al-Dīn 164
al-Dimyāṭī al-Dimashqī, Aḥmad b. Ibrāhīm 157
al-Dimyāṭī al-Qāhirī, Khalīl b. Ibrāhīm Abū l-Jūd *see* Imām Manṣūr
Dioscurides 118–9, 141–143
Dirāya 158

Empedocles 16
Enoch 203, 207
Euclid 84, 95, 98, 100, 119, 123, 129, 146–147, 156, 166, 171, 192

al-Fāʾiz ibn al-ʿĀdil 51n3
al-Fanarī, Shams al-Dīn 89, 95, 161
al-Fārābī, Abū Naṣr 14, 74, 183, 192, 194
al-Faraḍī 167
Faraj ʿAlī ibn Muḥammad, Nūr ad-Dīn ibn ash-Shāhid 179
al-Fāraskūrī al-Ḥarīrī, Muḥammad b. Yūsuf 47
al-Fārisī, ʿUmar b. Dāʾūd b. Sulaymān 123
al-Fāsī, Muḥammad b. Muḥammad b. Sulaymān 166
Fātiḥ ʿAlī Tippu 89, 95, 129
al-Fattāl, Ibrāhīm 166
al-Fawī, Muḥammad 159
Fayżī 125

Galen 16, 16n58, 203, 206–208, 210–211, 213, 215–218, 216n20
al-Gharraqī, Muḥammad b. Muḥammad 47
al-Gharraqī, al-Shams 47
al-Ghazālī, Abū Ḥāmid 10, 12–13, 20, 30, 84, 89, 102
Ghāzzān 89, 94
al-Ghazzī, Taqī al-Din 22, 26
al-Ghumārī, Abū ʿAbdallāh 35, 159
González de Clavijo, Ruy 244

Ḥabash al-Ḥāsib 223–224, 224n9, 226–230, 233
al-Ḥaḍramī, ʿAbd Allāh ibn Walī 9
al-Ḥāfiẓ Arslanshāh ibn Abī Bakr 51n3
al-Ḥājj Qalandar 162
al-Ḥalabī al-ʿAyntābī al-Kutubī, Qāsim b. Aḥmad 160
al-Ḥalabī al-Dimashqī, Yaḥyā b. Taqī al-Dīn *see* al-Faraḍī
al-Ḥalabī al-Ḥanafī, Muḥammad b. Muḥammad al-Shams *see* Ibn Amīr Ḥajj
al-Hamadānī, Najm ad-Dīn Asʿad 183
al-Hamāwī al-Qāhirī al-Muwaqqit, ʿAbd al-Raḥīm b. Muḥammad 45
al-Ḥārithī al-ʿĀmilī al-Hamadānī, Muḥammad b. Ḥusayn, called Bahāʾ al-Dīn ʿIzz al-Dīn 166
Ḥasan 18, 31, 114
al-Ḥasanī al-Malikī al-Ṣaliḥī, Naṣir al-Dīn Muḥammad b. Tulak 128
al-Ḥāsib al-Iskandarī, ʿAlī b. Aḥmad 34
al-Ḥāsib al-Malikī al-Manṣūrī, Ibrāhīm 115
al-Ḥāsib al-Malikī al-Nāṣirī, Ibrāhīm 30
al-Ḥaṣkafī, Aḥmad b. Muḥammad*see* Ibn Munlā
al-Ḥasnāwī, Abū l-Ḥasan ʿAlī b. Ibrāhīm 163
al-Ḥasnāwī, Mūsā b. Ibrāhīm 163
al-Haythamī al-Tibnawī al-Qāhirī, ʿAlī b. Muḥammad 46
Henricus Agrippus 237–238
Hermes (Trismegistos) 16, 203, 207–208
al-Ḥijāzī, al-Shams 35, 158–159
al-Hilālī al-Ḥamawī al-ʿAnbarī, ʿUmar b. Aḥmad al-Siraj *see* Ibn al-Khadar
Hippocrates 16, 16n58, 203, 213
Ḥubaysh b. al-Ḥasan 214
al-Ḥuḍarī, Muḥammad b. ʿAbdallāh 196
Humāyūn 127
Ḥunayn b. Isḥāq 129, 213–217
Ḥusām al-Dīn Sālār 130–131
Ḥusayn 125
Hülägü 89, 94, 143

Ibn Abī l-Bayyān, Sadīd al-Dīn 71, 73
Ibn Abī l-Munan, Abū Sulaymān Dāwūd 52n7, 61
Ibn Abī Saʿīd, Yūsuf Muhadhdhab al-Dīn 62
Ibn Abī Shukr al-Maghribī, Muḥyī l-Dīn 70, 71n109
Ibn Abī Sulaymān, al-Fāris Abū l-Khayr 64
Ibn Abī Sulaymān, Abū Shākir 65
Ibn Abī Sulaymān, Abū Saʿīd 64
Ibn Abī Ṭāhir Ṭayfūr 227, 233
Ibn Abī Uṣaybiʿa 48–74, 50n1, 51n3, 71n109, 79, 86, 89, 120–121, 135–137, 183, 185, 191–193, 193n105, 203–218
Ibn Afshūsh 163
Ibn al-Akfānī 89, 102
Ibn al-Aʿlam, al-Sharīf 89–90
Ibn al-ʿAmīd 89, 91, 100
Ibn Amīr Ḥājj 35, 47
Ibn ʿArabī 30n9, 195
Ibn ʿArabshāh 244
Ibn Asad 43
Ibn al-ʿAṭṭār, al-Muhibb 159
Ibn al-Bannāʾ 162
Ibn Barrī 160
Ibn al-Burhān 184
Ibn al-Dahhān al-Thuʿaylab 69
Ibn al-Daybaʿ 160
Ibn al-Durayhim 120, 120n35
Ibn Fahīd 70
Ibn Fallūs 191
Ibn Fāris 67
Ibn al-Furāt, al-ʿIzz 35
Ibn al-Fuwatī 89, 93

Ibn al-Gharābīlī 156
Ibn al-Ghazāl al-Ḥimṣi 16n58
Ibn Ghazāl ibn Abī Sa'īd, Amīn al-Dawla Abū l-Ḥasan 62
Ibn Ghurrāb 137
Ibn al-Hā'im, Shihāb al-Dīn 145, 151, 155, 158–159
Ibn Ḥajar *see* al-'Asqalānī
Ibn al-Ḥājib, Jamāl al-Dīn Abū 'Amr 8, 13, 161–162
Ibn al-Ḥājib, Muhadhdhab al-Dīn 68–69
Ibn al-Ḥājj, Muḥammad 7
Ibn Ḥalīqa, Rashīd al-Dīn 67
Ibn Ḥanbal, Aḥmad 21
Ibn al-Ḥanbalī, al-Rāḍī 165
Ibn Ḥawqal 126
Ibn al-Haytham 86, 89, 101, 194
Ibn al-Ḥimṣī 23
Ibn Ḥusayn, Rajab 166
Ibn al-'Īd, Taqī ad-Dīn 197
Ibn al-'Imād 179–180
Ibn Imām al-Kāmiliyya 162
Ibn Ismā'īl, Abū Ya'qūb Yūsuf 163
Ibn al-Ja'dī, Abū Muḥammad 68n99
Ibn Jamā'a, 'Izz al-Dīn 15, 15n53, 19, 179, 196
Ibn Jazla 125
Ibn Juljul 208
Ibn al-Kayyāl, 'Abd al-Laṭīf 165
Ibn al-Khadar 46
Ibn Khaldūn 8, 12–13, 35, 147, 154
Ibn Khālid 37
Ibn Khalīfa, Rashīd al-Dīn 53, 61, 68
Ibn Khallikān 184–186, 189–190
Ibn al-Khashshāb, al-Sharaf 32, 35, 37, 46, 159
Ibn Khiḍr 30
Ibn al-Kuways 139
Ibn al-Majdī 14, 15, 19, 29, 31–33, 35–36, 40, 46–47, 89, 95, 148, 151, 155–161, 171
Ibn Mālik 162
Ibn Marzūq, Muḥammad 158, 160, 163
Ibn Maẓhar 139
Ibn Minfākh, Najm al-Dīn 62
Ibn Munlā 165
Ibn al-Munlā *see* Ibn Munlā
Ibn al-Muṭrān, Muwaffaq al-Dīn As'ad b. Ilyās 59–61, 206, 210–213, 217–218
Ibn al-Muwaqqit *see* Ibn Amīr Ḥajj
Ibn al-Nadīm 89, 91, 111
Ibn al-Nafīs 130
Ibn al-Najjār, Abū 'Abdallāh Muḥammad 163
Ibn al-Naqqāsh, Muhadhdhab al-Dīn 52n7
Ibn al-Naqqāsh, Nūr al-Dīn 31, 33, 45–47, 89, 96, 159
Ibn Qāḍī Shuhba 6–12, 9n16–17, 11n22, 14, 14n46, 21n81, 25
Ibn Qāsim, Abū l-Ḥasan 'Alī 163
Ibn al-Qattān 162
Ibn Qayyim al-Jawziyya 171, 176, 181, 194
Ibn Qāyt Bāy, Ibrāhīm 36
Ibn al-Qifṭī 48, 74–75, 78, 120, 120n38, 125, 185, 185n70, 189
Ibn Qurqmās, Nāṣir al-Dīn Muḥammad 159
Ibn Qurqmas al-Hamzāwī al-Ḥanafī, Yūsuf 35–36, 159
Ibn Qutlūbughā, Qāsim 36
Ibn al-Rawandī 21
Ibn Raqīqa, Sadīd al-Dīn 52n7
Ibn Rāzin, 'Abd al-Raḥīm 33, 44
Ibn al-Ṣalaḥ, Taqī ad-Dīn 20, 30, 121, 190n93
Ibn Sharafshāh, al-Ḥasan 8
Ibn ash-Shāhid, Faraj 'Alī ibn Muḥammad Nūr ad-Dīn 179
Ibn al-Shāṭir 37, 40, 89, 101,
Ibn Shū'a, al-Muwaffaq 65
Ibn Sīnā, Abū 'Alī al- Ḥusayn 12, 14, 30n9, 71, 73–74, 88–89, 119, 121, 130–131, 181–184, 182n54, 188, 191–194, 196–197
Ibn Sīrāfī 43, 157
Ibn al-Tansī, Nāṣir al-Dīn 44
Ibn Ṭaybughā, Shihāb al-Dīn Aḥmad b. Rajab *see* Ibn al-Majdī
Ibn Taghrī Birdī, Abū l-Maḥāsin 173n22, 174n23
Ibn Tūmart 21
Ibn 'Ulayf 159
Ibn 'Unayn 76, 188
Ibn Walī al-Dīn 159
Ibn Wāṣil, Jamāl al-Dīn 70, 74–77, 89, 93, 95, 95n4, 137, 186–189
Ibn Yūnus 224, 226, 231, 233
Ibn Yūnus, Kamāl al-Dīn 11, 121, 182n56, 183, 190, 192
Ibn al-Za'frān 64, 64n77
Ibn Zāghū, Abū l-'Abbās Aḥmad 160, 163
Ibn Zuhayra, al-Shihāb 39, 159
Ibn Zuqqā'a 30
Ibrāhīm ibn Majāhid ibn Asad al-Dīn Shīrkūh 51n3
Ibrāhīm Sulṭān 235–236, 241–242
Idrīs 207
al-Idrīsī 222n5, 236–241, 236n35, 240n46
al-Ījī, 'Aḍud al-Dīn 8, 40
Ilḥāq b. Abī Isḥāq 123–124
Imām Manṣūr 34, 34

Imru' al-Qays 162
Īnāl 32, 45, 160
ʿĪsā b. ʿAlī, Abū ʿAlī 194
ʿĪsā b. Amizzabān 160
ʿĪsā b. Yaḥyā 214
al-Iṣbahānī, Salām Allāh 41
al-Iṣfahānī, Shams al-Dīn Maḥmūd 11–12, 15, 130, 183–184
Isḥāq b. Ḥunayn 214, 217
Iskandar ibn Qarā Yūsuf 174, 174n23
Iskandar Sulṭān 119, 122, 127, 235, 238, 241
al-Iskandarī, Ibrāhīm b. ʿAbd al-Razzāq *see* Ibn Ghurrāb
Ismāʿīl b. Bulbul 86, 89
al-Isrāʾīlī, Abū l-Faḍl 69, 131
al-Isrāʾīlī, ʿImrān 53, 56
al-Iṣṭakhrī, Abū Isḥāq Ibrāhīm b. Muḥammad 119–120, 128
al-ʿIzz Abū Naṣr Beg Arslān 89, 92
ʿIzz ad-Dīn Aybak 172, 178
ʿIzz al-Dīn Farrukhshāh 51n3, 62
ʿIzz al-Dīn al-Ḥasan 89, 95

al-Jaʿbartī 195
Jaʿfar 159–160
Jaʿfar b. al-Muktafī 235
al-Jaghmīnī *see* Čaghmīnī
Jahāngīr 125
Jamāl al-Dīn Ārqūsh 172–173
al-Jamāl b. Mūsā 33
Jamshīd 203
Jānibāk al-Dawādār 16, 16n54
Jaqmaq, al-Ẓāhir 138, 173n22
al-Jāwulī, Alṭunbughā 195
Jawza 65
al-Jazarī, ʿAbd ar-Raḥmān b. ʿAbdallāh 179
al-Jazarī, ʿAbd al-Razzāq 69, 113, 119, 128, 141, 158
al-Jīlī, Rafīʿ al-Dīn 71, 73
al-Jilyānī, ʿAbd al-Munʿim 61
Johannes Philoponnus 206–207, 209
Junaybat 35, 156
al-Jurjānī, Ismāʿīl b. Ḥusayn 130
al-Jurjānī, ʿAlī b. Muḥammad *see* al-Sayyid al-Sharīf
al-Juwaynī, Imām al-Ḥaramayn Abū l-Maʿālī 12

al-Kāfiyājī 14, 17, 40, 148–149, 160–162, 183
al-Kāfiyajī, Muḥammad ibn Sulaymān Muḥyī ad-Dīn 14, 17, 25, 40, 148–149, 160–162, 173

Kamāl ad-Dīn 179
Kamāl al-Dīn b. Yūnus *see* Ibn Yūnus
al-Kāmil 51, 51n3, 54n22, 55, 60–61, 64, 66–67, 72, 75, 121, 185, 187, 191–192
al-Karajī, Abū Bakr Muḥammad b. al-Ḥasan 89, 100–101
al-Kāshī, Ghiyāth al-Dīn 89, 95, 103, 116, 122, 129
al-Kāshifī, Ḥusayn Waʿiẓ 130
al-Kashshī, Zayn ad-Dīn 182
al-Kātibī, Najm ad-Dīn 197
al-Kawm al-Rishī (Kawmrīshī) al-Qāhirī al-Mīqātī, Aḥmad b. Ghulām Allāh 33, 44, 176
al-Khāfī, al-Burhān Amīr Ḥaydar 161
al-Khafrī, Shams al-Dīn 101, 119, 142
Khalīfa Sulṭān 125
Khalīl Bahādur Sulṭān 123
al-Khalīlī, Shams al-Dīn 37, 47
al-Kharaqī, Bahāʾ al-Dīn Abū Bakr Muḥammad 82, 89
al-Khawāfī, al-Zayn 161
al-Khawāṣṣ, al-Shihāb 40, 43, 46, 157
al-Khayyām, ʿUmar 86, 89
al-Khāzin, Abū Jaʿfar 89, 91
al-Khāzinī, ʿAbd al-Raḥmān 88–89, 110–111
al-Khilāṭī ash-Sharīf, Ibrāhīm b. ʿAbdallāh 195
al-Khirāzī 162
al-Khūnjī, Afḍal al-Dīn = Khwāja Fażl Allāh b. Rūzbehān 71, 123, 162, 182, 193
al-Khuttalī, Muḥammad b. Yaʿqūb b. Khazzām 119n33
al-Khūyʾī Shams al-Dīn 71, 73, 187, 193
al-Khusrawshāhī, Shams al-Dīn 71–73, 75, 77, 121, 182, 187, 193
al-Khwārazmī al-Qāhirī al-Ḥanafī, ʿAbd al-Ḥayy b. Mubarakshāh 37, 45
al-Khwārazmī, ʿAlī Shāh Muḥammad 130
Kibrīt, Sayyid Muḥammad 9
al-Kīlānī al-Sharīf al-Ḥusaynī, Khān Aḥmad 165
al-Kindī, Yaʿqūb b. Isḥāq 115, 234
al-Kirmānī, Burhān al-Dīn Nafīs 123
al-Kūhī, Abū Sahl 86, 89
al-Kūrānī, al-Jamāl 148–149, 156
al-Kurdī, al-ʿAlāʾ ʿAlī 162, 166
al-Kūtāʾī, ʿAbd al-Wāḥid 161
al-Kūtāʾī, Wājid 161

Lājīn 113
al-Lārī, Mullā Muṣliḥ al-Dīn 164

al-Lubūdī, Najm al-Dīn 62, 73, 89, 94
al-Lubūdī, Shams ad-Dīn 183

al-Maʿarrī, Abū al-ʿAlāʾ 21
al-Maghribī, Abū Jābir 211, 218
al-Maghribī, ʿAlāmat al-Dīn ʿĪsā ibn Muḥammad al-Jaʿfarī 10n21
al-Mahallī 46
Maḥmūd b. Sebügtegīn 89
Maḥmūd ibn Mamdūd 177
Maḥmūd al-Ustādār 138
al-Makhzūmī, Aḥmad ibn Muḥammad 15, 25
al-Makhzūmī al-Qāhirī, Muḥammad b. Aḥmad *see* Ibn al-Khashshāb
al-Makkī al-Ḥillawī, Ḥusayn b. Muḥammad *see* Ibn ʿUlayf
al-Makkī al-Qāhirī, Aḥmad b. Ṣadaqa *see* Ibn Sīrāfī
al-Malik al-Muẓaffar *see* Qutuz
al-Mālikī, ʿAlī al-Zayn Ṭāhir 157, 167
al-Maʾmūn 21–22, 87, 89, 99, 202, 223–235
Manṣūr b. Muḥammad b. Aḥmad 122
Manṣūr b. Ṣafī 44
al-Manṣūr ibn Taqī al-Dīn 51n3, 75
al-Manṣūr Qalāwūn 113, 115
al-Manṣūrī al-Malikī, Ibrāhīm Ḥāsib 115
Manūchihr Khān, Abū l-Fatḥ 119, 142–143
al-Maqdisī, Abū l-Taqī Ṣāliḥ ibn Aḥmad 68n99
al-Maqdisī, al-Rāḍī b. Abī l-Luṭf 166
al-Maqdisī al-Ḥaṣkafī, Abū l-Luṭf Muḥammad b. ʿAlī *see* Ibn al-Ḥimṣī
al-Maqrīzī, Taqī al-Dīn Aḥmad ibn ʿAlī 35, 156, 172–177, 173n22, 179–181
al-Marʿashī, Muḥammad 10n20
al-Māridānī, Badr al-Dīn *see* Sibṭ al-Māridānī
al-Māridānī, Jamāl al-Dīn 37, 39–40, 44–46, 89, 95, 145, 148–149, 155–156, 159
al-Māridīnī, Ismāʿīl b. Ibrāhīm *see* Ibn Fallūs
al-Marwarrudhi, Khalid b. ʿAbd al-Malik 223–224, 227–229, 233–234
al-Marwazī, ʿAyn al-Zamān Abū ʿAlī b. ʿAlī 128
Māshāʾallāh 124
al-Mashdāllī, Abū l-Faḍl Muḥammad b. Muḥammad 162
al-Maṣʿūd 75, 187
al-Maṣʿūd b. Maḥmūd 128
al-Masʿūd Aqsīs [sic] ibn al-Kāmil 51n3
Mawlānāzāde 30n9
al-Mawṣilī, al-Ḥājj Zayn al-Dīn Ṭāhir b. Qāḍī 162
al-Maybudī, Qāżī Ḥusayn 130–131
Mehmet III 119
Mehmet Fātiḥ 89, 92, 95, 116, 126–128
al-Mibyawī al-Rūmī, Muḥammad b. Sulaymān *see* al-Kāfiyājī
Mīrza Muḥammad Ibrāhīm 142
al-Miṣrī al-Samannūdī, Muḥammad b. ʿAlī *see* Ibn al-Qattān
al-Miṣrī, Quṭb ad-Dīn 182
al-Mizzī, Muḥammad b. Aḥmad 36
Moshe ben Maimon, or Maimonides 52
al-Muʾayyad Najm al-Dīn Masʿūd ibn Ṣalāḥ al-Dīn 51n3
al-Muʾayyad Shaykh 30–31, 33, 113
al-Muʿaẓẓam ʿIsā 51, 51n3, 54, 54n22, 60–64, 66, 68, 72–73, 75–76, 120–121, 121n23, 187–189, 191, 193
al-Muʿaẓẓam Turanshāh [sic] 51n3
al-Muʿaẓẓamī, ʿIzz al-Dīn 50n1
Mubāriz al-Dīn Muḥammad 124
Mubashshir b. Fātik 208
Muhadhdhab al-Dīn *see* ʿAbd al-Raḥīm b. ʿAlī
Muḥammad 82, 177, 204, 216
Muḥammad ʿĀdil Shāh = Muḥammad b. Armaghān 89–90, 95
Muḥammad b. Ibrāhīm Ṣalāḥ ad-Dīn *see* Ibn al-Burhān
Muḥammad b. Khalaf 90–91, 112
Muḥammad b. Manṣūr 123
al-Muḥibb 138
al-Muḥibbbī, Muḥammad Amīn b. Faḍl Allāh 3, 9, 10n21, 14, 16n58, 19, 25, 150–153, 164–165
al-Muʿīnī, Jawhar 44
al-Muʿizz 35, 156
Mullā ʿAlī 18
Mullā Khiżr Bey 127–128
al-Munāwī, Ṣadr ad-Dīn Abū Maʿālī Muḥammad ibn Ibrāhīm 179
al-Munlā Sharīf al-Kurdī 166
al-Munshī, Idrīs b. Ḥusām al-Dīn 123
Mūsā b. Muḥammad al-Sharaf al-Muwaqqit 37, 47
al-Mustanṣir 93
al-Mustaʿrib, al-Fāris Aqṭaʿī 177
al-Muʿtaḍid 86, 87n3, 90
al-Muʿtamid 86–87, 87n3, 90
Muʿtazz b. Ṭāhir, Tāj al-Dīn 142

al-Mutawakkil ʿAlāʾ Allāh 44
al-Muwaffaq 86–87, 87n3, 90
Muwaffaq al-Dīn ʿAbd al-ʿAzīz 58, 69, 135
Muwaffaq al-Dīn Yaʿqūb ibn Saqlāb 54n22, 61–63, 69
al-Muẓaffar Taqī ad-Dīn ʿUmar b. al-Malik al-Amjad 193

al-Nābigha al-Dhubyānī 162
Nādir Shāh 129, 131
al-Naḥrīrī, Muḥammad 159
al-Najdī al-Ḥanbalī al-Faraḍī, Muḥammad 165
an-Najm an-Naḥḥās 178
Najm al-Dīn Ayyūb ibn al-Kāmil 51n3
al-Naqqāsh al-Mīqātī, ʿAlī b. ʿAbd al-Qādir 32, 45
al-Nāṣir Aḥmad 30
al-Nāṣir Dāwūd 54n22, 62, 72, 75, 77, 121
al-Nāṣir Faraj 30–31, 138
al-Nāṣir Muḥammad b. Qalāʾūn 30
al-Nāṣir Ṣalāḥ al-Dīn Yūsuf 70
al-Nāṣir Yūsuf ibn Muḥammad 51
al-Nisābūrī, Niẓām al-Dīn 41
al-Nuʿaymī, ʿAbd al-Qādir 78, 186
Nūr al-Dīn ibn Jamāl al-Dīn ibn Artuq
Nūr al-Dīn Maḥmūd b. Zangī 52, 52n7, 55, 68, 113
al-Nuwayrī, Abū l-Qāsim 162

Pīr Muḥammad Bahādur Khān 122
Plato 16, 16n58, 74, 211, 217, 237
Pythagoras 16

al-Qabīṣī, Abū l-Ṣaqr 41
al-Qāhirī, ʿAbd al-Raḥīm b. Ibrāhīm 160
al-Qāhirī, Aḥmad b. ʿAlī 138
al-Qāhirī, Aḥmad ibn Muḥammad Shihāb al-Dīn, called Ibn al-Qurdāḥ 15n53, 139, 155
al-Qāhirī, ʿAlī b. Muḥammad al-Nūr 46
al-Qāhirī al-Azharī al-Bilqāsī, Aḥmad b. Sulaymān 157
al-Qāhirī al-Azharī al-Qāḍī, Zakariyyāʾ b. Muḥammad 158
al-Qāhirī al-Bījūrī, Aḥmad b. Muḥammad 158
al-Qāhirī al-Ghazzī, Muḥammad b. Qāsim *see* Ibn al-Gharābīlī
al-Qāhirī al-Ḥanafī, ʿAbd al-Razzāq b. Aḥmad 37, 45
al-Qāhirī al-Ḥanafī, Aḥmad b. ʿAlī 138
al-Qāhirī al-Lakhmī al-Santarāwī, Muḥammad b. Muḥammad 162
al-Qāhirī al-Maqsī, ʿAlī b. ʿUmar 31, 46
al-Qāhirī al-Mīqātī, ʿAlī b. ʿAbd al-Qādir 160
al-Qāhirī al-Sarāyʾī al-ʿAjamī, Muḥammad b. Aḥmad 161
al-Qāhirī al-Shāfiʿī, Aḥmad b. Muḥammad 139
al-Qāhirī al-Sikandarī, Aḥmad b. Asad *see* Ibn Asad
al-Qāhirī al-Sikandarī, Muḥammad b. Muḥammad 155
al-Qāhirī al-Sīwāsī, Muḥammad b. ʿAbd al-Wāḥid 156
al-Qāhirī al-Ṭalkhāwī, Ḥasan b. ʿAlī 32, 158
al-Qāʾinī, Ḥasan b. Saʿd 119
al-Qalaṣādī 160
al-Qalaṣāwī [sic] *see* al-Qalaṣādī
al-Qayatī 35
al-Qaymarī, Ḥasan b. ʿAlī 37, 44–45, 89, 95
Qāytbāy 123
Qāżī Sirāj al-Dīn 130
Qāżī Ḥusayn Maybudī *see* al- Maybudī
Qāżīzāde Rūmī 115–116, 122, 128, 146–147, 156
al-Qazvīnī, Zakariyyāʾ b. Muḥammad 117–178n25, 118–119, 126
Qujmas 31, 46
al-Qurashī al-Andalusī al-Bastī, ʿAlī b. Muḥammad*see* al-Qalaṣādī
al-Qurashī, Ibn al-Nafīs *see* Ibn al-Nafīs
al-Qurashī, Ibrāhīm b. Muḥammad *see* Ibn Zuqqāʿa
al-Qurashī al-Qalqashandī al-Qāhirī, ʿAlī b. Aḥmad 155
al-Qushjī, Alī 94, 101, 116, 122, 126, 165, 165n86
al-Qusantīnī, Aḥmad b. Muḥammad 158
al-Qusṭurlī, Muḥammad 160
Quṭuz, Sayf ad-Dīn 177–178

al-Raḥbī, Raḍī al-Dīn 52n7, 54, 54n22
al-Rashīdī 45
al-Rāzī, Abū Yūsuf Yaʿqūb b. Muḥammad 90, 100
al-Rāzī, Fahkr al-Dīn 8, 10–13, 72–74, 76, 12n34, 121, 130, 182, 182n54, 187–188, 133, 197

al-Rāzī, Muḥammad b. Zakariyyā' 129
al-Rāzī, Qutb al-Dīn 16n58, 30n9, 130–131
al-Rīfī, Abū Ya'qūb Yūsuf 162
Roger II 235–241
Rukn al-Dawla 90–91, 100
Rukn ad-Dīn b. Quba' 184

al-Sā'ātī, Fakhr al-Dīn Riḍwān ibn Rustam 62, 66, 68
Sa'd b. al-Sharīf Zayd al-A'lam 151, 164
Sadīd al-Dīn Abū Manṣūr 52n7, 62
Ṣadr al-Sharī'a 90, 101
aṣ-Ṣafadī, Khalīl b. Aybak 183, 197
Ṣafī 125
Ṣafī al-Dīn ibn Shukr 136
al-Saḥyūnī, Abū Bakr Taqī al-Dīn 166
al-Sakhāwī, Shams al-Dīn 9n19, 14–16, 15n53, 16n58, 26, 28–41, 30n9, 41n81, 43, 137–138, 144–151, 145n4, 155
Ṣalāḥ al-Dīn ibn Ayyūb 50, 50n2, 51n3, 52, 52n7, 54, 54n22, 59, 61, 64–67, 69, 72, 120, 211
Ṣalāḥ al-Dīn ibn Yāghīsān 53
al-Ṣāliḥ lsma'īl ibn al-'Ādil 51n3, 61
aṣ-Ṣāliḥ Ṣalāḥ ad-Dīn 174n23
al-Ṣāliḥī, 'Abd al-Khāliq 31–32, 45
al-Samarqandī, Najīb al-Dīn Muḥammad b. 'Alī 123
al-Samarqandī, Shams al-Dīn 146–147, 150
as-Sāmarī, Ṣadaqa 191
al-Sāmarrī, Muwaffaq al-Dīn Ya'qūb 73
al-Samaw'al, Abū Naṣr 90, 101
al-Samnūdī al-Shāfi'ī, 'Umar b. 'Īsā 34
Ṣamṣām al-Dawla 90, 100
Sanad b. 'Ali 223–224, 224n11
Sanjar 88, 90, 110
Sarah bint Jamā'a 35
al-Sarmīnī al-Ḥalabī, Aḥmad b. Ibrāhīm 33
al-Saṭḥī, al-Shihāb 46
Sayf ad-Dīn Abū Bakr 178
al-Sayyid al-Faraḍī, 'Alī b. 'Abd al-Qāḍir [. . .] al-Shāmī al-Qāhirī al-Azharī, called Tilmīdh Ibn al-Majdī 16, 148, 155, 160
al-Sayyid al-Sharīf al-Jurjānī, 'Alī b. Muḥammad 8, 30n9, 40–41, 115, 124, 130, 155–156, 197
Sebügtegīn 88, 90, 111
ash-Sha'bānī, Qurqmās 173–174, 173n22
Shāh Jahān 90, 95, 125–126, 129
Shāh Shujā' 124
al-Shāhid, al-Shams 158
al-Shahrastānī, Abū l-Fatḥ 129
al-Shahrazūrī, Ibn al-Ṣalāḥ 121
al-Shahrazūrī, Shams al-Dīn 86, 87n3, 90
Shāhrukh 123, 241, 243
Shajarat ad-Durr 172
al-Shambārī al-Makkī, Ibrāhīm b. 'Alī *see* al-Zamzāmī, Ibrāhīm b. 'Alī
al-Shāmī, Niẓām al-Dīn 'Alī 241
al-Shams Muḥammad b. Ayyūb 40
Sharaf ad-Dīn *see* Ibn 'Unayn
al-Sharfī b. al-Jay'ānī 159
al-Shāṭir Shūmān 158
al-Shawwā al-Qāhirī, 'Abd al-Wahhāb b. Muḥammad 45
al-Shaybānī al-Zabīdī, 'Abd al-Raḥmān b. 'Alī *see* Ibn al-Dayba'
Shaykhzāde Khuzbānī 17, 25
al-Shifā'ī al-Ḥasanī, Muẓaffar b. Muḥammad 143
al-Shīrāzī, Ibrāhīm b. Aḥmad 33
al-Shīrāzī, Qiwām al-Dīn 162
al-Shīrāzī, Quṭb al-Dīn 8, 14, 90, 94, 101, 133, 142–143
al-Shirwānī, Shams ad-Dīn = Muḥammad b. Marāhim [sic] al-Dīn al-Shams al-Shirwānī, then al-Qāhirī 35, 156, 159, 160, 183
Sibṭ al-Abharī 179
Sibṭ ibn al-Jawzī 21, 70, 74–75, 78, 121, 121–122n43, 187, 189
Sibṭ al-Māridānī *see* Sibṭ al-Māridānī
Sibṭ al-Māridānī, Muḥammad b. Muḥammad Badr al-Dīn 44–45, 47, 90, 95–96, 145, 158–159
Sijistānī, Abū Sulaymān 90–91
Sijzī, Abū Sa'īd 115, 124
al-Sijīnī al-Qāhirī al-Azharī al-Faraḍī, Aḥmad b. 'Ubayd Allāh al-Shihāb 40, 43, 157, 159
al-Sikandarī al-Shāfi'ī, Muḥammad b. 'Alī 37
al-Sikandarī, Muḥammad b. 'Awad *see* Junaybat
al-Sindī, Sayyid Ṣibghat Allāh ibn Rūḥ Allāh 9n19
Sirāj al-Dīn wa-l-Dunyā 224, 226–227, 231, 233–234
Sirāj al-Rūmī 162
Sitt al-Shām 53, 53n16, 62, 191
al-Subkī, Tāj al-Dīn 21–22, 22n83, 32, 186
al-Subkī, Taqī al-Dīn 10–11
al-Ṣūfī, 'Abd al-Raḥmān 90, 94, 117–118n25, 118–119, 125, 128, 142–143, 142n25
al-Suhrawardī, Shihāb al-Dīn 30n9, 48, 71–72, 184

al-Sulamī, Sa'd al-Dīn ibn 'Abd al-'Azīz 54n22
Sulaymān 125
al-Suyūṭī, Jalāl al-Dīn 16–17, 20, 148

al-Ṭabarī, Muḥammad b. Jarīr 233
al-Ṭabarī, al-Muḥibb 152
al-Tabrīzī 158
al-Taftāzānī, Ḥāfiz al-Dīn 30n9
al-Taftāzānī, Sa'd al-Dīn 8, 30n9, 161
Tāj b. al-Ẓarīf 44
al-Takhawī al-Qāhirī, Ḥasan b. 'Alī 32
aṭ-Ṭanāḥī, Nāṣir ad-Dīn 196
Ṭarafa b. al-'Abd 162
Taqī al-Dīn 'Umar 51n3, 73
Tashköprüzade 10n20, 13, 116, 128
al-Tawḥīdī, Abū Ḥayyān 14, 21
aṭ-Ṭaybarsī, Muḥammad b. 'Alī b. 'Abdallāh 196
Thābit b. Qurra 86–87, 90, 234
Themistios 74
al-Tibranī *see* al-Haythamī
al-Tilimsānī, Abū Bakr 162
Timur 86, 90, 111, 115, 117, 122, 241–244
al-Tīrūnī, Ya'qūb 162
aṭ-Ṭughrā'ī, Mu'ayyad ad-Dīn 196
al-Ṭūlūni, Muḥammad ibn Aḥmad Shams al-Dīn 18
al-Tūnisī, Muḥammad b. Aḥmad *see* al-Wānnawghī
al-Tūnisī al-Bijāyī, Aḥmad b. Muḥammad*see* Abū l-'Abbās b. Kuḥayl
al-Ṭuntadā'ī, al-Shams 44, 158
al-Ṭūsī, Aṣīl al-Dīn 90
al-Ṭūsī, Naṣīr al-Dīn 8, 14, 30n9, 41, 90, 93–95, 101, 123–124, 126, 129–130, 142–143, 143n30, 146, 155–156, 184, 192, 197
al-Ṭūsī, Salmān 118
al-Ṭūsī, Sharaf al-Dīn 69, 71n109, 182n56

Ulugh Beg 1, 90, 94–95, 103, 115–116, 119, 122–123, 126–129, 142, 202
al-'Umarī, Sayf ad-Dīn Bazlār 178
al-'Uqbānī, Abū l-Qāsim b. Sa'īd 158, 163
al-'Uqbānī, Qāsim 160
Uqlīdis 16, 35, 155–156; *see also* Euclid
al-'Urḍī, Mu'ayyad al-Dīn 70, 71n109, 89–90, 92–93, 101
al-Urmawī, al-Ṣafī 'Abd al-Mu'min 162
al-Urmawī, Sirāj ad-Dīn 182
al-Urmawī, Tāj al-Dīn 11, 13, 182, 187
'Uthmān Qarā Yūluk 174
al-'Uthmānī al-Iṣfūnī, Ḥasan b. Muḥammad 183
Ūzūn Ḥasan, Ḥasan Bahādur Khān 116, 123–124

al-Wafā'ī, al-'Izz 'Abd al-'Azīz 37, 45–46
al-Warā'ī, al-'Izz 158
al-Wānnawghī 160
al-Wānsharīsī, Abū Bakr b. 'Īsā 162
al-Wardīrī 35, 159
al-Wāsiṭī, Ḥusayn b. Bistām 130
William I 235, 237–238

Yaḥyā b. Abī Manṣūr 224n9, 224n11, 229
Yaḥyā b. Aktham 223–227, 224n9, 229, 231, 233–234
al-Yamanī, Muḥammad b. 'Abd al-Qādir 166
al-Yanabī, Abū l-Jūd 35, 159
Ya'qūb Bahādur Khān 123
Yashbak min Mahdī al-Dawādār 123
al-Yazdī, Sharaf al-Dīn 'Alī 238n41, 241
al-Yūnīnī, Mūsā b. Muḥammad 177–178
al-Yushbughāwī, 'Alī b. Sūdūn al-'Alā' 35

al-Ẓāhir (Ayyubid prince) 51n3, 72
al-Ẓāhir *see* Baybars
Zakī ad-Dīn b. az-Zakī 191
al-Zamzāmī, Abū l-Fatḥ 39
al-Zamzāmī, al-Badr Ḥusayn 39–40, 159
al-Zamzāmī, Ibrāhīm b. 'Alī Burhān al-Dīn 33, 39
al-Zamzāmī, Ismā'īl b. Nābit 40
al-Zamzāmī, Muḥammad b. 'Abd al-'Azīz al-Jamāl 29–30, 39, 161
al-Zamzāmī, Muḥammad b. Abī l-Fatḥ b. Ismā'īl 39
al-Zamzāmī, Nābīt b. Ismā'īl 39
al-Zamzāmī, Nūr al-Dīn 'Alī b. Muḥammad 39–40, 161
al-Zawāwī al-Mīqātī, 'Umar b. 'Abd al-Raḥmān 29
Zayn b. Ḥājjī 123
al-Zayn b. al-Ṣā'igh 158
Zuhayr b. Abī Sulmā 162

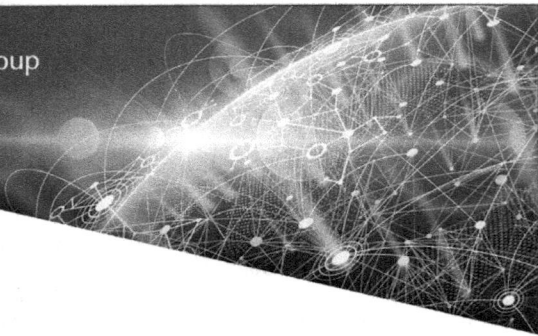